Titles in This Series

11 **Larry Finkelstein and William M. Kantor, Editors,** Groups and Computation
10 **Joel Friedman, Editor,** Expanding Graphs
9 **William T. Trotter, Editor,** Planar Graphs
8 **Simon Gindikin, Editor,** Mathematical Methods of Analysis of Biopolymer Sequences
7 **Lyle A. McGeoch and Daniel D. Sleator, Editors,** On-Line Algorithms
6 **Jacob E. Goodman, Richard Pollack, and William Steiger, Editors,** Discrete and Computational Geometry: Papers from the DIMACS Special Year
5 **Frank Hwang, Clyde Monma, and Fred Roberts, Editors,** Reliability of Computer and Communication Networks
4 **Peter Gritzmann and Bernd Sturmfels, Editors,** Applied Geometry and Discrete Mathematics, The Victor Klee Festschrift
3 **E. M. Clarke and R. P. Kurshan, Editors,** Computer-Aided Verification '90
2 **Joan Feigenbaum and Michael Merritt, Editors,** Distributed Computing and Cryptography
1 **William Cook and Paul D. Seymour, Editors,** Polyhedral Combinatorics

Groups and Computation

DIMACS

Series in Discrete Mathematics
and Theoretical Computer Science

Volume 11

Groups and Computation

Workshop on Groups and Computation
October 7–10, 1991

Larry Finkelstein
William M. Kantor
Editors

NSF Science and Technology Center
in Discrete Mathematics and Theoretical Computer Science
A consortium of Rutgers University, Princeton University,
AT&T Bell Labs, Bellcore

American Mathematical Society

This DIMACS workshop on Groups and Computation was held at DIMACS, October 7–10, 1991.

1991 *Mathematics Subject Classification.* Primary 20B40, 20C40.

Library of Congress Cataloging-in-Publication Data

Groups and computation/Larry Finkelstein, William M. Kantor, editors.
p. cm.—(DIMACS series in discrete mathematics and theoretical computer science; v. 11)
Papers from the workshop held at DIMACS, Rutgers University, Oct. 7–10, 1991.
Includes bibliographical references.
ISBN 0-8218-6599-4
1. Group theory—Data processing—Congresses. 2. Algebra—Data processing—Congresses. I. Finkelstein, Larry. II. Kantor, W. M. (William M.), 1944– . III. Series.
QA174.7.D36G76 1993 93-4968
512′.2–dc20 CIP

Printed in the United States of America.
The paper used in this book is acid-free and falls within the guidelines established to ensure permanence and durability. ♾
Printed on recycled paper. ♻

This volume was prepared by the authors using AMS-TEX and AMS-LATEX, the American Mathematical Society's TEX macro system.

10 9 8 7 6 5 4 3 2 1 98 97 96 95 94 93

In memory of
Daniel Gorenstein
(1923–1992)

Contents

Foreword

This DIMACS volume on Groups and Computation contains papers from talks at a workshop held at DIMACS, October 7–10, 1991.

We would like to thank the workshop organizers, Daniel Gorenstein, Charles Sims, Larry Finkelstein, and William Kantor for organizing this workshop that brought together so many outstanding speakers within this field.

Our special thanks go to Larry Finkelstein and William Kantor for all their work in editing this volume.

Fred Roberts, Acting Director
Robert Tarjan, Co-Director
Diane Souvaine, Acting Associate Director

Preface

The Workshop "Groups and Computation" took place at DIMACS, the Center for Discrete Mathematics and Theoretical Computer Science on the campus of Rutgers University, on October 7–10, 1991. The Workshop explored and fostered the interplay of a number of research areas:

- *Symbolic algebra and computer algebra*, which have developed algorithms for group–theoretic computation and have built large integrated software packages (such as Cayley and GAP).
- *Theoretical computer science*, which has studied the complexity of computation with groups.
- *Group theory*, which has provided new tools for understanding the structure of groups, many of which involve the classification and properties of finite simple groups.
- *Applications of group computation* within mathematics or computer science, which have dealt with such diverse subjects as simple groups, combinatorial search, networks and the analysis of data.

An important Workshop theme was the comparison between algorithms that have had effective implementations and recently developed algorithms that have improved worst–case asymptotic times. Many algorithms of these two types depend heavily on properties of simple groups, both for motivation and correctness proofs, while other algorithms have depended more on novel data structures than on group theory.

The scientific program consisted of invited lectures and short research announcements, as well as informal discussions and software demonstrations. The four extended length talks (by C. C. Sims, J. Neubüser, E. M. Luks and L. Babai) laid out part of the groundwork for discussions of relationships between implementation and complexity questions; this was continued in further talks and in informal settings. The present Proceedings, which contain most, but not all, of the talks given at the Workshop, are dominated by papers amplifying these relationships. However, there were further subjects discussed in the Workshop and represented here: parallel algorithms for groups, computation in associative algebras, asymptotic behavior of permutation groups, studying finite groups using infinite reflection groups, combinatorial searching, computing with representations (including applications to statistics), and Cayley graphs as models for interconnection networks. Some of these subjects naturally led to practical computational questions. In fact, the difficulty encountered in attemping to separate or pigeon–hole the papers in this volume is indicative of the fact that there are no clear dividing lines between practical and theoretical aspects of the subject of the Workshop, nor between the needs of users and those more interested in the algorithms themselves.

It was our belief that a full discussion of implementation experiences would lead to a deeper understanding of the mathematical issues that underlie group computations, leading both to new algorithms and to improvements of existing ones. In order to provide some background for this, software provided by the participants was available during the entire Workshop; it was discussed in a number of talks, and one evening was devoted to informal software presentations. In addition, there was a panel discussion whose theme was "Problems whose current algorithmic solutions are, in practice, too time– or space–consuming." The panelists were selected for their broad experience in building software for group computation; the discussions provided insight into implementation problems and goals.

We are grateful for the assistance of our co–organizers, Daniel Gorenstein and Charles Sims. The possibility of arranging a Workshop arose accidentally during lunch with Danny, a mere thirteen months before the Workshop took place. His encouragement and enthusiasm helped us to overcome many of the usual difficulties encountered when organizing an international workshop. We also thank the staff at DIMACS for their efforts in helping the Workshop to function smoothly. Donna Harmon and Christine Thivierge of the American Mathematical Society provided invaluable assistance in the production of these Proceedings. Finally, we thank DIMACS, the National Science Foundation, the National Security Agency and Rutgers University for their financial support of the Workshop.

Larry Finkelstein
William M. Kantor

Workshop Program

Talks

L. Babai, *Combinatorial methods for algorithms in finite groups*

R. Beals, *Structure forests in nearly linear time for small base groups*

K. D. Blaha, *Minimum bases for permutation groups: the Greedy Algorithm*

M. Clausen, *The FFT on nonabelian groups*

J. H. Conway, *Infinite reflection groups as a tool for computing*

G. Cooperman, *New tools for computational group theory*

J. D. Dixon, *Computing ordinary representations of finite groups*

L. Finkelstein, *New tools for computational group theory*

D. F. Holt, *Computing quotients of finitely presented groups*

A. J. Goodman, *Subdirectly reducible groups and edge minimal graphs with a given automorphism group*

A. A. Ivanov, *Package for computing with coherent configurations*

W. M. Kantor, *Orders of elements*

C. W. H. Lam, *Application of group theory to combinatorial searches*

E. M. Luks, *Polynomial time computation and permutation groups*

P. D. Mark, *Parallel algorithms for Sylow theorems*

J. Neubüser, *Remarks on Coxeter groups and GAP*

P. M. Neumann, *Algorithms for computing in matrix groups*

C. E. Praeger, *Algorithms for computing in matrix groups*

L. Pyber, *Generation and enumeration of permutation groups*

D. Rockmore, *Computing isotypic projections for data on homogeneous spaces*

A. L. Rosenberg, *Almost all Cayley graphs are almost direct product graphs and why we care*

N. Sarawagi, *Group membership for primitive groups*

G. Schneider, *Computing Loewy series of* $Sz(8)$

Á. Seress, *Composition factors in primitive groups*

C. C. Sims, *Complexity issues in infinite group theory*

L. H. Soicher, *Computing with graphs and groups*

I. Weisz, *PERM – A program for computing strong generating sets*

B. W. York, *Implications of parallel architectures for group computations*

Panel Discussion on Implementations

G. Butler, G. Cooperman, L. Finkelstein (*Moderator*), J. S. Leon, G. Havas, D. F. Holt, J. Neubüser, C. C. Sims.

Software Demonstrations

A. M. Cohen, *LIE: A system for computing with Lie groups*

G. Havas, *Algorithms for computing with finitely presented groups*

D. F. Holt and S. Rees, *Computing with quotients of finitely presented groups*

A. A. Ivanov, *Computing with coherent configurations*

C. W. H. Lam, *ISOM*

J. Leon, *Partition backtrack algorithms for permutation groups*

J. Neubüser and F. Celler, *GAP*

Á. Seress and I. Weisz, *PERM: Computing strong generating sets for permutation groups*

L. H. Soicher, *Computing with graphs and groups*

Participants

László Babai *Department of Computer Science, University of Chicago, Chicago, IL 60637.* laci@gargoyle.uchicago.edu

Ranan B. Banerjee *Department of Mathematics and Computer Science, St. Joseph's University, Philadelphia, PA 19131.*

Ulrich Baum *Institut für Informatik V, Rhein. Friedrich-Wilhelms-Universität, 5300 Bonn, Germany.* uli@leon.informatik.uni-bonn.de

Kenneth D. Blaha *Department of Computer Science, Pacific-Lutheran University, Tacoma, WA 98447.* blaha_k@plu.bitnet

Greg Butler *Department of Computer Science, Concordia University, Montreal, Quebec H3G 1M8.* gregb@maxwell.concordia.ca

Peter J. Cameron *Department of Mathematics, Queen Mary College, London E1 4NS, England.* pjc@maths.qmw.ac.uk

Frank Celler *Lehrstuhl D für Mathematik, RWTH Aachen, 5100 Aachen, Germany.* fceller@samson.math.rwth-aachen.de

Michael Clausen *Institut für Informatik V, Rhein. Friedrich Wilhelms Universität, 5300 Bonn, Germany.* clausen@leon.informatik.uni-bonn.de

Peter Clote *Department of Computer Science, Boston College, Boston, MA 02167.* clote@bcuxs1.bc.edu

Arjeh M. Cohen *Centrum voor Wiskunde en Informatica, 1098SJ Amsterdam, The Netherlands.* marc@cwi.nl

John H. Conway *Department of Mathematics, Princeton University, Princeton, NJ 08540.* conway@fine.princeton.edu

Gene Cooperman *College of Computer Science, Northeastern University, Boston, MA 02115.* gene@ccs.northeastern.edu

John D. Dixon *Department of Mathematics, Carleton University, Ottawa, Ontario K15 5B6.* jdixon@carleton.ca

Wayne Eberly *Department of Computer Science, University of Calgary, Calgary T2N 1N4, Alberta.* eberly@cpsc.ucalgary.ca

Charles M. Fiduccia *Supercomputing Research Center, Bowie, MD 20715.* fiduccia@super.org

Larry Finkelstein *College of Computer Science, Northeastern University, Boston, MA 02115.* laf@ccs.northeastern.edu

David E. Flesner *Department of Mathematics, Gettysburg College, Gettysburg, PA 17325.* flesner@cc.gettysburg.edu

Daniel E. Frohardt *Department of Mathematics, Wayne State University, Detroit, MI 48202.* danf@math.wayne.edu

Robert Gilman *Department of Mathematics, Stevens Institute of Technology, Hoboken, NJ 07030.* rgilman@sitult.stevens-tech.edu

Albert J. Goodman *Department of Mathematics, University of Chicago, Chicago, IL 60637.* albert@gargoyle.uchicago.edu

Daniel Gorenstein *Department of Mathematics, Rutgers University, New Brunswick, NJ 08903.*

Jonathan I. Hall *Department of Mathematics, Michigan State University, East Lansing, MI 48824.* jhall@nsf1.mth.msu.edu

George Havas *Department of Computer Science, University of Queensland–St. Lucia, Brisbane, Queensland 4067, Australia.* havas@cs.uq.oz.au

Frederick Hoffman *Department of Mathematics, Florida Atlantic University, Boca Raton, FL 33431.* hoffman@acc.fau.edu

Derek F. Holt *Mathematics Institute, University of Warwick, Coventry CV4 7AL, England.* dfh@maths.warwick.ac.uk

Alexander A. Ivanov *Institute for System Studies, Academy of Sciences, 117312 Moscow, Russia.*

William M. Kantor *Department of Mathematics, University of Oregon, Eugene, OR 97403.* kantor@bright.math.uoregon.edu

Peter Kenne *Faculty of Information Science and Engineering, University of Canberra, Belconnen ACT 2616, Australia.*

Clement W. H. Lam *Department of Computer Science, Concordia University, Montreal, Quebec H3G 1M8.* lam@vax2.concordia.ca

Jeffrey S. Leon *Department of Mathematics, University of Illinois, Chicago, IL 60680.* U10407@UICVM.uic.edu

Eugene M. Luks *Department of Computer and Information Science, University of Oregon, Eugene, OR 97403.* luks@cs.uoregon.edu

Richard N. Lyons *Department of Mathematics, Rutgers University, New Brunswick, NJ 08903.* lyons@math.rutgers.edu

Peter D. Mark *Department of Computer and Information Science, University of Oregon, Eugene, OR 97403.* pdm@cs.uoregon.edu

John McKay *Department of Computer Science, Concordia University, Montreal, Quebec H3G 1M8.* mckay@vax2.concordia.ca

Joachim Neubüser *Lehrstuhl D für Mathematik, RWTH Aachen, 5100 Aachen, Germany.* neubueser@math.rwth-aachen.de

Peter M. Neumann *Queens College, Oxford University, Oxford OX1 4AW, England.* neumann@vax.ox.ac.uk

Michael F. Newman *Department of Mathematics, Australian National University, Canberra 2601, Australia.* newman@pell.anu.edu.au

Michael E. O'Nan *Department of Mathematics, Rutgers University, New Brunswick, NJ 08903.*

Cheryl E. Praeger *Department of Mathematics, University of Western Australia, Nedlands, Western Australia 6009.* praeger@madvax.maths.uwa.oz.au

László Pyber *Mathematical Institute of the Hungarian Academy of Sciences, Budapest, 1364 Hungary.* h1130pyb@ella.hu

Sarah Rees *Department of Mathematics and Statistics, The University, Newcastle upon Tyne NE1 7RU, England.* Sarah.Rees@newcastle.ac.uk

Daniel Rockmore *Department of Mathematics, Dartmouth College, Hanover, NH 03755.* rockmore@gilligans.dartmouth.edu

Lajos Ronyai *Computer and Automation Institute, Hungarian Academy of Sciences, H1502 Budapest, Hungary.* h631ron@ella.hu

Arnold L. Rosenberg *Department of Computer and Information Science, University of Massachusetts, Amherst, MA 01003.* rsnbrg@elys.cs.umass.edu

Namita Sarawagi *College of Computer Science, Northeastern University, Boston, MA 02115.* namita@corwin.ccs.northeastern.edu

Stephen Schibell *National Security Agency, Ft. George G. Meade, MD 20755.* stschib@afterlife.ncsc.mil

Gerhard Schneider *Fachbereich Mathematik der Universität, D-4300 Essen 1, Germany.* mat420@de0hrz1a.bitnet

Ákos Seress *Department of Mathematics, Ohio State University, Columbus, OH 43210.* akos@function.mps.ohio-state.edu

Charles C. Sims *Department of Mathematics, Rutgers University, New Brunswick, NJ 08903.* sims@math.rutgers.edu

Michael Slattery *Department of Mathematics, Marquette University, Milwaukee, WI 53233.* mikes@sylow.mscs.mu.edu

Neil J. A. Sloane *Mathematical Sciences Research Center, AT&T Bell Laboratories, Murray Hill, NJ 07974*

Stephen D. Smith *Department of Mathematics, University of Illinois, Chicago, IL 60680.* smiths@math.uic.edu

Leonard H. Soicher *Department of Mathematics, Queen Mary College, London E1 4NS, England.* leonard@maths.qmw.ac.uk

Iván Weisz *Department of Mathematics, Ohio State University, Columbus, OH 43210.* weisz@function.mps.ohio-state.edu

Bryant W. York *College of Computer Science, Northeastern University, Boston, MA 02115.* york@ccs.northeastern.edu

DIMACS Series in Discrete Mathematics
and Theoretical Computer Science
Volume 11, 1993

Computing Composition Series in Primitive Groups

LÁSZLÓ BABAI, EUGENE M. LUKS, AND ÁKOS SERESS

ABSTRACT. We give a fast polynomial-time algorithm for computing a composition series in a primitive permutation group given by a list of generators. Permutation representations for the composition factors are also obtained.

The algorithm will be a key procedure in an $O(sn^3 \log^c n)$ algorithm that will the solve the same problems for an arbitrary subgroup of S_n given by s generators. The general algorithm will be described in a forthcoming paper.

The first polynomial-time algorithm for this problem was given by Luks. Our procedure follows the overall architecture of that original algorithm, while replacing the subroutines by much faster ones. New combinatorial estimates of suborbit sizes in primitive groups guarantee the improved performance.

1. Introduction

In recent years, polynomial-time algorithms have been found for a great number of permutation-group problems. These range from constructing a strong generating set (SGS) [**24**], which is the fundamental data structure for membership testing and other basic tasks (finding the order, constructing normal closures, etc.), to obtaining structural information, such as composition factors [**20**] and Sylow subgroups [**14**]. (See [**15**] and [**21**] for other examples.) In this paper, we develop a more efficient procedure for the Composition-Series Problem: *Given generators for a permutation group G, find generators for the subgroups in a composition series $G = N_0 \rhd N_1 \rhd \ldots \rhd N_m = 1$ and permutation representations for the composition factors N_i/N_{i+1}.*

1991 *Mathematics Subject Classification.* Primary 20B40,68Q40; Secondary 20B15.

The first author was supported in part by NSF Grant CCR-9014562 and "OTKA" Grant (Hungary) #2581. The second author was supported in part by NSF Grant CCR-9013410. The third author was supported in part by NSF Grant CCR-9201303 and NSA Grant MDA904-92-H-3046.

This is the final version of this paper.

Our main result is

THEOREM 1.1. *Given s permutations generating a primitive group $G \leq S_n$, a composition series for G, together with permutation representations for the composition factors, can be found in $O^{\sim}(n^3 + sn^2)$ time.*

In the above, as in other stated results, we use the "soft" version ([**5**]) of the big-O notation: for two functions $f(n), g(n)$, we write $f(n) = O^{\sim}(g(n))$ (read "soft-oh") if, for some constants $c, C > 0$, $f(n) \leq Cg(n)\log^c n$. Note, in this notation, that we ignore the possible dependency of f and g on s. Indeed, the main impact of our present algorithms is in the (most typical) situation where s is small relative to n.

For possible implementations, it may cause concern that the exponent of $\log n$, hidden in the tilde notation, is rather large (greater than 20 in some lower order terms $n\log^c n$). However, for the present, our objective is to establish an asymptotic performance guarantee that is several orders of magnitude better than those previously known and we do not attempt to minimize the exponent of $\log n$. Nevertheless, we do point out that, in the "practical" range of n, say $n < 10^6$, one can eliminate the procedures that contribute these large exponents.

The first polynomial-time, $O(n^8 + sn^2)$, algorithm for finding a composition series was given by Luks [**20**]. Although Luks's goal was only to demonstrate that the problem is solvable in polynomial time, he also established the requisite structure, group-theoretic and algorithmic, upon which to build improvements. Thus, the overall architecture of our algorithm still follows Luks's original procedure, while the subroutines have been replaced by much faster ones, taking advantage of new, mostly combinatorial estimates.

The case of primitive groups is a crucial instance of the composition-series problem for general groups. The algorithm described herein will play a key role in a forthcoming paper where the following general result will be proved.

THEOREM 1.2. [**6**] *Given s permutations generating $G \leq S_n$, a composition series for G, together with permutation representations for the composition factors, can be found in $O^{\sim}(sn^3)$ time.*

Following [**20**], we remark that it is trivial to reduce the composition-series problem to the primitive case in polynomial time. However, the naive reduction will not yield the strong timing estimates stated in Theorem 1.2; indeed, it is highly nontrivial to perform even the rudimentary task of finding strong generators within the time bound stated. The latter accomplishment is the main result of [**6**] and the results of the present paper are used as a subroutine in the SGS construction given there. By the time the SGS has been constructed, a composition series is a byproduct.

En route to Theorem 1.1, the present paper also resolves the following subcase of Theorem 1.2, which involves "small-base" groups (see Section 2.2).

THEOREM 1.3. *Let c be fixed. Given s permutations generating $G \leq S_n$, where $\log|G| = O(\log^c n)$, a composition series for G, together with permutation representations for the composition factors, can be found in $O^\sim(n^3 + sn)$ time.*

We also point out that our algorithm for this problem requires only $O^\sim(sn)$ space.

While no deterministic algorithm is currently known to beat the asymptotic timing of Theorem 1.3, Beals and Seress [**8**] have recently constructed a randomized (Monte Carlo) algorithm for finding a composition series in small-base groups, running in *nearly linear,* $O^\sim(ns)$ time. That algorithm follows the outline of the present paper and uses novel randomization techniques to overcome the n^2 and n^3 bottlenecks (cf. Section 4).

To handle the primitive groups of large order, our algorithm is strongly guided by their structure theory (cf. Theorems 2.1, 2.7, and 2.8 below).

The timing analysis of our algorithm depends on the classification of finite simple groups via estimates of the orders of primitive groups (Theorems 2.1, 2.2) and Schreier's Conjecture (Theorem 2.3). We shall also make use of the Odd Order Theorem [**11**].

2. Notation, Definitions and Background

2.1. Group Theory. For basic definitions and facts about groups and permutation groups, we refer the reader to [**12**] and [**26**].

For $H \leq G$, H^G denotes the *normal closure* of H and $\mathrm{N}_G(H)$ is the *normalizer* of H. The *automorphism group* and *outer automorphism group* of G are denoted by $\mathrm{Aut}(G)$ and $\mathrm{Out}(G)$, respectively. The *socle* of G, $\mathrm{Soc}(G)$, is the subgroup of G generated by all minimal normal subgroups. If $G = T_1 \times \cdots \times T_r$ is the direct product of groups each isomorphic to a fixed group T, the *diagonal* $\mathrm{Diag}(T_1 \times \cdots \times T_r)$ of G is the subgroup comprised of $\{f_1(t)f_2(t)\cdots f_r(t) \mid t \in T\}$ where, for each i, $f_i : T \to T_i$ is a fixed isomorphism. For $H \leq G$, G/H denotes the set of right cosets of H in G.

The full symmetric group acting on the domain Ω is denoted by $\mathrm{Sym}(\Omega)$, while A_n and S_n denote the alternating and symmetric groups of degree n, respectively. We say that G *acts* on a set Ω if a homomorphism $G \to \mathrm{Sym}(\Omega)$ is given; the action is *faithful* if its kernel is the identity. If G acts on Ω and $\omega \in \Omega$, we denote by ω^G the *orbit* of ω under G, namely $\{\omega^g \mid g \in G\}$; we say G acts *transitively* on Ω if Ω consists of one orbit. If G acts transitively on Ω, a nonempty subset $\Delta \subseteq \Omega$ is called a *block of imprimitivity* for G if, for all $g \in G$, $\Delta^g = \Delta$ or $\Delta^g \cap \Delta = \emptyset$; the singletons and Ω itself are the *trivial* blocks; G is said to act *primitively*, and is said to be *primitive* in the case $G \leq \mathrm{Sym}(\Omega)$, if it has no nontrivial blocks of imprimitivity. A G-invariant partition $\mathcal{B}$ of Ω is *minimal* if G acts primitively on $\mathcal{B}$. For $B \subseteq \Omega$, G_B denotes the *pointwise stabilizer* of B in G. A transitive group $G \leq \mathrm{Sym}(\Omega)$ is called *regular* if $G_\omega = 1$ for any (all) $\omega \in \Omega$.

When speaking of subgroups of S_n, we call S_n and A_n the *giants*. They are the two largest subgroups and, by far, the two largest primitive subgroups of S_n.

For the analysis of our algorithm, we need several consequences of the classification of finite simple groups. The first one originated with Cameron [**9**] who, based on results of Kantor [**13**], classified the primitive groups of degree n and order $> n^{c \log \log n}$. We state a simplified version due to Liebeck [**17**], suitable for algorithmic purposes. (Unless otherwise indicated, logarithms in this paper are to base 2.)

THEOREM 2.1. [**17**] *Let $G \leq \mathrm{Sym}(\Omega)$ be primitive and let $n = |\Omega|$. Then either $|G| \leq n^{9 \log n}$ or the socle of G is the product of isomorphic alternating groups.*

We also need an estimate for the order of non-giant primitive groups $G \leq S_n$. The following can be deduced from results of Cooperstein [**10**] and Liebeck, Saxl [**19**].

THEOREM 2.2. *Let $G \leq \mathrm{Sym}(\Omega)$ be a non-giant primitive group, and let $n = |\Omega|$. If $n \notin \{8, 11, 12, 24\}$ then $|G| \leq 5^{n/2}$. In the exceptional cases, only $AGL(3,2)$ (for $n = 8$) and the Mathieu groups (for $n = 11, 12, 24$) violate this bound. In any case, $|G| \leq 7(5^{n/2})$.*

Remark. Elementary estimates $|G| < 4^n$ and $|G| \leq n^{4\sqrt{n} \log n}$ have been obtained by Praeger, Saxl [**22**] and Babai [**2, 3**], respectively. While Babai's bound is asymptotically sharper than $7(5^{n/2})$, the latter is better, and more convenient, for small n.

We cite the following additional consequence of the classification, traditionally termed Schreier's "Conjecture".

THEOREM 2.3. *If G is simple then $\mathrm{Out}(G)$ is solvable. Moreover, if G has a permutation representation of degree n, then $|\mathrm{Out}(G)|$ has no prime factor $> \sqrt{n}$.*

2.2. Algorithmic Concepts and Results. It is assumed that permutation groups are input or output via a list of generators.

In his pioneering work [**24**],[**25**], Sims introduced the notions of a base and strong generating set as the fundamental data structures for computing with permutation groups. A *base* for a permutation group $G \leq \mathrm{Sym}(\Omega)$ of degree n is a subset $B = \{b_1, b_2, .., b_M\}$ of Ω such that $G_B = 1$. Viewing B as a linearly-ordered set, the *point-stabilizer chain* of G relative to B is the chain of subgroups

$$G = G^{(1)} \geq G^{(2)} \geq \ldots \geq G^{(M+1)} = 1,$$

where $G^{(i)} = G_{\{b_1,..,b_{i-1}\}}$. The base B is called *nonredundant* if there is strict inclusion $G^i > G^{i+1}$ for all $1 \le i \le M$. A *strong generating set* (SGS) for G relative to B is a set S of generators of G with the property that

$$\langle S \cap G^{(i)} \rangle = G^{(i)}, \text{ for } 1 \le i \le M+1.$$

With reference to some fixed $c > 0$, an (infinite) family $\mathcal{G}$ of groups is called a family of *small-base groups* if all members G of degree n admit bases of size $O(\log^c n)$. Equivalently, $\log |G| = O(\log^{c'} n)$ for a fixed constant c' and each $G \in \mathcal{G}$ of degree n.

Let $B = \{b_1, ..., b_M\}$ be a base of the group G and let $G = G^{(1)} \ge G^{(2)} \ge ... \ge G^{(M+1)} = 1$ be the corresponding point-stabilizer chain. Moreover, let R_i denote a *transversal* (complete set of right coset representatives) for $G^{(i+1)}$ in $G^{(i)}$, $1 \le i \le M$. Each $g \in G$ can be written uniquely in the form $g = r_M r_{M-1} \cdots r_2 r_1$, $r_i \in R_i$. The process of factoring g in this form is called *sifting* or *stripping*.

Knuth [**16**] has offered a version of the Schreier–Sims SGS construction that is particularly suitable for small-base groups. Although he did not explicitly analyze the timing in this fashion, it is straightforward to check the following.

THEOREM 2.4 ([**16**]). *Given $G = \langle S \rangle \le S_n$, $|S| = s$, the following can be computed in $O(n^2 \log^3 |G| + sn \log |G|)$ time.*

(i) *A base of size $\le \log |G|)$ and an SGS of size $\le \log^2 |G|$ relative to this base. The SGS allows for membership-testing in the group in time $O(n \log |G|)$.*
(ii) *Strong generators for the kernel of an action $G \to \mathrm{Sym}(\Delta)$ on a set Δ of size $O(n)$.*

Remark. The space requirement of Knuth's procedure can be expressed as $O(n^2 \log |G|)$. In [**4**, Lemma 2.2], Babai, Cooperman, Finkelstein, and Seress gave a version of Sims's Schreier vector data structure [**24**] for the storage of the (partial) transversals. This allows a time-space trade-off, adjusting the space demand to $O(n \log^2 |G|)$ while increasing the time for SGS construction to $O(n^2 \log^4 |G| + sn \log^2 |G|)$ and membership-testing to $O(n \log^2 |G|)$. Note, however, that for small-base groups this trade-off improves the space requirement by essentially an order of magnitude, from $O^\sim(n^2)$ to $O^\sim(n)$, while maintaining a time of the form $O^\sim(n^2 + sn)$ for SGS construction and $O^\sim(n)$ for membership-testing.

If a base and an SGS for a small-base group are already in hand, certain further elementary operations can be performed in nearly linear time. Standard methods, augmented by [**4**, Lemma 2.2], yield the following theorem. (Note: although [**4**] deals with probabilistic methods, the cited lemma is deterministic.)

THEOREM 2.5. *If a base B of size $O(\log |G|)$ and an SGS of size $O(\log^2 |G|)$ are known for $G \le \mathrm{Sym}(\Omega)$ then, given s' generators for $H \le G$, the following can be computed in time $O(n \log^5 |G| + s' log^3 |G|)$.*

(i) *An SGS of size* $O(\log^2 |G|)$ *for* H *(relative to the base* B*).*
(ii) *An SGS of size* $O(\log^2 |G|)$ *for the normal closure of* H*.*

Atkinson's algorithm [**1**] shows that blocks of imprimitivity can be found in $O(sn^2)$ time. While this bound suffices for Theorem 1.1, we observe that, for small-base groups, Beals [**7**] has the following improvement.

THEOREM 2.6 ([**7**]). *Given* $G = \langle S \rangle \leq S_n$, $|S| = s$, *it can be decided in time* $O(n \log^3 |G| + sn \log |G|)$ *whether or not* G *is primitive. If not, the procedure finds a nontrivial block of imprimitivity.*

Finally, we quote two results from [**5**] which enable us to handle the large primitive groups.

THEOREM 2.7 ([**5**]). *Given a primitive group* $G = \langle S \rangle \leq S_n$, $|S| = s$, *it can be recognized in* $O^{\sim}(n^3 + sn^2)$ *time whether or not* G *contains* A_n. *If it does, then, within the same time bound, a strong generating set of size* n *can also be constructed from* S.

Remark. The phrase "constructed from S" in the above signals the construction of a *straight-line program*, that is, a sequence of group elements starting with elements of S, in which each successive element is determined as a product or power of some predecessor(s). Thus, for example, having identified a group as A_n, we do not allow the luxury of writing down arbitrary even permutations as members. The reason for such restriction is that we must deal with groups that are alternating only on some orbit or some induced representation, but permutations must be specified globally.

THEOREM 2.8 (NATURAL ACTION [**5**]). *If the order of the primitive group* $G = \langle S \rangle \leq S_n$ *exceeds* $n^{9 \log n \log \log n}$ *and* $\mathrm{Soc}(G) \simeq (A_k)^r$ *then these properties can be recognized and the natural imprimitive action of* G *on a set of size* kr, *with the factors of the socle acting on blocks of size* k *as* A_k, *can be constructed in* $O^{\sim}(sn^2)$ *time.*

Remark. The discrepancy between the bounds $n^{9 \log n}$ in Theorem 2.1 and $n^{9 \log n \log \log n}$ in Theorem 2.8 is due to the fact that, for small values of n, our algorithm may not recognize the imprimitive action if $|G| \leq n^{9 \log n \log \log n}$. Asymptotically, the cutoff point can be pushed down to $n^{5 \log n \log \log n}$ [**5**].

3. Primitive Permutation Groups

In this section, we state the results about primitive groups that are necessary to prove the correctness and timing of the composition-series algorithm.

A central ingredient is the O'Nan-Scott theorem, which classifies primitive groups according to their socles [**23**],[**9**],[**18**]. The statement below represents a regrouping of the subcases to suit our purposes. Note that a primitive group may belong to more than one subcase.

THEOREM 3.1. *Let G be a primitive permutation group of degree n. Then at least one of the following holds:*

(i) *G is cyclic of prime order.*
(ii) *G has a proper normal subgroup of index $\leq 7\sqrt{n}$.*
(iii) *G has one or two regular normal subgroups.*
(iv) *$N = \mathrm{Soc}(G)$ is the unique minimal normal subgroup of G, and it admits a decomposition $N = N_1 \times \ldots \times N_m$, with $m!/2 > 7\sqrt{n}$ (and, consequently, $m \geq 1 + \log n/(2\log\log n)$), where the N_i are conjugate under the action of G. Moreover, G acts on $\{N_i\}_{1\leq i\leq m}$ as the full alternating group A_m. In addition, one of the following holds:*
   (a) *$N_\omega = (N_1)_\omega \times \ldots \times (N_m)_\omega$, $1 \neq (N_i)_\omega \neq N_i$.*
   (b) *$N = T_1 \times \cdots \times T_r$ for some $m|r$, where the T_i are isomorphic nonabelian simple groups and $N_\omega = \mathrm{Diag}(T_1 \times \ldots \times T_r)$.*
(v) *G is nonabelian simple.*

PROOF. Let G be a primitive group. If G has a regular normal subgroup then G falls in case (i) or (iii). Otherwise, G has a unique minimal normal subgroup, $\mathrm{Soc}(G) = T_1 \times ... \times T_r$ where the T_i are isomorphic nonabelian simple groups, and $r \leq \log_5 n$. If $r = 1$ then $T_1 \leq G \leq \mathrm{Aut}(T_1)$ and, by Theorem 2.3, G falls in case (ii) or (v).

If $r > 1$ then G acts, by conjugation, transitively on $\{T_1, ..., T_r\}$. Let M be the pointwise stabilizer of a minimal block system in this transitive action. Then $M \lhd G$ and $G/M \simeq L$ for some primitive group L, $L \leq S_m$ for some $m|r$. If L does not contain A_m then, by Theorem 2.2, $|L| \leq 7(5^{r/2}) \leq 7\sqrt{n}$ and G falls in case (ii). If $L = S_m$ then G has a normal subgroup of index 2. Finally, if $L = A_m$ then either $m!/2 \leq 7\sqrt{n}$ and G falls in case (ii) or $m!/2 > 7\sqrt{n}$ and G is in case (iv). Each N_i is the direct product of the T_j in a block of the minimal block system. The two subcases of case (iv) involve an elementary rearrangement (as in [**20**, pp. 96-97]) of cases in the O'Nan-Scott Theorem. □

Remark. To establish Theorem 1.1, it would suffice to work with the coarser bound n in case (ii) of Theorem 3.1, eliminating the need for Theorem 2.2. However, the $7\sqrt{n}$ bound eliminates one n^3 bottleneck in the algorithm.

DEFINITION. We say that primitive groups of the types described in Theorem 3.1 (iv)(a) and (iv)(b) are of *type (iv)a* and *type (iv)b* respectively.

In order to speed up the handling of primitive groups of types (iv)a and (iv)b, we must strengthen some of the results of [**20**]. We start by examining the connection between the conjugation action of G on $\{N_1, ..., N_m\}$ and the orbits of G_ω, for $\omega \in \Omega$.

We have $N = \mathrm{Soc}(G) = T_1 \times \ldots \times T_r$ where the T_i are isomorphic nonabelian simple groups and G acts (via conjugation) as A_m on a minimal block system in $\{T_j\}_j$ for some $m|r$; the blocks in $\{T_j\}_j$ correspond to the N_i in case (iv) of Theorem 3.1. Furthermore, since $G/N \simeq G_\omega/N_\omega$, G_ω also acts as A_m on

$\{N_1, \ldots, N_m\}$. Let $K \lhd G_\omega$ be the kernel of this action. Then $G_\omega/K \simeq A_m$, and conjugation by K fixes each N_i. Note that $N_\omega \le K$.

Since N is transitive, Ω can be identified with N/N_ω via $\omega^x \leftrightarrow N_\omega x$; considering the induced conjugacy-action of G_ω on N/N_ω, this identification is a G_ω-map. In particular, the orbits of G_ω correspond to the G_ω conjugacy classes in N/N_ω. Also, if G is of type (iv)a, we have

$$\Omega \simeq N/N_\omega \simeq N_1/(N_1)_\omega \times \cdots \times N_m/(N_m)_\omega. \tag{1}$$

In this situation, let $\psi \in \Omega$ and take $x \in N$ such that $\psi = \omega^x$. Then $x = x_1 \cdots x_m$ with $x_i \in N_i$. We call $(N_i)_\omega x_i$ the ith *coordinate* of ψ and call this coordinate *nontrivial* if $x_i \notin (N_i)_\omega$.

We also need the following general observation.

FACT 1. *Let $\bar{G} \le \mathrm{Sym}(\bar{\Omega})$ be primitive and $\bar{N}$ be a nontrivial, nonregular normal subgroup of $\bar{G}$. For $\bar{\omega} \in \bar{\Omega}$, the only point fixed by the stabilizer $\bar{N}_{\bar{\omega}}$ is $\bar{\omega}$.*

Indeed, the equivalence relation on $\bar{\Omega}$ defined by

$$\bar{\omega} \sim \bar{\psi} \text{ iff } \bar{N}_{\bar{\omega}} = \bar{N}_{\bar{\psi}}$$

is clearly $\bar{G}$-invariant. □

LEMMA 3.2. *Let $G \le \mathrm{Sym}(\Omega)$ be a primitive group of type (iv)(a). Then, for any $\omega \in \Omega$, the size of the smallest orbit of G_ω on $\Omega \setminus \{\omega\}$ is $< \log^3 n$. Furthermore, the elements of such a smallest orbit have precisely one nontrivial coordinate.*

PROOF. By (1), $n = |N_1 : (N_1)_\omega|^m$ and therefore $|N_1 : (N_1)_\omega| \le \log^2 n$, since $m \ge 1 + \log n/(2 \log\log n)$.

As above, let $K \lhd G_\omega$ be the kernel of the conjugation action of G_ω on $\{N_1, \ldots, N_m\}$. Let q be the size of the smallest K-orbit in $N_1/(N_1)_\omega$. We claim that if $\psi \in \Omega$ has k nontrivial coordinates, then the G_ω-orbit of ψ has at least $\binom{m}{k} q 2^{k-1}$ conjugates under G_ω. To show this, we first note that the $\binom{m}{k}$ term accounts for the action of $A_m \cong G_\omega/K$. We may now assume that the k nontrivial coordinates of ψ occur in $N_i/(N_i)_\omega$, $i = 1, \ldots, k$. The term q derives from the action of K on $N_1/(N_1)_\omega$. Finally, we see that there is an additional independent factor of 2 contributed by the conjugacy action of $(N_i)_\omega$ on $N_i/(N_i)_\omega$, for each i, $2 \le i \le k$; for this, observe, if a coset $(N_i)_\omega v$ were fixed by $(N_i)_\omega$ then $N_\omega v$ would be fixed by N_ω, contrary to Fact 1.

On the other hand, there is an element $\psi = \omega^u \in \Omega$ with one nontrivial coordinate, such that $N_\omega u$ has q conjugates under K, so that the G_ω-orbit of ψ has cardinality mq.

Hence coset representatives in the smallest G_ω-orbit have only one nontrivial coordinate and the size of the smallest G_ω-orbit is $mq \le (\log_5 n)|N_1 \colon (N_1)_\omega| < \log^3 n$. □

LEMMA 3.3. *Let $G \leq \mathrm{Sym}(\Omega)$ be a primitive group of type (iv)(a) and let $\omega \in \Omega$. Then there exist points $\psi, \chi, \phi \in \Omega$ such that*

(i) *ψ, χ belong to the smallest orbit of G_ω in $\Omega \setminus \{\omega\}$.*

(ii) *ϕ belongs to the smallest orbit of G_χ in $\Omega \setminus \{\chi\}$.*

(iii) *The subgroup $H = \langle G_{\omega\psi}, G_{\chi\phi} \rangle \neq G$, and the action of G on the cosets of any maximal subgroup containing H is either nonfaithful or of degree $\leq n/2$.*

PROOF. By Lemma 3.2, for any i, $1 \leq i \leq m$, there is an element in the smallest G_ω-orbit in $\Omega \setminus \{\omega\}$ that is nontrivial only in the ith coordinate. Let ψ, χ belong to that smallest orbit, with ψ nontrivial in the second coordinate and with χ nontrivial in the first coordinate. There exist $t \in N_1$, $u \in N_2$, such that $\omega^t = \chi$, $\omega^u = \psi$. Let $\phi = \chi^u = \psi^t$ so that ϕ belongs to the smallest orbit of G_χ $(= G_\omega^t)$ in $\Omega \setminus \{\chi\}$. Thus ψ, χ, ϕ satisfy (i) and (ii).

To show that ω, ψ, χ, ϕ satisfy (iii), we first show that $H = \langle G_{\omega\psi}, G_{\chi\phi} \rangle \leq \mathrm{N}_G(N_2) < G$. Suppose $x \in G_{\omega\psi}$. If $N_2^x = N_j$ then $u^x u^{-1} \in N_\omega \cap N_j N_2 = (N_j)_\omega (N_2)_\omega$. Since $u \notin (N_2)_\omega$, we must have $j = 2$. Thus, $G_{\omega\psi}$ normalizes N_2. Similarly for $G_{\chi\phi}$.

Let M be a maximal subgroup of G containing H. Suppose that the action of G on the right cosets of M is faithful. We show that $|G : M| \leq n/2$.

Since $t \in N_1$, $(N_i)_\omega = (N_i)_\chi$ for $i > 1$, but, by Fact 1, $N_\omega \neq N_\chi$; thus, $(N_1)_\omega \neq (N_1)_\chi$. Since u centralizes N_1, $(N_1)_\omega \leq G_{\omega\psi}$; similarly, $(N_1)_\chi \leq G_{\chi\phi}$. We have

$$(N_1)_\omega < \langle (N_1)_\omega, (N_1)_\chi \rangle \leq (N_1)_H \leq (N_1)_M$$

(the latter two being "point" stabilizers in the actions of N_1 on G/H and G/M, respectively). In particular, $|(N_1)_M| \geq 2|(N_1)_\omega|$. But, since G_M acts on $\{N_i\}_{1 \leq i \leq m}$ as does G, we know that $(N_i)_M$ is conjugate under G_M to $(N_1)_M$. (Of course, G_M is just M but it is more useful to emphasize the point-stabilizer interpretation.) Thus $|N_M| \geq (2|(N_1)_\omega|)^m$. Since N is transitive in the (primitive) action of G on G/M,

$$|G \colon M| = |N \colon N_M| \leq (|N_1 : (N_1)_\omega)|/2)^m = n 2^{-m}. \qquad \square$$

The next two lemmas give us the machinery to deal with primitive groups of type (iv)b.

Let $\Omega^{(2)}$ denote the set of unordered pairs of elements of Ω. Then G acts naturally on $\Omega^{(2)}$.

LEMMA 3.4. *Let $G \leq \mathrm{Sym}(\Omega)$ be a primitive group of type (iv)(b) and let $\omega \in \Omega$. Then there is a point $\psi \neq \omega \in \Omega$ such that*

(i) *ψ belongs to a G_ω-orbit of size $< \log^3 n$.*

(ii) *If $\mathcal{B}$ is a minimal block system in the G-orbit of the pair $\{\omega, \psi\} \in \Omega^{(2)}$ then the action of G on $\mathcal{B}$ is either nonfaithful or $|\mathcal{B}| \leq n/2$.*

PROOF. We have $N = \mathrm{Soc}(G) = T_1 \times \ldots \times T_r$, $n = |T_1|^{r-1}$, and $|N_1| = |T_1^{r/m}|$. Therefore $n = |N_1|^{m(1-1/r)}$. On the other hand, $r \geq m \geq 1 + \log n/(2 \log\log n)$, hence $|T_1| \leq |N_1| < \log^2 n$. Also, since $r - 1 < \log_{|T_1|} n$, we have $r < \log n$.

Let $t \in T_1$ be an involution (guaranteed to exist by the Odd Order Theorem[**11**]) and set $\psi = \omega^t$. To see that (i) holds, note that any $t \in T_1$ has at most $|T_1|r < \log^3 n$ conjugates under G_ω so that the G_ω-orbit of $N_\omega t$ in N/N_ω is also bounded by $\log^3 n$. The result follows now by the G-set identification $N/N_\omega \simeq \Omega$.

Suppose that the action of G on $\mathcal{B}$, as in (ii), is faithful. We show that $|\mathcal{B}| \leq n/2$.

Let b denote the block in $\mathcal{B}$ containing $\{\omega, \psi\}$. Since t fixes $\{\omega, \psi\}$, it fixes b in the action on $\mathcal{B}$. It follows from $(T_1)_b \neq 1$ that $|\mathcal{B}| = |T_1 \colon (T_1)_b|^r$ (e.g., by consideration of the cases of the O'Nan-Scott Theorem [**9**]). Since $\sqrt{n} < r!/2$, we have $|T_1| = n^{1/(r-1)} < (r!/2)^{2/(r-1)}$ and the latter is less than 2^{r-1} for all $r \geq 2$. Therefore, $|\mathcal{B}| = |T_1 \colon (T_1)_b|^r \leq |T_1|^r/2^r = n|T_1|/2^r < n2^{r-1}/2^r$. □

In our application of Lemma 3.4, we need to locate such a ψ without having to search through many points. For this, we have

LEMMA 3.5. *Let $G \leq \mathrm{Sym}(\Omega)$ be a primitive group of type (iv)(b) and let $\omega \in \Omega$. The union of all G_ω-orbits that have $< \log^3 n$ points has cardinality $O^\sim(1)$.*

PROOF. We consider points in Ω as equivalence classes of N, each class of size $|N_\omega| = |\mathrm{Diag}(T_1 \times \ldots \times T_r| = |T_1| \leq \log^2 n$. It suffices then to show that there are only $O^\sim(1)$ elements of N with fewer than $\log^5 n$ conjugates under G_ω, for if an an element has at least $\log^5 n$ conjugates, then the number of equivalence classes among these conjugates is at least $\log^5 n/|N_\omega| \geq \log^3 n$.

Let k^* be the smallest integer such that $\binom{m}{k^*} > \log^5 n$. (For large n, we have $k^* = 6$). Let $u \in N$ be arbitrary and write u in the form $u = u_1 u_2 \cdots u_m$, $u_i \in N_i$. We note that the diagonal subgroup $N_\omega = \mathrm{Diag}(T_1 \times \cdots \times T_r)$ defines a G_ω-invariant identification (specific isomorphism) between the T_i; this in turn determines an identification between the N_i. So we can compare $u_1, u_2, \ldots$. If a particular element of N_1 occurs exactly ℓ times among the u_i where $k^* \leq \ell \leq m - k^*$ then u has at least $\binom{m}{\ell} > \log^5 n$ conjugates, due to the A_m action. If there are at least $k^* + 2$ different values among the u_i then u has at least $m^{k^*} > \log^5 n$ conjugates. Finally, we observe that there are $< \binom{m}{k^*}^{k^*+1} |N_1|^{k^*+1} = O^\sim(1)$ elements $u \in N$ which belong to neither of the above cases. □

We shall also need the following simple lemma.

LEMMA 3.6. *Suppose that the primitive group $G \leq \mathrm{Sym}(\Omega)$ has a regular normal subgroup N, $\omega, \psi \in \Omega$, and let M be a maximal subgroup containing $G_{\omega\psi}$. Then $|G : M| \leq |\Omega|$.*

PROOF. Let $|\Omega| = n$. We know that $|G : G_{\omega\psi}| \leq n(n-1)$ and $G_{\omega\psi} \cap N = 1$. We distinguish two cases:

(i) M contains N. In this case, $|M| \geq |G_{\omega\psi}||N|$, so $|G : M| \leq n - 1$.

(ii) M does not contain N. In this case, since M is maximal, $MN = G$. Hence $|G| = |MN| \leq |M||N|$ and $|G : M| \leq n$. □

4. The Algorithm

We reduce Theorem 1.1 to Theorem 1.3 as follows. Given a primitive G, then by Theorem 2.8, in $O^{\sim}(sn^2)$ time it is possible to detect whether or not $|G| \geq n^{9 \log n \log \log n}$ and by constructing the imprimitive natural action of G, to reduce the composition-series problem to the giant and the small-base case. However, for giant G, we can, by Theorem 2.7, identify G/A_n and construct an SGS for A_n.

Thus, it remains to prove Theorem 1.3.

The core of the algorithm used for the problem in Theorem 1.3 is a procedure that solves the following problem and is especially efficient for small-base groups.

PROBLEM. GENERALIZED PROPER NORMAL SUBGROUP (GPNS)
INPUT: $G = \langle S \rangle \leq \mathrm{Sym}(\Omega)$, $|\Omega| = n$, $|S| = s$.
OUTPUT: *One of the following.*

(1) *The report "G is simple".*
(2) *Generators for a proper normal subgroup of G.*
(3) *A faithful action of G on a domain of size at most n/2.*

Our procedure for GPNS follows the outline of the algorithm given in [**20**, Section 5]; the novel parts are the handling of primitive groups with a regular normal subgroup (see Step 4) and the improvement in the handling of primitive groups with a unique nonabelian normal subgroup (see Steps 5 and 6).

In most cases, outputs of types (2) and (3) are discovered in induced actions of G. Each constructed action gives rise to a primitive action in which we test the kernel and the size of the domain. The procedure TEST_ACTION formalizes these steps. The input, Δ, is a set of size > 1, on which G acts transitively.

procedure TEST_ACTION(Δ)
  begin
    $\mathcal{B}$:= a minimal G-block system on Δ;
    N:= the kernel of the G-action on $\mathcal{B}$;
    if $N \neq 1$ **then** (**output** N; **halt**);
    if $|\mathcal{B}| \leq |\Delta|/2$ **then** (**output** $\{G \to \mathrm{Sym}(\mathcal{B})\}$; **halt**)
  end.

TEST_ACTION is called at several points in procedure for GPNS. If it does not detect a proper normal subgroup (output of type (2)) or a smaller domain (output of type (3)) then control is passed back to the calling procedure.

The algorithm for the GPNS problem:

procedure GPNS

Step 0. Construct a nonredundant base and an SGS for G

Step 1. $\Psi :=$ any nontrivial orbit of G;
TEST_ACTION(Ψ).

Step 2. **if** $|G| = |\Psi|$ **then** (**output** "G is simple of prime order"; **halt**).

Step 3. $A :=$ any subset of G of size $7\sqrt{n}+1$;
for each pair $\{a, b\} \in A$ **do**
$N := \langle ab^{-1}\rangle^G$;
if $N \neq G$ **then** (**output** N; **halt**).

Step 4. $\omega, \psi :=$ any two points of Ψ;
$H := \mathrm{N}_G(G_{\omega\psi})$;
$M :=$ a maximal subgroup of G containing H;
if $|G : M| \leq n$ **then** TEST_ACTION(G/M).

Step 5. $\omega :=$ any point in Ψ;
for all ψ, χ in smallest G_ω-orbit **do**
for all ϕ in smallest G_χ-orbit **do**
$H := \langle G_{\omega\psi}, G_{\chi\phi}\rangle$;
if $H \neq G$ **then** TEST_ACTION(G/H).

Step 6. $\omega :=$ any point in Ψ;
for all ψ in G_ω-orbits of size $< \log^3 n$ **do**
$\Gamma_\psi :=$ the G-orbit of $\{\omega, \psi\}$ in set of unordered pairs;
TEST_ACTION(Γ_ψ).

Step 7. **output** "G is nonabelian simple".

LEMMA 4.1. *The output of the procedure GPNS is correct.*

PROOF. It is clear that the output is correct if the algorithm encounters a proper normal subgroup or a smaller domain. We have to prove that G is simple if the output says so.

If Step 1 is passed then G acts primitively on Ψ. A primitive regular group is cyclic of prime order, so the output of Step 2 is correct. Suppose that the algorithm passed Step 2; it is sufficient to show that, if it reaches Step 7, then G is simple.

If G falls in case (ii) of Theorem 3.1 then the algorithm halts at Step 3.

We claim, if G falls in case (iii), that the algorithm halts at Step 4. To see this, let N be a regular normal subgroup of G. By Lemma 3.6, $|G : M| \leq n$, so TEST_ACTION will be called. The unique $x \in N$ carrying ω to ψ centralizes $G_{\omega\psi}$, so $x \in M$. This means that N cannot act regularly on the cosets of M; so, if the action on the cosets is faithful then the number of cosets is $\leq |N|/2 = |\Psi|/2$

(Since N is normal in G, it acts transitively in any primitive action of G, whence the size of the domain is the index of a point-stabilizer in N.)

If G falls in case (iv)(a) of Theorem 3.1 then Lemma 3.3 implies that the algorithm halts at Step 5.

If G falls in case (iv)(b) then, by Lemma 3.4, the algorithm halts at Step 6.

So, if Step 7 is reached then G is simple. □

As in Theorems 2.4 and 2.5, we express the timing of GPNS as a function of n and $|G|$. At this point, however, it is convenient to introduce an extension of our "soft O" notation. In considering numerical functions on the pairs (G, n), where $G \leq S_n$, we write $f(G,n) = O^{\approx}(g(G,n))$ if, for some constants $c, c', C > 0$, $f(G,n) \leq Cg(G,n)\log^c n \log^{c'} |G|$. If we were dealing exclusively with small-base groups $G \leq S_n$, the $\log^{c'} |G|$ term would be superfluous. However, we ultimately need to use the GPNS procedure in induced representations of a small-base group. With respect to these, possibly smaller, domains the image of G need not be a small-base group.

LEMMA 4.2. *The procedure GPNS runs in* $O^{\approx}(n^3 + sn)$ *time.*

PROOF. First, we point out that a call of TEST_ACTION takes $O^{\approx}(n^2)$ time on a set of size $O^{\approx}(n)$ since, by Theorem 2.6, a minimal block system $\mathcal{B}$ can be computed in $O^{\approx}(n)$ time and, by Theorem 2.4, the kernel of the action on $\mathcal{B}$ can be obtained in $O^{\approx}(n^2)$.

By Theorem 2.4, Step 0 runs in $O^{\approx}(n^2 + sn)$ and, as observed above, Step 1 runs in $O^{\approx}(n^2)$. Theorem 2.5 shows that Step 3 takes $O^{\approx}(n^2)$ time.

Step 4 contains the n^3 bottleneck of the algorithm. One can choose ω, ψ to be the first two points of the base returned in Step 0. Let $\{u_i\}$ and $\{v_j\}$ denote the set of coset representatives for G_ω in G and for $G_{\omega\psi}$ in G_ω, respectively and let $S_{\omega\psi}$ be the set of strong generators for $G_{\omega\psi}$. Then $|S_{\omega\psi}| = O^{\approx}(1)$. The subgroup $H = \mathrm{N}_G(G_{\omega\psi})$ can be computed in $O^{\approx}(n^2)$ time since for each of the $< n^2$ coset representatives $v_j u_i$ for $G_{\omega\psi}$ in G, it is enough to trace the image of ω and ψ at the conjugates $y^{v_j u_i}$, for $y \in S_{\omega\psi}$. Computing M can be done by setting $M := H$ and then, for all coset representatives of $G_{\omega\psi}$ in G, computing $M^* = \langle M, v_j u_i \rangle$. If $M^* = G$ then discard $v_j u_i$; otherwise, set $M := M^*$. By Theorem 2.5, one computation of M^* takes $O^{\approx}(n)$ time. Eventually, a maximal subgroup M is obtained in $O^{\approx}(n^3)$ time. The action of each of the $O^{\approx}(1)$ (strong) generators y of G on the cosets of M can be constructed in $O^{\approx}(n^2)$ time since for each coset representative z of M in G, the coset of zy can be found in $O^{\approx}(n)$ time, by testing which of the ratios $zy\bar{z}^{-1}$, $\bar{z} \in G/M$, are in M. Testing one $\bar{z}$ amounts to a sifting through an SGS of M. This can be done in $O^{\approx}(1)$ time, since it is enough to follow the images of the (common) base of G and M.

Step 5 runs in $O^{\approx}(n^2)$ time since, by Lemma 3.2, we have to test only $O^{\approx}(1)$ triples ψ, χ, ϕ. By Theorem 2.5, H can be computed in $O^{\approx}(n)$ time. Since $H = \langle G_{\omega\psi}, G_{\chi\phi} \rangle$ contains $G_{\omega\psi}$, the action of G on the cosets of H is the action of G on a block system of cosets of $G_{\omega\psi}$. The action of G on the cosets of $G_{\omega\psi}$ is the action of G on $(\omega, \psi)^G$, and can be found in $O^{\approx}(n)$ time since the index of $G_{\omega\psi}$ in G is $O^{\approx}(n)$. The block system corresponding to H can be found in $O^{\approx}(n)$ time since we can compute generators for H. Finally, calling TEST_ACTION on G/H costs $O^{\approx}(n^2)$.

Step 6 also runs in $O^{\approx}(n^2)$ time since there are only $O^{\approx}(1)$ choices for ψ (by Lemma 3.5), for each ψ, the images of the unordered pair $\{\omega, \psi\}$ can be found as a block system of cosets of $G_{\omega\psi}$, and, as above, this block system is obtained in $O^{\approx}(n)$. Calling TEST_ACTION on Γ_ψ costs $O^{\approx}(n^2)$. □

As indicated, we apply the above algorithm to $\phi(G)$ where $\phi : G \to S_m$ is some induced representation of G and always $m = O^{\sim}(n)$. Although not essential for the asymptotic result, it seems most reasonable to carry out such procedure calls "locally", e.g. elementary operations, such as permutation multiplications, are performed only in the m element domain. To facilitate the "lifting" of the answer back to $G \le S_n$, in an initial construction of the base and SGS for $\phi(G)$, preimages in G of the SGS are kept; for small-base G, doing this via the method of Theorem 2.4 costs $O^{\sim}(nm+sn)$ time. The lifting of elements (e.g., generators of targeted subgroup) is done at a per-element cost of $O^{\sim}(n)$ by sifting through the base of $\phi(G)$, copying the computation in the lifting of the SGS. The complete lifting of a subgroup also includes the kernel of the action (Theorem 2.4(ii)).

We complete, finally, the

PROOF OF THEOREM 1.3. It suffices to indicate how one obtains a maximal normal subgroup M together with a permutation representation for G/M, for this may be repeated $O^{\sim}(1)$ times to obtain a composition series of G. We outline the construction of M, which follows that in [**20**].

To find a maximal normal subgroup M, we apply the procedure for GPNS to G. If the output of GPNS is of type (3), we rerun GPNS with the induced action. Thus, for nonsimple groups, at most $\log n$ applications of GPNS will produce a proper normal subgroup N (output of type (2)). In such case, we can construct a nontrivial action $\phi : G \to \mathrm{Sym}(\Psi)$ with $N \le \ker(\phi)$ and $|\Psi| \le |\Omega|$, namely, take $\Psi = G/G^{(i)}N$, where i is minimal such that $G > G^{(i)}N$ (by Theorems 2.4 and 2.5), this action is constructible in $O^{\sim}(n^2)$ time). Again, we may assume the output is of type (1) or (2). If $\phi(G)$ is simple then $M = \ker(\phi)$ is maximal normal in G and ϕ induces a faithful representation of G/M. Otherwise, if $\bar{N}$ is a proper normal subgroup of $\phi(G)$, we have $N < \phi^{-1}(\bar{N}) < G$. Replace N by $\phi^{-1}(\bar{N})$ and repeat, i.e., construct an action whose kernel includes N, etc. □

Acknowledgment. The authors wish to thank the referees for their effort, their interest, and their suggestions.

References

1. M.D. Atkinson, *An algorithm for finding the blocks of a permutation group*, Math. Comp. **29** (1975), pp. 911–913.
2. L. Babai, *On the order of uniprimitive permutation groups*, Ann. of Math. **113** (1981), pp. 553–568.
3. L. Babai, *On the order of doubly transitive permutation groups,* Inventiones Math. **65** (1982), pp. 473-484.
4. L. Babai, G. Cooperman, L. Finkelstein, and Á. Seress, *Nearly linear time algorithms for permutation groups with a small base*, Proc. ISSAC'91 (Internat. Symp. on Symbolic and Algebraic Computation), Bonn 1991, pp. 200–209.
5. L. Babai, E.M. Luks, and Á. Seress, *Fast management of permutation groups*, Proc. 29th IEEE FOCS (1988), pp. 272–282.
6. L. Babai, E.M. Luks, and Á. Seress, *Fast deterministic management of permutation groups*, in preparation.
7. R. Beals, *Constructing blocks of imprimitivity in nearly linear time for small-base groups*, in these Proceedings.
8. R. Beals and Á. Seress, *Structure forest and composition factors for small base groups in nearly linear time*, Proc. 24th ACM STOC (1992), pp. 116–125.
9. P. J. Cameron, *Finite permutation groups and finite simple groups*, Bull. London Math. Soc. **13** (1981), pp. 1–22.
10. B.N. Cooperstein, *Minimal degree for a permutation representation of a classical group*, Israel J. Math. **30** (1978), pp. 213–235.
11. W. Feit and J.G. Thompson, *Solvability of groups of odd order*, Pacific J. Math. **13** (1963), pp. 775–1029.
12. M. Hall, Jr., The Theory of Groups, Macmillan, New York 1959.
13. W.M. Kantor, *Permutation representations of the finite classical groups of small degree or rank*, J. Algebra **60** (1979), pp. 158–168.
14. W.M. Kantor, *Sylow's theorem in polynomial time*, J. Comp. and Syst. Sci. **30** (1985), pp. 359–394.
15. W.M. Kantor and E.M. Luks, *Computing in quotient groups*, Proc. 22nd ACM STOC, 1990, pp. 524–563.
16. D.E. Knuth, *Notes on efficient representation of perm groups*, Combinatorica **11** (1991), pp. 57–68 (preliminary version circulated since 1981).
17. M.W. Liebeck, *On minimal degrees and base sizes of primitive groups*, Arch. Math. **43** (1984), pp. 11–15.
18. M.W. Liebeck, C.E. Praeger, and J. Saxl, *On the O'Nan-Scott theorem for finite primitive permutation groups*, J. Australian Math. Soc. **44** (1988), pp. 389-396.
19. M.W. Liebeck and J. Saxl, *The orders of maximal subgroups of the finite exceptional groups of Lie type*, Proc. London Math. Soc. (3) **55** (1987), pp. 299–330.
20. E. M. Luks, *Computing the composition factors of a permutation group in polynomial time*, Combinatorica **7** (1987), pp. 87–99.
21. E. M. Luks, *Permutation groups and polynomial-time computation*, in these Proceedings.
22. C. E. Praeger and J. Saxl, *On the orders of primitive permutation groups*, Bull. London Math Soc. **12** (1980), pp. 303–307.
23. L. L. Scott, *Representations in characteristic p*, Proc Santa Cruz Conf. on Finite Groups, AMS (1980), pp. 319–322.
24. C.C. Sims, *Computational methods in the study of permutation groups*, in: Computational Problems in Abstract Algebra, J. Leech, ed., Pergamon Press 1970, pp. 169–183.
25. C.C. Sims, *Computation with Permutation Groups*, in Proc. Second Symposium on Symbolic and Algebraic Manipulation, S.R. Petrick, ed., ACM, New York, 1971, pp. 23–28.
26. H. Wielandt, Finite Permutation Groups , Acad. Press, New York 1964.

Department of Computer Science, University of Chicago, Chicago, IL 60637-1504, Department of Algebra, Eötvös University, Budapest, Hungary H-1088
E-mail address: laci@cs.uchicago.edu

Department of Computer and Information Science, University of Oregon, Eugene, OR 97403
E-mail address: luks@cs.uoregon.edu

Department of Mathematics, The Ohio State University, Columbus, OH 43210
E-mail address: akos@function.mps.ohio-state.edu

DIMACS Series in Discrete Mathematics
and Theoretical Computer Science
Volume **11**, 1993

Computing blocks of imprimitivity for small-base groups in nearly linear time

ROBERT BEALS

ABSTRACT. A *base* of a permutation group G is a subset B of the permutation domain such that only the identity of G fixes B pointwise. The permutation representations of important classes of groups, including all finite simple groups other than the alternating groups, admit $O(\log n)$ size bases, where n is the size of the permutation domain. Groups with very small bases occur frequently in work on permutation groups within computational group theory.

We address the problem of finding blocks of imprimitivity and a structure forest for permutation groups with small (polylogarithmic size) bases. Previous methods of Atkinson work for general permutation groups in $O(n^2)$ time. We give an $O(n \log^3 |G|)$ algorithm, which is $O(n \log^c n)$ in the small-base case.

In another paper, with Á. Seress, we use this algorithm to find composition factors for small-base groups in $O(n \log^c n)$ time [**6**].

1. Introduction

Let $\Omega = \{1, \ldots, n\}$ and let $\mathrm{Sym}(\Omega)$ denote the symmetric group acting on Ω, i.e. the group of all permutations of Ω. A *structure forest* of the permutation group $G \leq \mathrm{Sym}(\Omega)$ admits G as automorphisms and describes the imprimitivity structure in an algorithmically useful way. Its nodes correspond to blocks of imprimitivity, and the block corresponding to a node is maximal in the block corresponding to its parent. (The structure forest is usually not uniquely associated with G.) A structure forest can serve to guide divide-and-conquer algorithms for permutation groups (see e.g. [**7**]). The results of this paper have been used by Beals and Seress [**6**] as a crucial step in finding composition factors in nearly linear time for small-base groups. To obtain a structure forest for G, we need to

1991 *Mathematics Subject Classification.* Primary 20B40; Secondary 68Q20.
Supported by NSF Grant CCR 9014562.
This paper is in final form and no version of it will be submitted for publication elsewhere.

find (independently) a structure tree (see below) for each transitive constituent of G. Therefore we may henceforth *assume that G is transitive.*

For $i \in \Omega$, we denote by G_i the stabilizer of i. If $\Delta \subseteq \Omega$ we denote by G_Δ the pointwise stabilizer of Δ and by $G_{\{\Delta\}}$ the *setwise* stabilizer of Δ.

In [**9**], Sims introduced the notion of a *base*. A base for G is a subset $B = \{x_1, x_2, .., x_M\}$ of Ω with the property $G_B = 1$. If B is a non-redundant base, i.e. no proper subset of B is a base, then the following relation is immediate:

$$2^M \leq |G| \leq n^M.$$

Having fixed a constant $c > 0$, we call G a *small-base group* if G has a base of size $M \leq (\log n)^c$. By the inequality above, this is equivalent to the condition that $\log |G| \leq \log^{c'} n$ for some constant c'. (We refer to infinite families of groups; the constants should not depend on how large a member of the family is chosen.)

$\Delta \subseteq \Omega$ is a *block of imprimitivity* if $\forall \sigma \in G$ either $\Delta^\sigma = \Delta$ or $\Delta^\sigma \bigcap \Delta = \emptyset$. G is *primitive* if the only blocks of imprimitivity are Ω and the singletons.

If $\Delta \subseteq \Omega$ is a block of imprimitivity, then its images under the group action partition Ω; this partition is called a *block system.* We may think of a block system as corresponding to a depth two tree on which G acts as automorphisms: the leaves of the tree are the points of Ω, each leaf has as its parent an interior node corresponding to the block in which the leaf lies, and the interior nodes have the root as their parent.

DEFINITION 1.1. *A structure tree of the transitive group $G \leq \mathrm{Sym}(\Omega)$ is a rooted tree T with Ω as leaves admitting G as automorphisms such that for each node $v \in T$, the stabilizer G_v acts primitively on the children of v. Each level of T corresponds to a block system which is minimally coarser (as a partition) than the block system one level below.*

THEOREM 1.1. *Suppose the transitive group $G \leq \mathrm{Sym}(\Omega)$ is given by a generating set S. Let $s = |S|$. One can find a structure tree for G in time $O(n \log |G|(s + \log^2 |G|))$.*

This is nearly linear time ($O(ns \log^c n)$) if G is a small-base group (ns is the length of the input). If in addition we are given the orbits of G_1 together with a complete set of coset representatives of G_1 in G written as short words, we can find a structure tree faster:

THEOREM 1.2. *Assume orbits of G_1 are given along with coset representatives for G_1 in G written as words of length ℓ. Then the total time to calculate a structure tree is $O(n \log n(\ell + \log n))$.*

This would be true, for example, if we had previously run the nearly linear time Monte Carlo algorithm of Babai, Cooperman, Finkelstein, and Seress [**3**] for constructing strong generating sets for small-base groups. Note that $\ell = O(\log |G|)$ can be achieved by the "double the cube" method of [**3**]. We sketch this method in Section 3.

A previous algorithm due to Atkinson [**2**] works for all permutation groups in time $O(n^2)$, and can actually require quadratic time even for groups of base size 2. Atkinson's algorithm is based on the following principle: for $i \in \Omega$, the smallest block of imprimitivity containing 1 and i is the connected component of 1 in the graph on Ω with edge set $\{1, i\}^G$. The algorithm works by calculating this block, which for some groups is frequently all of Ω, successively for $i = 2, 3, \dots, n$. We achieve savings over this by not calculating the whole block at once (which may take $\Omega(n)$ time), but only calculating a small part of it before moving on to a new i. By doing this carefully, we avoid spending too long on i for which the block is too big. The effectiveness of our data structure is partly due to the use of the linear time algorithm of R. E. Tarjan [**10**] to find the strongly connected components of a directed graph [**1**].

In Section 3 we present a deterministic algorithm for the special case that the orbits of G_1 are known. In Section 5 we show how to extend the algorithm of Section 3 (deterministically) to the general case, i.e. when only generators for G are known.

The results of this paper have been previously announced in [**6**].

2. Notation

For $X \subseteq \Omega$, let X^* denote $X \cup \{1\}$, and let $\overline{X}$ denote the smallest block of imprimitivity of G containing X^*. For $i \in \Omega$, let $\bar{i}$ denote $\overline{\{i\}}$. We introduce the following equivalence relation on Ω: for $i, j \in \Omega$, we write $i \sim j$ to mean that $\bar{i} = \bar{j}$. We also introduce a transitive relation on subsets of Ω: for $X, Y \subseteq \Omega$, we write $X \preceq Y$ to mean $\overline{X} \subseteq \overline{Y}$. Observe that $\preceq$ restricted to equivalence classes of $\sim$ is a partial order. The following are immediate:

PROPOSITION 2.1. *For all* $X \subseteq \Omega$, $X \subseteq \overline{X} = \overline{\overline{X}}$. □

PROPOSITION 2.2. *Let X be an equivalence class of $\sim$, and Δ be a block of imprimitivity containing 1. If $X \cap \Delta \neq \emptyset$, then $\overline{X} \subseteq \Delta$.* □

The algorithm maintains an equivalence relation P on $\Omega \setminus \{1\}$ with the invariant property that for $i, j \in \Omega \setminus \{1\}$, $i\, P\, j \Rightarrow i \sim j$. Let $\widetilde{P}$ denote the set of equivalence classes of P. We will maintain a partial order R on $\widetilde{P}$ with the invariant property that for $X, Y \in \widetilde{P}$, $X\, R\, Y \Rightarrow X \preceq Y$.

3. Algorithm for the case that the orbits of G_1 are known

The algorithm makes use of an arbitrary permutation $\pi \in \mathrm{Sym}(\Omega)$ such that the cycles of π are the orbits of G_1. Note that we do not require that π be an element of G, so it may be calculated in linear time from the orbits of G_1.

A *transversal* for $H \leq G$ is a complete set of coset representatives for H in G.

The first step of the algorithm is to calculate a transversal for G_1 in the compact form of a shallow Schreier tree, i.e. a rooted tree with edges labeled by elements of G such that each leaf corresponds to a coset, and the product of the elements along the path from the root to the leaf gives a representative of the corresponding coset [**9**].

One of the novelties of [**3**] is the use of the "doubling the cube" technique (originally introduced by Babai and Szemerédi [**5**] in another context) to find *shallow* Schreier trees:

LEMMA 3.1. *Let $G \leq \mathrm{Sym}(\Omega)$ be given by a set S of generators. One can deterministically in time $O(sn\log|G| + n\log^2|G|)$ calculate $\ell \leq 2\log|G|$ group elements $\{g_1,\dots,g_\ell\}$, arranged in a Schreier tree of depth ℓ, where $n = |\Omega|$ and $s = |S|$.*

PROOF. We sketch the method of [**3**]: For $g_1,\dots,g_k \in G$, let

$$C_k = C(g_1,\dots,g_k)$$

denote the *cube*

$$\{g_1^{\epsilon_1}\cdot g_2^{\epsilon_2}\cdot\dots\cdot g_k^{\epsilon_k} \mid \epsilon_i \in \{0,1\}\}.$$

We say that C_k is *non-degenerate* if $|C_k| = 2^k$. Note that $|C_k| = 2|C_{k-1}|$ iff $g_k \notin {C_{k-1}}^{-1}C_{k-1}$.

We will construct a non-degenerate cube inductively as follows: Let $g_1 \in S$ be arbitrary, so C_1 is non-degenerate. Suppose that we have calculated $g_1,\dots,g_i$ so that C_i is non-degenerate. Let $X_i = 1^{{C_i}^{-1}C_i}$. If $X_i \neq \Omega$, then for some $x \in X_i, g \in S, x^g \notin X_i$. We can calculate $\omega_x \in G$ with $1^{\omega_x} = x$. Let $g_{i+1} = \omega_x g \notin {C_i}^{-1}C_i$, and repeat.

If $X_i = \Omega$, then the elements $g_1,\dots,g_i$ and their inverses can be arranged in a Schreier tree of depth at most $2i \leq 2\log|G|$ (since C_i is non-degenerate) and we are done. For each j, it takes time $O(nj)$ to calculate X_j, and time $O(sn)$ to calculate x and g. Calculating ω_x takes time $O(nj)$ since it is a product of length $\leq 2j$ of the $g_1,\dots,g_j$ and their inverses. So the time to calculate g_j is $O(sn + nj)$, and the total time to calculate $g_1,\dots,g_i$ is $O(sni + ni^2) = O(sn\log|G| + n\log^2|G|)$. □

Consequently we obtain n words $\{w_1,\dots w_n\}$ over $\{g_1,\dots,g_\ell\}$ such that for $i \in \Omega$, $|w_i| \leq \ell$ and $1^{w_i} = i$. Once these have been calculated, we can for any $i,j \in \Omega$ calculate i^{w_j} in time ℓ.

We use the subroutine GROW described below:

GROW(X)
INPUT: $X \in \widetilde{P}$.
OUTPUT: A set Y containing at least $\min(|X|, |\overline{X} \setminus X^*|)$ points of $\overline{X} \setminus X^*$.

First we need a lemma:

LEMMA 3.2. *For any $X \in \widetilde{P}$, for any $i \in X$, $G_{\{\overline{X}\}} = \langle G_1, w_i\rangle$.*

PROOF. Let K denote $\langle G_1, w_i\rangle$. Observe that $\overline{X} = \bar{i}$ since X is a subset of an equivalence class of $\sim$. Since $\bar{i}$ is a block of imprimitivity containing 1 and i, both G_1 and w_i stabilize it, so $K \leq G_{\{\bar{i}\}}$. Let $\Delta = 1^K$. Both K and $G_{\{\bar{i}\}}$ are unions of cosets of G_1, so to prove $K = G_{\{\bar{i}\}}$ it suffices to show that $\overline{X} \subseteq \Delta$. First observe that Δ is a block of imprimitivity: suppose $g \in G$ with $j \in \Delta^g \cap \Delta \neq \Delta$. By multiplying g by an element of K we may assume that $j = 1$. Then g belongs to a coset of G_1 which is contained in K, so $g \in K$ and $\Delta^g = \Delta$, a contradiction. Since Δ contains 1 and i, by Proposition 2.2 we have $\overline{X} \subseteq \Delta$. Therefore $G_{\{\overline{X}\}} = K$. □

OBSERVATION 3.3. *$\langle G_1, w_i\rangle$ and $\langle \pi, w_i\rangle$ have the same orbits.* □

Even though we do not know $\overline{X}$ explicitly, we know that $\overline{X} = 1^{\langle \pi, w_i\rangle}$. GROW uses an array which stores for all $j \in \Omega$ the last time j was located by GROW, and uses standard search techniques to find the orbit of 1 under the action of $\langle \pi, w_i\rangle$, but halts as soon as $\min(2|X| + 1, |\overline{X}|)$ points are found. The set Y of points found in $\overline{X} \setminus X^*$ is returned. Note that $|Y| \geq \min(|X|, |\overline{X} \setminus X^*|)$ as desired. Since ℓ steps are required to find the image of a point under w_i, GROW takes $O(|X|\ell)$ steps to run.

The equivalence relation P is initialized to correspond to the partition of $\Omega \setminus \{1\}$ into orbits of G_1. Since any block of imprimitivity containing 1 is a union of orbits of G_1, P is a refinement of $\sim$. The ordering R is initialized to be an antichain (no two elements are comparable). This is certainly weaker than $\preceq$. In the course of the algorithm, we add comparisons to R, always making sure that it remains weaker than $\preceq$. For instance, if $Y = \text{GROW}(X)$, then for any $Z \in \widetilde{P}$ such that $Z \cap Y \neq \emptyset$, we know that $Z \preceq X$, so we may add the relation $Z\,R\,X$ to R. This may create cycles, so we periodically need to run a procedure MERGE, which uses Tarjan's algorithm [**10, 1**], to find strongly connected components of R and merges them into single equivalence classes of $\widetilde{P}$. This maintains the invariant property that P is a refinement of $\sim$.

Below we present pseudocode for the main part of the algorithm:

BLOCK$(G, \widetilde{P})$
INPUT: $G \leq \text{Sym}(\Omega)$, $\widetilde{P}$ is the partition of $\Omega \setminus \{1\}$ into orbits of G_1.
OUTPUT: A minimal block of imprimitivity properly containing $\{1\}$.
/* *Note that G is primitive if and only if the output is Ω.* */
STEP 1: Calculate $\ell \leq \log|G|$ group elements $\{g_1, \ldots, g_\ell\}$ and n words $\{w_1, \ldots, w_n\}$ over $\{g_1, \ldots, g_\ell\}$ such that for $i \in \Omega$, $|w_i| \leq \ell$ and $1^{w_i} = i$.
/* *Use the "doubling the cube" technique of Lemma* 3.1.*/
STEP 2: Fix π, an arbitrary permutation such that the cycles of π are the orbits of G_1.
STEP 3: Initialize R to be an antichain.
STEP 4: Let $M = \{X \in \widetilde{P} \mid X \text{ is } R\text{-minimal}\}$.

STEP 5: FOR all $X \in M$ DO

(i) Let $Y = \text{GROW}(X)$.
(ii) IF $Y = \emptyset$ THEN HALT and OUTPUT X^*.
(iii) FOR all $Z \in \widetilde{P}$ such that $Z \cap Y \neq \emptyset$, DO add $Z\,R\,X$ to R.

STEP 6: MERGE.
STEP 7: GOTO STEP 4.

The following lemma establishes the correctness of the algorithm:

LEMMA 3.4. *For any* $X \in \widetilde{P}$, *if* $\text{GROW}(X) = \emptyset$ *then* X^* *is a minimal block of imprimitivity properly containing* $\{1\}$.

PROOF. Since $\text{GROW}(X) = \emptyset$ it follows that $X^* = \overline{X}$. Since X is a subset of a $\sim$-equivalence class, we know by proposition 2.2 that no block of imprimitivity lies strictly between $\{1\}$ and $\overline{X}$. □

If we wish to use the algorithm to find a structure tree, then once we find a minimal block X, we remove X from $\widetilde{P}$ (i.e. make P an equivalence relation on $\Omega \setminus \overline{X}$). Rewrite the generating set S and the permutations $g_1, \ldots, g_\ell$ as permutations acting on the set Ω' of images of $\overline{X}$. Redefine π to be an arbitrary permutation of Ω' such that the cycles of π are the orbits of the action of $G_{\{\overline{X}\}}$ on Ω'. If an orbit of $G_{\{\overline{X}\}}$ or an image of $\overline{X}$ intersects two classes of $\widetilde{P}$, then merge them into a single class. Replace Ω by Ω', and continue the algorithm starting at STEP 4. It will give us a minimal block of imprimitivity properly containing $\overline{X}$. Continuing in this way we get all the nodes of a structure tree.

4. Analysis of the algorithm

Standard techniques for UNION-FIND (cf. [**1**]) can be used to maintain $\widetilde{P}$ in total time less than $O(n \log\log n)$. Most of the time of our algorithm is spent in calls to MERGE and GROW.

LEMMA 4.1. *The total number of calls to* MERGE *is* $O(\log n)$, (*even if we use the algorithm to find the entire structure tree*).

PROOF. Let $\Delta_1 \subsetneq \Delta_2 \subsetneq \ldots \subsetneq \Delta_k = \Omega$ be the sequence of minimal blocks constructed. Since for all i, $|\Delta_i|$ divides Δ_{i+1}, we have $k \leq \log n$. After every call to MERGE until the last one before Δ_1 is output, the cardinality of R-minimal elements of $\widetilde{P}$ which are subsets of Δ_1 at least doubles (since they have all been GROWn). So at most $\lceil \log|\Delta_1| \rceil$ calls are made before Δ_1 is found. Once Δ_1 is found, Ω is replaced by a set of size $n/|\Delta_1|$, so at most $\lceil \log(|\Delta_1|/|\Delta_2|) \rceil$ calls are made before Δ_2 is found. Continuing in this way, we see that the total number of calls is at most $k + \log n \leq 2\log n$. □

LEMMA 4.2. *The total amount of time spent on* GROW *is* $O(n\ell \log n)$.

PROOF. Let $k \leq \log n$ be the depth of the structure tree obtained. Each element $i \in \Omega$ is in an R-minimal set at most $k + \log n$ times, by an argument similar to that in the proof of the previous lemma. Each call to GROW takes at most ℓ steps per element of the set being GROWn, so the total time is $O(n\ell \log n)$. □

THEOREM 4.3. *Assume orbits of G_1 are given along with a depth ℓ Schreier tree representing a transversal for G_1. Then the total time to calculate a structure tree (omitting* STEP 1) *is $O(n \log n(\ell + \log n))$.*

PROOF. By Lemma 4.1, the algorithm makes $O(\log n)$ calls to MERGE. Since at most n arcs are added to R between calls to MERGE, each call is made on an input of size at most $n \log n$. Since Tarjan's algorithm is linear time, the total time spent in calls to MERGE, is $O(n \log^2 n)$. By Lemma 4.2 the amount of time spent in calls to GROW is $O(n\ell \log n)$. Combining, we get the desired time bound. □

5. Algorithm for the general case

The general algorithm uses a group $H = \langle S_1 \rangle$ with the invariant property that $H \leq G_1$. Initially, $S_1 = \emptyset$, $H = 1$, and $\pi = 1$. We will run the BLOCK algorithm as if H and G_1 had the same orbits. However, whenever GROW is about to return fewer than $|X|$ points of $\overline{X} \setminus X^*$ (it has found all of the points of $\Delta = 1^{\langle H, w_i \rangle}$), it calls a procedure VERIFY-BLOCK. This procedure will either verify that Δ is a block, in which case the BLOCK algorithm will continue to search for a minimal block properly containing $\{1\}$, but the search will be restricted to Δ. Otherwise, VERIFY-BLOCK will find an element $g \in G_1 \setminus H$. In the latter case it replaces S_1 by $S_1 \cup \{g\}$, H by $\langle H, g \rangle$, and replaces π so that the cycles of the new π are the orbits of the new H. Then it returns to GROW, which can now continue finding points of $\overline{X}$. Note the timing analysis of Section 4 still applies to the time used by MERGE and GROW. The extra time we need to expend in the general case is spent in calls to VERIFY-BLOCK.

LEMMA 5.1. *In time $O(n(\ell + s))$ we can either verify that Δ is a block, or calculate $g_1, g_2 \in G$ such that $\emptyset \neq \Delta^{g_1} \cap \Delta^{g_2} \neq \Delta^{g_1}$.*

PROOF. To verify that Δ is a block, we successively take coset representatives $w_{i_1}, w_{i_2}, \ldots$ with the property that for all j:

$$i_j \notin \Delta \cup \bigcup_{k<j} \Delta^{w_{i_k}}$$

We stop if we find a j such that $\Delta^{w_{i_j}} \cap \Delta^{w_{i_k}} \neq \emptyset$ for some $k < j$, or when Ω is partitioned by the $\Delta^{w_{i_j}}$. This take time $O(n\ell)$. If we have partitioned Ω into images of Δ, we can in time sn verify that the partition is preserved by the generators of G (in which case Δ is a block) or find $g_1, g_2 \in G$ such that

$\emptyset \neq \Delta^{g_1} \cap \Delta^{g_2} \neq \Delta^{g_1}$. If we stopped before partitioning Ω into images of Δ, it is because we have already found such a g_1, g_2. $\square$

OBSERVATION 5.2. *We may assume* $g_2 = 1$.

PROOF. Replace g_1 by $g_1 {g_2}^{-1}$. $\square$

LEMMA 5.3. *Given* g_1 *such that* $\emptyset \neq \Delta^{g_1} \cap \Delta \neq \Delta$, *we may calculate an element* $g \in G_1 \setminus H$ *in time* $O(s_1 n \log |G| + n \log^2 |G|)$ *(where* $s_1 = |S_1|$*).*

PROOF. Let $K = \langle H, w_i \rangle$. Take $j, k \in \Delta$ such that $j^{g_1} = k$. Use the "cube doubling" of [**5, 3**] to find ${w_j}', {w_k}' \in K$ such that $1^{{w_j}'} = j$ and $1^{{w_k}'} = k$ in time $O(s_1 n \log |K| + n \log^2 |K|)$. Let $g = {w_j}' g_1 {w_k}'^{-1} \in G_1$. Since ${w_j}'$ and ${w_k}'$ stabilize Δ setwise, g does not. Therefore, $g \in G_1 \setminus H$. $\square$

REMARK. It is worth trying $g = w_j g_1 {w_k}^{-1}$ first. There is unfortunately no guarantee that g will not stabilize the set Δ, but if $\Delta^g \neq \Delta$, then there is no need to calculate ${w_j}'$ and ${w_k}'$.

Combining the above two Lemmas, and observing that $s_1 \leq \log |G|$, we get:

LEMMA 5.4. *The procedure* VERIFY-BLOCK *runs in time at most* $O(n(\ell + s + \log^2 |G|))$. $\square$

Below we give pseudocode for the VERIFY-BLOCK procedure:

VERIFY-BLOCK(Δ_0, i)
INPUT: $\Delta_0 \subseteq \Omega, i \in \Delta_0, \Delta_0 = 1^{\langle H, w_i \rangle}$.
OUTPUT: TRUE if Δ_0 is a block of imprimitivity for G, FALSE otherwise.
/ If* Δ_0 *is not a block, this procedure has the following side effects: some* $g \in G_1 \setminus H$ *is calculated and added to* S_1, π *is updated so that its cycles are the orbits of the new* $\langle S_1 \rangle$. **/*
STEP 1: Let $j = 1$.
STEP 2: WHILE $\cup_{k<j} \Delta_k \neq \Omega$ DO
  / First we try to partition* Ω *into images of* Δ_0. **/*
  (i) Select $i_j \in \Omega \setminus (\cup_{k<j} \Delta_k)$; let $\Delta_j = \Delta^{w_{i_j}}$.
  (ii) IF $\Delta_j \cap \Delta_k \neq \emptyset$ for some $k < j$ THEN let $g_1 = w_{i_j} {w_{i_k}}^{-1}$; BREAK and GOTO STEP 5.
  (iii) Let $j = j + 1$.

STEP 3: For all $\sigma \in S, k < j$ DO
  / We verify that the partition is G-invariant. */*
  (i) Calculate ${\Delta_k}^{\sigma}$
  (ii) IF for some $k' \neq k$ we have $\emptyset \neq {\Delta_k}^{\sigma} \cap \Delta_{k'} \neq \Delta_{k'}$ THEN let $g_1 = w_{i_k} \sigma w_{i_{k'}}$; BREAK and GOTO STEP 5.

STEP 4: HALT and OUTPUT TRUE.
/ In this case Δ_0 is a block. */*

STEP 5: Take $j, k \in \Delta$ such that $j^{g_1} = k$.
/ g_1 has the property that $\emptyset \neq \Delta_0^{g_1} \cap \Delta_0 \neq \Delta_0$. */*

STEP 6: Calculate $w_j', w_k' \in \langle H, w_i \rangle$ such that $1^{w_j'} = j$ and $1^{w_k'} = k$.
/ Use the "cube doubling" technique of* [**5, 3**]. */*

STEP 7: Let $g = w_j' g_1 {w_k'}^{-1}$.
/ $g \in G_1 \setminus H$. */*

STEP 8: Let $S_1 = S_1 \cup \{g\}$.

STEP 9: Let $\pi \in \mathrm{Sym}(\Omega)$ be such that the cycles of π are the orbits of $\langle S_1 \rangle$.

STEP 10: OUTPUT FALSE.
*/** GROW *will continue finding points of $\overline{X}$. */*

If we wish to calculate a structure tree, the Δ_k along with the action of the $\sigma \in S$ on $\{\Delta_k \mid k < j\}$ calculated by VERIFY-BLOCK should be stored in global variables or otherwise passed to the rest of the program.

THEOREM 5.5. *The general algorithm runs in time*

$$O(n \log |G|(\ell + s + \log^2 |G|)).$$

PROOF. Since H can be increased at most $\log |G|$ times, VERIFY-BLOCK is called at most $\log |G|$ times with a Δ that is not a block. If Δ is a block, VERIFY-BLOCK only takes time $O(n(\ell + s))$. Since at most $\log n$ blocks will be found by the algorithm, the total time for calls to VERIFY-BLOCK with Δ a block is $O(n(\ell + s))$. Therefore $O(n \log |G|(\ell + s + \log^2 |G|))$ time is spent in calls to VERIFY-BLOCK. By Theorem 4.3, this dominates the running time. □

Since $\ell \leq 2 \log |G|$, this implies Theorem 1.1.

ACKNOWLEDGEMENT: I wish to thank professor László Babai for suggesting this problem to me and for helpful discussions.

REFERENCES

1. A.V. Aho, J.E. Hopcroft, J.D. Ullman: *The Design and Analysis of Computer Algorithms*, Addison-Wesley, Reading 1974, pp. 189–195.
2. M.D. Atkinson: An algorithm for finding the blocks of a permutation group, *Math. Comp.* **29** (1975), pp. 911–913.
3. L. Babai, G. Cooperman, L. Finkelstein, Á. Seress: Nearly linear time algorithms for permutation groups with a small base, *Proc. ISSAC'91* (Internat. Symp. on Symbolic and Algebraic Computation), Bonn 1991, pp. 200–209.
4. L. Babai, G. Cooperman, L. Finkelstein, E. M. Luks, Á. Seress: Fast Monte Carlo algorithms for permutation groups, *Proc. 23rd ACM STOC* (1991), pp. 90–100.
5. L. Babai, E. Szemerédi: On the complexity of matrix group problems I, *Proc 25th IEEE FOCS* (1984), pp. 229–240.
6. R. Beals, Á. Seress: Structure forest and composition factors in nearly linear time for small base groups, *Proc 24th ACM STOC* (1992), pp. 116–125.

7. Z. Galil, C. M. Hoffmann, E. M. Luks, C. P. Schnorr, A. Weber: An $O(n^3 \log n)$ deterministic and an $O(n^3)$ Las Vegas isomorphism test for trivalent graphs, *J. ACM* **34** (1987), pp. 513–531.
8. D. E. Knuth: Efficient representation of perm groups, *Combinatorica* **11** (1991), pp. 33–44 (preliminary version circulated since 1981).
9. C.C. Sims: Computation with Permutation Groups, in: *Proc. Second Symposium on Symbolic and Algebraic Manipulation*, (S.R. Petrick,ed.), ACM, New York, 1971, pp. 23–28.
10. R. E. Tarjan: Depth first search and linear graph algorithms, *SIAM J. Comp.* **1** (1972), pp. 146–160.
11. H. Wielandt: *Finite Permutation Groups*, Acad. Press, New York 1964.

DEPARTMENT OF COMPUTER SCIENCE, UNIVERSITY OF CHICAGO, CHICAGO, ILLINOIS 60637
E-mail address: beals@cs.uchicago.edu

DIMACS Series in Discrete Mathematics
and Theoretical Computer Science
Volume **11**, 1993

Fast Fourier Transforms for Symmetric Groups

MICHAEL CLAUSEN AND ULRICH BAUM

March 1, 1992

ABSTRACT. Recently, it has been shown that a Fourier transform for the symmetric group S_n can be evaluated in less than $0.5(n^3+n^2)n!$ arithmetic operations. We look at this algorithm in more detail and show that it allows an efficient software implementation using appropriate data structures. We give running times of our program and compare them to other existing implementations.

1. Introduction

In 1965, Cooley and Tukey [**6**] published a fast algorithm for the evaluation of Discrete Fourier Transforms. Since then, the DFT and its variants have become extremely important tools in digital signal processing and related areas. In terms of representation theory, the DFT describes an algebra isomorphism from the complex group algebra $\mathbb{C}C_n$ of the cyclic group C_n (the signal domain) onto the algebra of n-square diagonal matrices over $\mathbb{C}$ (the spectral domain).

Wedderburn's theorem allows us to generalize the DFT to arbitrary finite groups: Let K be a splitting field of the finite group G with $charK \not| |G|$. Then the group algebra KG is isomorphic to an algebra of block diagonal matrices: $KG \simeq \oplus_{i=1}^{h} K^{d_i \times d_i}$, where the blocks correspond to the equivalence classes of irreducible representations of KG and h is the number of conjugacy classes of G. Every algebra isomorphism $D = \oplus_{i=1}^{h} D_i : KG \rightarrow \oplus_{i=1}^{h} K^{d_i \times d_i}$ is called a *Fourier transform* for KG. The constituents $D_1, \ldots, D_h$ of D form a transversal of irreducible representations of KG. With respect to natural bases in KG and

1991 *Mathematics Subject Classification.* Primary 20C30,20C40; Secondary 68Q40,68R05.
The final version of this paper will be submitted for publication elsewhere.

$\oplus_i K^{d_i \times d_i}$, every Fourier transform of KG can (and will) be viewed as a matrix $D \in K^{|G| \times |G|}$ and every $a \in KG$ as a column vector from $K^{|G|}$.

This gives rise to three closely related computational problems: Efficient *construction* of a suitably encoded Fourier transform matrix for a given group, fast *evaluation* of such a transform, which amounts to computing a matrix-vector product and finally fast *interpolation*, which means fast evaluation of the inverse Fourier transform. Usually, the construction is a precomputation step that has to be done only once for multiple evaluations.

The simplest way to do this would be to precompute and store the whole $|G|$-square Fourier matrix and then to evaluate it using the standard matrix-vector multiplication formula. But this procedure is very inefficient: The precomputation step requires computing the representing matrices $D_i(g)$ for all i and all $g \in G$. Obviously, this takes a lot of time and one has to store $|G|^2$ numbers. The evaluation would take order $|G|^2$ arithmetic operations. Both time and memory restrictions prohibit using this approach for large groups. To obtain practically feasible algorithms, one takes advantage of the group structure and the properties of the chosen transversal of irreducible representations. Non-abelian groups have irreducible representations of degree > 1 which have different matrix forms depending on the choice of bases in the corresponding simple KG-modules. Hence there are essentially different Fourier transforms which can also widely differ in their construction and evaluation complexities.

We are going to define a computational model for the evaluation of Fourier transforms and their inverses: The *K-linear complexity* of a matrix $A \in K^{n \times n}$ is the minimal number of K-linear operations (i.e. additions, subtractions and scalar multiplications) sufficient to evaluate A at a generic input vector x. As non-abelian groups have many Fourier transforms, we define the *K-linear complexity $L_K(G)$ of a finite group G* as the minimum of the K-linear complexities of all Fourier transforms for KG.

Obviously, $|G|-1 \leq L_K(G) < 2|G|^2$. The classical FFT algorithms [**6, 3, 16**] show that $L_K(G) = O(|G| \log |G|)$ for abelian groups G. This has recently been extended to a much larger class containing the supersolvable groups [**1**] using monomial and symmetry adapted representations which are also surprisingly simple to generate. In a restricted computational model, one can prove that these algorithms are asymptotically optimal [**2**].

This paper is concerned with fast Fourier transforms for symmetric groups and their implementation. Some interesting applications of these transforms in the statistical analysis of ranked data have been investigated by Diaconis [**7, 8**]. In the 1930's, Alfred Young has found simple explicit formulae for two transversals of irreducibles of S_n, the so-called seminormal and orthogonal forms. Independently, Diaconis/Rockmore [**9, 14**] and Clausen [**5**] have used Young's seminormal form as a basis of a more efficient algorithm for the evaluation of the corresponding Fourier transform of S_n. Although their methods are quite similar, the resulting upper bounds differ substantially: Diaconis and Rockmore show that

$$L_K(S_n) = O((n!)^{3/2}n),$$

while Clausen proves the explicit upper bound

$$L_K(S_n) < \frac{1}{2}(n^3 + n^2)n!. \tag{1.1}$$

Recently, Linton/Michler/Olsson [12] suggested a completely different approach to compute Fourier transforms for symmetric groups, having an asymptotic complexity of

$$L_K(S_n) = O((n!)^{3/2}e^{\sqrt{n}}n^{-1/2}).$$

Diaconis/Rockmore and Linton/Michler/Olsson sketch implementations of their algorithms. Both implementations seem to be rather time and space consuming, see the tables in section 4. In particular, the Fourier transform for S_{10} does not appear feasible using either of these algorithms.

This has motivated us to implement Clausen's algorithm and compare its time and memory requirements to those of the other solutions mentioned above. Our 1000-line C program proved superior to the other implementations in precomputation time, running time and memory. For example, for $n = 9$, our program runs 400 times faster than the Linton/Michler/Olsson program (although we use a slower workstation) and takes about 90 milliseconds for precomputations. Computing the Fourier transform for S_{10} is no problem with our program, even on a small workstation.

In this paper, we look at Clausen's algorithm and the necessary precomputations in more detail and show how to turn it into an efficient program using appropriate data structures. Finally, we present some data on the performance of our program and compare it to the other two implementations.

In the full version, we will give more detailed comparisons of the three algorithms. We will also describe a similarly efficient implementation of the inverse Fourier transform.

2. Fast Prccomputation

A uniform approach to design efficient DFT algorithms is based on adapting the irreducible representations to a chain of subgroups: Suppose we want to evaluate a Fourier transform D of a finite group G at a generic input vector $a = \sum_{g\in G} a_g g \in KG$. Any algorithm based on the formula $D(a) = \sum_{g\in G} a_g D(g)$ has to specify the order of summation. This can be done in a natural way using subgroups: If U is a subgroup of G, one rewrites the input $a = \sum_{g\in G} a_g g \in KG$ according to the partition $G = \cup_j h_j U$ of G into left cosets of U: $a = \sum_j h_j a_j$ with $a_j := \sum_{u\in U} a_{h_j u} u \in KU$. Then $D(a) = \sum_j D(h_j)D(a_j)$ reduces the original problem to several evaluations of D at elements of KU. Now if the restriction of D to KU is itself (up to multiplicities) *equal* to a Fourier transform of the subgroup U, we can apply this method recursively without performing a base change. This idea can be formalized as follows.

DEFINITION 2.1. *Let $\mathcal{T} = (G = G_n > \ldots > G_0 = \{1\})$ be a tower of subgroups of the finite group G and K a splitting field of all G_i with $\mathrm{char} K \not| |G|$.*[1] *A matrix representation D of KG is called $\mathcal{T}$-*adapted *if for all j, $0 \leq j \leq n$, the following conditions hold:*

(a) *The restriction $D \downarrow KG_j$ of D to KG_j is* equal *to a direct sum of irreducible matrix representations of KG_j.*
(b) *Equivalent irreducible constituents of $D \downarrow KG_j$ are* equal.

As copying is free in our computational model, condition (b) allows us to use intermediate results several times, saving arithmetic operations. This will lead to more efficient DFT algorithms. The concept of symmetry adapted representations has also been successfully applied to various mathematical and physical problems, see e.g. [**11, 13**].

An easy induction argument shows that every representation of KG is equivalent to a $\mathcal{T}$-adapted representation. Moreover, $\mathcal{T}$-adapted representations are almost unique under certain conditions:

THEOREM 2.1. *For a $\mathcal{T}$-adapted representation D of KG, the following statements are equivalent:*

(a) *D is multiplicity-free and for all j, $1 \leq j \leq n$, if F is an irreducible constituent of $D \downarrow KG_j$, then $F \downarrow KG_{j-1}$ is multiplicity-free.*
(b) *If Δ is a $\mathcal{T}$-adapted representation equivalent to D, then there exists a monomial matrix X such that $D(a) = X^{-1}\Delta(a)X$ for all $a \in KG$. (In this case, we call D and Δ* monomially equivalent.*)*

A proof can be found in [**4**] or [**1**]. As we shall see later, the last theorem applies to the tower

$$\mathcal{T}_n := (S_n > S_{n-1} > \ldots > S_1)$$

of S_n. Next, we are going to describe Young's seminormal form, which is a $\mathcal{T}_n$-adapted Fourier transform for S_n. Although it is well-known, see e.g. the standard text by James/Kerber [**10**], we are going to revisit Young's construction from an implementation point of view.

It is known that every field is a splitting field for S_n. So let K be any field with *char*$K \not| n!$. The conjugacy classes of S_n as well as the equivalence classes of irreducible representations of KS_n are usually parametrized by the partitions of n. A *partition* $\alpha = (\alpha_1, \alpha_2, \ldots)$ of n, abbreviated $\alpha \vdash n$, is a non-increasing sequence of positive integers summing up to n. Partitions can be illustrated by the corresponding diagrams: The diagram of α is the set

$$\bigcup_i \{(i,j) \mid 1 \leq j \leq \alpha_i\},$$

[1] Unless otherwise specified, we will assume in this section that G, $\mathcal{T}$ and K are defined as here.

which can be visualized as a left-justified arrangement of α_i boxes in the i-th row. By abuse of notation we will make no difference between partitions and their diagrams. The following figure shows all diagrams for $n = 4$:

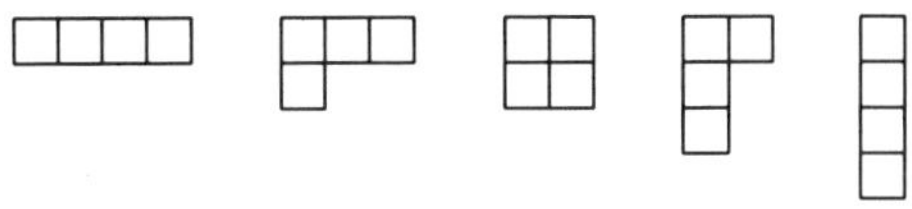

Hence there are 5 inequivalent irreducible representations of KS_4. For every partition α of n, we denote by $[\alpha]$ an irreducible representation of S_n of "type" α. Inclusion defines a partial ordering on the set of all diagrams, the so-called *Young lattice*:

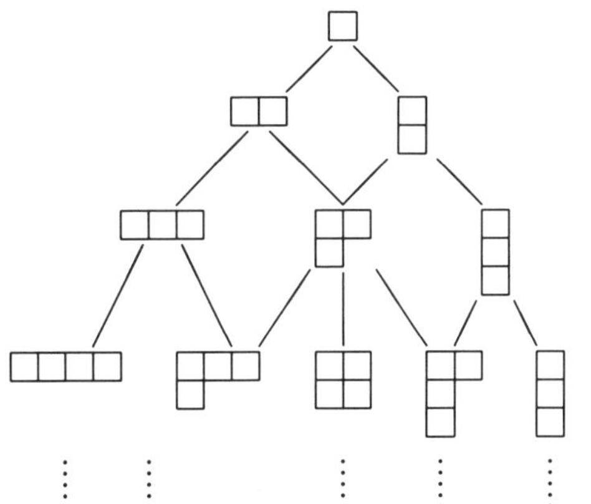

In fact, this has a representation theoretical meaning: The celebrated branching theorem tells us that $[\alpha]\downarrow S_{n-1}$ is multiplicity-free:

$$[\alpha]\downarrow S_{n-1} \sim \bigoplus_{\beta} [\beta],$$

where the sum is over all diagrams β of $n-1$ contained in α. In particular, the degree f_α of $[\alpha]$ equals the number of paths from (1) to α in the Young lattice. Every such path can be described by a standard α-tableau, which is an α-shaped matrix whose elements $1, 2, \ldots, n$ are arranged in such a way that the entries in each row and column are strictly increasing:

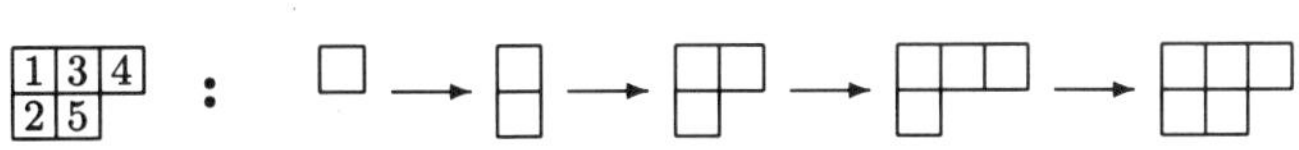

In 1930, Alfred Young has given surprisingly simple explicit formulae for $\mathcal{T}_n$-adapted irreducible representations of S_n. They are based on the so-called *last letter sequence*, which is a total ordering of the standard α-tableaux. It can be described by the leaves of a certain tree. We omit a formal definition and illustrate it by an

Example Last letter sequence of all standard $(3,2)$-tableaux:

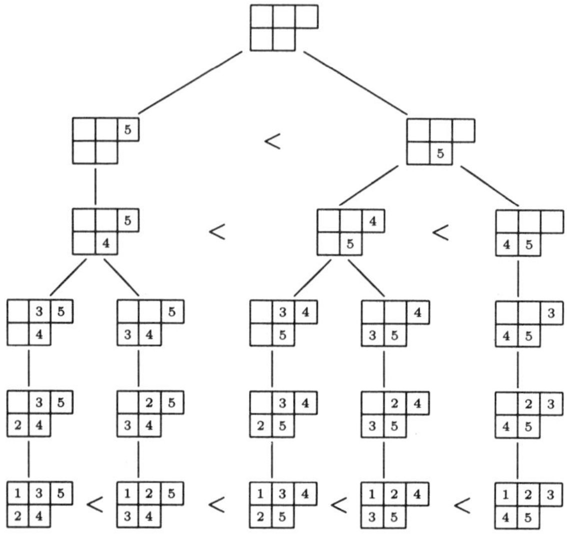

Since S_n is generated by all transpositions of consecutive numbers,

$$S_n = \langle (1,2), (2,3), \dots, (n-1,n) \rangle,$$

every representation D of KS_n is completely determined by all $D(i,i+1)$, $i < n$. Let α be a partition of n. We are going to describe Young's seminormal form σ^α, which is an irreducible representation of KS_n of "type α". The rows and columns are parametrized by the last letter sequence $T_1 < \dots < T_r$ of all standard α-tableaux, where $r := f_\alpha$. In order to describe for a fixed $i < n$

$$\sigma^\alpha(i,i+1) =: (\sigma_{kl})_{1\le k,l\le r} \tag{2.1}$$

we have to consider two cases.

Case 1 For $a \le r$ the numbers i and $i+1$ are in the same row (resp. column) of T_a: Then the only non-zero entry in the a-th row and a-th column of (σ_{kl}) is the diagonal position: $\sigma_{aa} = 1$, if i and $i+1$ are in the same row of T_a, whereas $\sigma_{aa} = -1$, if i and $i+1$ are in the same column of T_a.

Case 2 T_b results from T_a by interchanging i and $i+1$: Then, if $a < b$,

$$\begin{pmatrix} \sigma_{aa} & \sigma_{ab} \\ \sigma_{ba} & \sigma_{bb} \end{pmatrix} = \begin{pmatrix} d^{-1} & 1-d^{-2} \\ 1 & -d^{-1} \end{pmatrix} \tag{2.2}$$

where $d := |u-x| + |v-y|$ is the axial distance of the positions (u,v) of i and (x,y) of $i+1$ in T_a. All other entries are zero. Consequently, the matrix (σ_{kl}) is sparse with at most two non-zero entries in each row and in each column. The following table shows all $\sigma^\alpha(i,i+1)$ for $n = 4$.

last letter sequence	$\sigma^\alpha(1,2)$	$\sigma^\alpha(2,3)$	$\sigma^\alpha(3,4)$
$\begin{smallmatrix}1&2&3&4\end{smallmatrix}$	(1)	(1)	(1)
$\begin{smallmatrix}1&3&4\\2\end{smallmatrix}$ $\begin{smallmatrix}1&2&4\\3\end{smallmatrix}$ $\begin{smallmatrix}1&2&3\\4\end{smallmatrix}$	$\begin{pmatrix}-1&&\\&1&\\&&1\end{pmatrix}$	$\begin{pmatrix}\frac{1}{2}&\frac{3}{4}&\\1&-\frac{1}{2}&\\&&1\end{pmatrix}$	$\begin{pmatrix}1&&\\&\frac{1}{3}&\frac{8}{9}\\&1&-\frac{1}{3}\end{pmatrix}$
$\begin{smallmatrix}1&3\\2&4\end{smallmatrix}$ $\begin{smallmatrix}1&2\\3&4\end{smallmatrix}$	$\begin{pmatrix}-1&\\&1\end{pmatrix}$	$\begin{pmatrix}\frac{1}{2}&\frac{3}{4}\\1&-\frac{1}{2}\end{pmatrix}$	$\begin{pmatrix}-1&\\&1\end{pmatrix}$
$\begin{smallmatrix}1&4\\2\\3\end{smallmatrix}$ $\begin{smallmatrix}1&3\\2\\4\end{smallmatrix}$ $\begin{smallmatrix}1&2\\3\\4\end{smallmatrix}$	$\begin{pmatrix}-1&&\\&-1&\\&&1\end{pmatrix}$	$\begin{pmatrix}-1&&\\&\frac{1}{2}&\frac{3}{4}\\&1&-\frac{1}{2}\end{pmatrix}$	$\begin{pmatrix}\frac{1}{3}&\frac{8}{9}&\\1&-\frac{1}{3}&\\&&-1\end{pmatrix}$
$\begin{smallmatrix}1\\2\\3\\4\end{smallmatrix}$	(-1)	(-1)	(-1)

It can be shown that

$$\sigma_n := \bigoplus_{\alpha \vdash n} \sigma^\alpha$$

is a $\mathcal{T}_n$-adapted Fourier transform for KS_n, see e.g. Theorem 3.3.10 in [**10**]. More precisely,

$$\sigma^\alpha \downarrow S_{n-1} = \bigoplus_{\beta \vdash n-1 : \beta \subset \alpha} \sigma^\beta, \tag{2.3}$$

where the direct summands appear in the order given by the first level of the last letter sequence tree for α. E.g.,

$$\sigma^{(4,2,1)} \downarrow S_6 = \sigma^{(3,2,1)} \oplus \sigma^{(4,1,1)} \oplus \sigma^{(4,2)}.$$

As each $\sigma^\alpha \downarrow S_{n-1}$ is multiplicity-free, Theorem 2.2 tells us that σ_n is – up to monomial basis transforms – the unique $\mathcal{T}_n$-adapted Fourier transform for S_n. For example, Young's *orthogonal form*, which is another well-known $\mathcal{T}_n$-adapted DFT for S_n, is obtained by a monomial basis transform from σ_n. Our DFT algorithm for S_n requires that, for all $k \leq n$, the matrices $\sigma^\beta(i, i+1)$ for all $\beta \vdash k$ and $i < k$ be precomputed and stored. One might think that this takes too much memory. But this is not the case if we use the right data structure: The sparse matrix $(\sigma_{ab}) := \sigma^\beta(i, i+1)$ is represented by a table $(b_a, d_a)_{1\leq a \leq f_\beta}$ of pairs of integers satisfying $1 \leq b_a \leq f_\beta$, $-1 \leq d_a < k$. The pair $(b, d) := (b_a, d_a)$ describes row a of the matrix. If $a = b$, then $\sigma_{aa} = d \in \{\pm 1\}$ according to case 1 above. Otherwise, we are in case 2. If $a < b$, $\sigma_{aa} = d^{-1}$ and $\sigma_{ab} = 1 - d^{-2}$. Finally, if $a > b$, we have $\sigma_{ab} = 1$ and $\sigma_{aa} = -d^{-1}$. This encoding is slightly redundant but allows a more efficient multiplication by (σ_{ab}). Altogether, we have to store

$$2 \sum_{k \leq n} (k-1) \sum_{\beta \vdash k} f_\beta$$

small integers, a minor quantity compared to the input size $n! = \sum_{\alpha \vdash n} f_\alpha^2$. For example, for $n = 10$, this requires space for 227376 small integers, while the input consists of $10! = 3628800$ floating-point numbers in case $K = \mathbb{R}$.

Now we describe the procedure for computing all $\sigma^\beta(i, i+1)$. In a first step, we generate the Young lattice up to level n in an obvious way. This gives us all partitions of $k \leq n$ as well as the degrees and the branching behaviour of the corresponding representations in negligible time and space.

Next, we compute the $\sigma^\beta(i, i+1)$ "bottom-up" for $S_2, S_3, \ldots, S_n$. Suppose $\beta \vdash k \leq n$ and $i < k$. For $i < k-1$, we have

$$\sigma^\beta(i, i+1) = (\sigma^\beta \downarrow S_{k-1})(i, i+1) = \bigoplus_{\gamma \vdash k-1 : \gamma \subset \beta} \sigma^\gamma(i, i+1)$$

according to equation 2.3. Hence the matrix $\sigma^\beta(i, i+1)$ can be assembled without any further calculation from its direct summands already computed at stage $k-1$.

Only for $i = k-1$, the matrix has to be constructed by Young's formula using the last letter sequence of β. To do this, we only need to know the positions of $k-1$ and k in the standard β-tableaux $T_1, \ldots, T_{f_\beta}$. Hence we do not have to construct the complete last letter sequence tree, but merely the nodes $N_1 < \cdots < N_m$ of depth 2. (These are pictures containing only the two entries $k-1$ and k.) This makes the construction of $\sigma^\beta(k-1, k)$ substantially faster.

Each N_j defines an interval $I_j \subseteq \{T_1, \ldots, T_{f_\beta}\}$, which consists of all leaves of the subtree rooted at N_j. The size of I_j equals the degree of the representation corresponding to the partition of $k-2$ which is obtained by deleting the positions $k-1$ and k from N_j. (This degree has already been computed in the first step of the algorithm.) For each $1 \leq j \leq m$, exactly one of the following two cases occurs:

(i) $k-1$ and k are contained in the same row (resp. column) of N_j. Then case 1 of Young's construction applies to all elements T_a in I_j.

(ii) For some $l > j$, N_j is transformed into N_l by interchanging $k-1$ and k. Then case 2 of Young's construction applies to every pair (T_a, T_b), consisting of the u-th element of I_j and I_l, respectively. Obviously, the axial distance d is the same for all these pairs.

In this way, the matrix $\sigma^\beta(k-1,k)$ can be quickly constructed. For $n=10$, the construction of $\sigma^\beta(i,i+1)$ for all partitions $\beta \vdash k$ and all $1 \leq i < k$, $k = 2, \ldots, 10$, takes 270 milliseconds on a SUN SPARCstation 1.

3. Fast Evaluation

In this section, we will show how the Fourier transform σ_n can be efficiently evaluated. Given an element $a = \sum_{g \in S_n} a_g g$ of KS_n, we have to compute $\sigma_n(a)$. To begin with, we decompose S_n into left cosets of the subgroup S_{n-1}: $S_n = \dot{\bigcup}_{j \leq n} g_{jn} S_{n-1}$, where g_{jn} is any fixed permutation in S_n which maps n to j. The choice of g_{jn} is crucial for our algorithm's efficiency and will be discussed in a moment. According to this decomposition, we can write $a = \sum_{j=1}^n g_{jn} a_j$, where $a_j = \sum_{h \in S_{n-1}} a_{g_{jn}h} h \in KS_{n-1}$. As σ_n is a morphism of K-algebras, we get

$$\begin{aligned}
\sigma_n(a) &= \sum_{j=1}^n \sigma_n(g_{jn})\sigma_n(a_j) \\
&= \sum_{j=1}^n \bigoplus_{\alpha \vdash n} \sigma^\alpha(g_{jn})\sigma^\alpha(a_j) \\
&= \sum_{j=1}^n \bigoplus_{\alpha \vdash n} \sigma^\alpha(g_{jn})(\sigma^\alpha \downarrow KS_{n-1})(a_j).
\end{aligned}$$

Now we use the fact that σ_n is $\mathcal{T}_n$-adapted:

$$\sigma_n(a) = \sum_{j=1}^n \bigoplus_{\alpha \vdash n} \sigma^\alpha(g_{jn}) \bigoplus_{\alpha \supset \beta \vdash n-1} \sigma^\beta(a_j).$$

Thus every $\sigma^\beta(a_j)$, once computed, can be used in the evaluation of $\sigma^\alpha(a_j)$ for all partitions α of n containing β. Because copying is free in our computational model, this means a substantial reduction of the number of operations. The assumption that copying is free is realistic for practical software implementations, as copying an element of K is usually much faster than an arithmetic operation. This is the fundamental advantage of $\mathcal{T}$-adapted Fourier transforms. (Clausen **[5]** employs this idea to derive improved DFT algorithms for arbitrary finite groups.)

We still have to specify the coset representatives g_{jn}. We use the cycle $g_{jn} := (j, j+1, \ldots, n)$. Why is this a good choice? As $g_{jn} = (j, j+1)(j+1, j+2) \ldots (n-1, n)$, $\sigma^\alpha(g_{jn})$ is the product of $n-j$ sparse matrices:

$$\sigma^\alpha(g_{jn}) = \sigma^\alpha(j, j+1) \cdot \sigma^\alpha(j+1, j+2) \cdot \ldots \cdot \sigma^\alpha(n-1, n).$$

Recall that $\sigma^\alpha(i, i+1)$ has at most two nonzero entries in each row and at most 3/4 of its nonzero entries are $\neq \pm 1$. Hence we can multiply $\sigma^\alpha(i, i+1)$ with an arbitrary f_α-square matrix in at most $5/2 \cdot f_\alpha^2$ operations. Instead of directly multiplying the "twiddle factor" $\sigma^\alpha(g_{jn})$ (which is typically not sparse) by $\sigma^\alpha(a_j)$, we compute the product as indicated by the expression

(3.1)

$$\Big(\sigma^\alpha(j, j+1) \cdot \ldots \cdot \Big(\sigma^\alpha(n-2, n-1) \cdot \Big(\sigma^\alpha(n-1, n) \cdot \sigma^\alpha(a_j)\Big)\Big) \ldots\Big)$$

and can thus perform this multiplication with at most $5/2 \cdot (n-j) f_\alpha^2$ arithmetic operations, compared to order f_α^3 for direct multiplication.[2]

Finally, we have to sum up the n block diagonal matrices $\sigma_n(g_j)\sigma_n(a_j)$. This takes

$$(n-1) \sum_{\alpha \vdash n} f_\alpha^2 = (n-1) \cdot n!$$

operations.

Let L_n denote the arithmetic cost of our algorithm to evaluate σ_n. Then we have the recursion

$$L_n \leq n \cdot L_{n-1} + \left(\sum_{j=1}^{n} \sum_{m=j+1}^{n} \sum_{\alpha \vdash n} \frac{5}{2} \cdot f_\alpha^2 \right) + (n-1) \cdot |S_n|.$$

Obviously, $L_1 = 0$. Combining this with $\sum_{\alpha \vdash n} f_\alpha^2 = |S_n|$ and the well-known formula $\sum_{i=2}^{n} \binom{i}{2} = \binom{n+1}{3}$, induction yields the following small improvement of Clausen's original result (1.1).

THEOREM 3.1.

$$L_K(S_n) \leq L_n \leq \left(\frac{5}{12} n^3 + \frac{1}{2} n^2 - \frac{11}{12} n \right) n!$$

(By a closer look at the matrices in 3.1, this upper bound can be slightly improved. However, the gain is too small to justify the much more technical proof.)

So far, we have followed a top-down approach to describe and analyze our algorithm. However, a top-down implementation recursively computing the Fourier transform would require too much memory and a lot of bookkeeping, see the discussion in [**9**]. Our implementation works bottom-up to avoid these problems. It takes a global view of all computations at *layer k* of the algorithm, where the

[2] Diaconis/Rockmore also use σ_n, but take $g_{jn} = (j, n)$ and directly multiply $\sigma^\alpha(g_{jn})$ by $\sigma^\alpha(a_j)$. This gives an upper bound of order $(n!)^{3/2} \cdot n$ for $L_K(S_n)$.

Fourier transform σ_k of S_k is evaluated once for each coset of S_k in S_n. More precisely, define $g_{jk} := (j, j+1, \ldots, k)$ for $1 \leq j \leq k \leq n$. Then the words

$$g_{\underline{j}} := g_{j_n n} g_{j_{n-1} n-1} \cdots g_{j_{k+1} k+1}, 1 \leq j_i \leq i,$$

form a transversal of left cosets of S_k in S_n. The lexicographical ordering of the vectors $(j_n, \ldots, j_{k+1})$ thus induces a total ordering of the cosets.

The data in layer k of our algorithm consists of $n!/k!$ blocks $\sigma_k(a_{\underline{j}})$ of size $k!$, where the $a_{\underline{j}} \in KS_k$ are defined by decomposing the input $a \in KS_n$ along our transversal: $a = \sum_{\underline{j}} g_{\underline{j}} a_{\underline{j}}$. As we shall see below, it is favorable for our algorithm to arrange the blocks in each layer k according to the lexicographical ordering of the cosets of S_k in S_n. Each block diagonal matrix $\sigma_k(a_{\underline{j}})$ is represented by a vector of length $k!$ as follows: The block constituents are ordered according to the lexicographic order of the partitions of k. Within each block, the entries are stored row-wise. With respect to this arrangement, the data in layer k can be described as a vector $v_k \in K^{n!}$. In particular, the input vector v_1 consists of the coefficients of a enumerated in the lexicographical order of the group elements.

Given v_1, our program successively computes $v_2, \ldots, v_n$ and outputs v_n. In layer k, we compute v_k from v_{k-1}. More precisely, we compute each $\sigma_k(a_{\underline{j}})$ from the k blocks $\sigma_{k-1}(a_{\underline{l}})$ with $g_{\underline{l}} S_{k-1} \subset g_{\underline{j}} S_k$ as described above. As our ordering of the $\sigma_k(a_{\underline{j}})$ is compatible with the stepwise refinement of our coset decomposition, the same linear transformation has to be applied to each block of length $k!$ in v_{k-1} to obtain the corresponding block in v_k. In a (sequential) software implementation, this is easy to realize by moving pointers over v_k and v_{k-1}. On the other hand, it is obvious that a very regular *parallel* implementation is possible. The layered structure of the algorithm also allows pipelining.

What about memory requirements? Of course, we do not have to store $(v_1, \ldots, v_n)$, but only v_k and v_{k+1} at layer k. In addition, we need storage for $f_\alpha^2 + 2f_\alpha$ elements of K for evaluating the expression 3.1. Altogether, our algorithm requires memory for

$$M(n) := 2 \cdot n! + f_n^2 + 2 \cdot f_n$$

elements of K, where f_n denotes the maximal degree of an irreducible representation of KS_n. For $n = 10$, we have $f_{10} = 768$ and $M(10) < 2.17 \cdot 10!$. Using 32-bit floating point arithmetic ($K = \mathbb{R}$), this takes less than 30 megabytes of virtual memory. In addition, we need less than one megabyte for storing the precomputed tables (see section 2) and local variables. So the Fourier transform for S_{10} is already feasible on a medium-sized workstation.

4. Running Times

We have implemented our algorithm in C on a SUN SPARCstation 1. The C program is less than 1000 lines long. It works over $K = \mathbb{R}$, using 32-bit `float` arithmetic. For $6 \leq n \leq 10$, the following table shows the precomputation time (in CPU seconds), size of precomputed tables (in kilobytes), evaluation time (in CPU seconds), number of arithmetic operations and the theoretical upper bound of our FFT algorithm.

n	precomp. (s)	table size (kb)	eval. (s)	arith. ops	upper bound
6	0.01	2	0.08	55440	73800
7	0.02	7	0.79	623952	811440
8	0.03	28	8.67	7507836	9596160
9	0.08	110	105.68	96756840	121927680
10	0.26	444	1518.62	1333294380	1660176000

Let us now look at the performance of the other two algorithms mentioned. Linton/Michler/Olsson [**12**] give the following running times of their algorithm on an IBM RISC 6000/540, which is faster than our SPARCstation 1 (evaluation times in CPU seconds).

n	eval. (s)
6	10.27
7	35.81
8	643.94
9	44791.26

Unfortunately, Diaconis and Rockmore [**9, 14**] do not present any running times of their algorithm. The only explicit information they give is that, for $n = 9$, the best variant of their DFT algorithm takes $243,802,167$ multiplications. In 32-bit arithmetic, the precomputed tables would require about 12 megabytes of memory. On the other hand, our algorithm takes $40,510,440$ multiplications for $n = 9$ and only 110 kilobytes of memory for the precomputed data. To be able to compare the running times, we are currently implementing the Diaconis/Rockmore algorithm on our SPARCstation 1. The results will be given in the final version of this paper. The final version will also include more detailed comparisons of the three algorithms with respect to precomputation times, evaluation times and memory requirements.

References

1. U. Baum, Existence and Efficient Construction of Fast Fourier Transforms on Supersolvable Groups, *Computational Complexity*, **1/3** (1992), 235–256.
2. U. Baum, M. Clausen, Some Lower and Upper Complexity Bounds for Generalized Fourier Transforms and Their Inverses, *SIAM J. Comput.*, **2** (1991), 451–459.
3. L.I. Bluestein, A Linear Filtering Approach to the Computation of the Discrete Fourier Transform, *IEEE Trans.*, **AU-18** (1970), 451–455.
4. M. Clausen, *Beiträge zum Entwurf schneller Spektraltransformationen*, Habilitationsschrift, Universität Karlsruhe, 1988.
5. M. Clausen, Fast Generalized Fourier Transforms, *Theor. Comp. Sci.*, **67** (1989), 55–63.
6. J.W. Cooley, J.W. Tukey, An Algorithm for the Machine Calculation of Complex Fourier Series, *Math. Comp.*, **19** (1965), 297–301.

7. P. DIACONIS, A Generalization of Spectral Analysis with Application to Ranked Data, *Ann. Statistics*, **17** (1989), 949–979.
8. P. DIACONIS, *Group Representations in Probability and Statistics*, Institute of Mathematical Statistics Lecture Notes - Monograph Series, **11**, Hayward, CA, 1988.
9. P. DIACONIS, D. ROCKMORE, Efficient Computation of the Fourier Transform on Finite Groups, *J. AMS*, **3** (1990), 297–332.
10. G.D. JAMES, A. KERBER, *The Representation Theory of the Symmetric Group*, Addison-Wesley, 1981.
11. D.J. KLEIN, C.H. CARLISLE, F.A. MATSEN, Symmetry Adaptation to Sequences of Finite Groups, *Advances in Quantum Chemistry*, **5** (1970), Academic Press, 219–260.
12. S.A. LINTON, G.O. MICHLER, J.B. OLSSON, Fast Fourier Transforms on Symmetric Groups, *preprint*, Universität Essen, 1991.
13. K. MUROTA, K. IKEDA, Computational Use of Group Theory in Bifurcation Analysis of Symmetric Structures, *SIAM J. Sci. Stat. Comput.*, **12** (1991), 273–297.
14. D. ROCKMORE, Computation of Fourier Transforms on the Symmetric Group, in: E. KALTOFEN, S.M. WATT (eds.), *Computers and Mathematics*, Springer, 1989.
15. J.P. SERRE, *Linear Representations of Finite Groups*, Springer, 1977.
16. S. WINOGRAD, On Computing the Discrete Fourier Transform, *Math. Comp.*, **32** (1978), 175–199.

INSTITUT FÜR INFORMATIK, UNIVERSITÄT BONN, RÖMERSTR. 164, 5300 BONN, GERMANY
E-mail address: clausen@leon.cs.uni-bonn.de

INSTITUT FÜR INFORMATIK, UNIVERSITÄT BONN, RÖMERSTR. 164, 5300 BONN, GERMANY
E-mail address: uli@leon.cs.uni-bonn.de

DIMACS Series in Discrete Mathematics
and Theoretical Computer Science
Volume 11, 1993

From Hyperbolic Reflections to Finite Groups

J. H. CONWAY

More than 10 years ago, I conjectured that the Coxeter group defined by the Y_{555} diagram (see Figure 1) yields the so-called Bimonster group when subjected to just one further relation. This conjecture has recently been proved by Norton [2] and Ivanov [3] building on some previous work of Norton, Soicher, Pritchard and myself, described in [1] and [2].

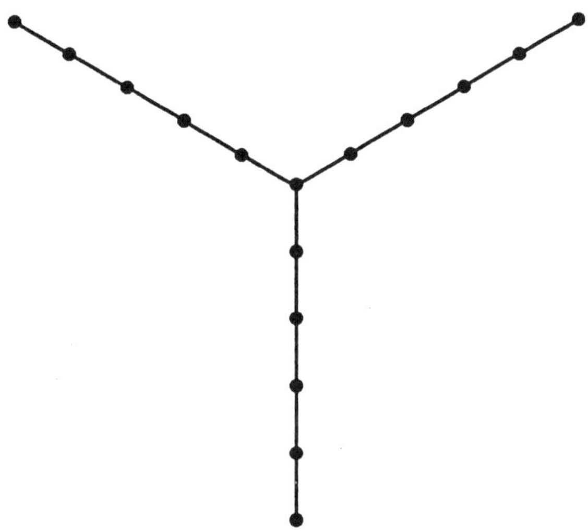

Figure 1. The Y_{555} diagram

Until the major new contribution provided by Ivanov knocked the wind out of our sails, Pritchard and I had been trying to prove the conjecture using hyperbolic geometry. This would have involved a massive amount of machine calculation as well as the geometry and group theory. Ivanov's ideas have made all

1991 *Mathematics Subject Classification.* Primary 20D08, 20F55.
This is the final version of this paper.

this unnecessary for the sake of proving the conjecture, and so probably nobody will find it worthwhile to force it through, which is in some ways a pity, since one result would almost certainly have been a practicable way of computing inside the Monster group.

Although it has failed for its original purpose, I think the "hyperbolic method" is still interesting for both theoretical and computational reasons. My aim in this note is to put it before a wider circle of readers, and in particular to show how, when it works, it can produce astonishingly short proofs of characterization theorems for groups.

THE $S(a, b)$ GROUPS.

The Monster group M contains a subgroup A isomorphic to the alternating group of degree 5 and whose centraliser is an alternating group of degree 12 in which the 7-point stabiliser is conjugate to A. Now symmetric groups arise naturally as reflection groups in a way that alternating ones do not. To overcome this difficulty, we introduce the wreathed square of the Monster, which we call the Bimonster.

This is the group obtained from the Cartesian square $M \times M$ by adjoining the automorphism t of order 2 for which $t(x, y)t = (y, x)$. Then the Bimonster contains a subgroup S isomorphic to the symmetric group of degree 5 and whose centraliser is a symmetric group of degree 12 in which the 7-point stabiliser is conjugate to S. In fact it follows from the Ivanov-Norton theorem that there are just two minimal groups with this property - the Bimonster and the symmetric group of degree 17.

In general we shall say that a group is an $S(a, b)$ group if it is minimal with the property that it contains a symmetric group S of degree a whose centraliser is a symmetric group of larger degree b, in which the stabiliser of $b - a$ points is conjugate to S. When a or b is 6 there are some special features which make this not quite natural or meaningful, so something more needs to be said in those cases - we shall ignore this problem in this paper.

$S(a, b)$ GROUPS FROM COXETER GROUPS.

Every $S(a, b)$ group can be exhibited as a homomorphic image of a Coxeter group. We shall consider as an example the case of $S(3, 8)$ groups, and will prove the theorem that any such group is isomorphic either to S_{11} or to a certain orthogonal group over the field of order 2. This is analogous to the Norton-Ivanov theorem that any $S(5, 12)$ group is isomorphic either to S_{17} or to the Bimonster.

Let Y be an $S(3, 8)$ group, and let the appropriate symmetric subgroup S of degree 3 be generated by the elements c, d playing the roles of the permutations (p, q), (q, r) respectively. Then we can suppose its centraliser in Y is generated by the elements

$$d',\ c',\ b',\ a,\ b'',\ c'',\ d''$$

playing the roles of the respective transpositions

$$(1,2),\ (2,3),\ (3,4),\ (4,5),\ (5,6),\ (6,7),\ (7,8)\ .$$

The elements mentioned so far generate the Coxeter group of Figure 2.

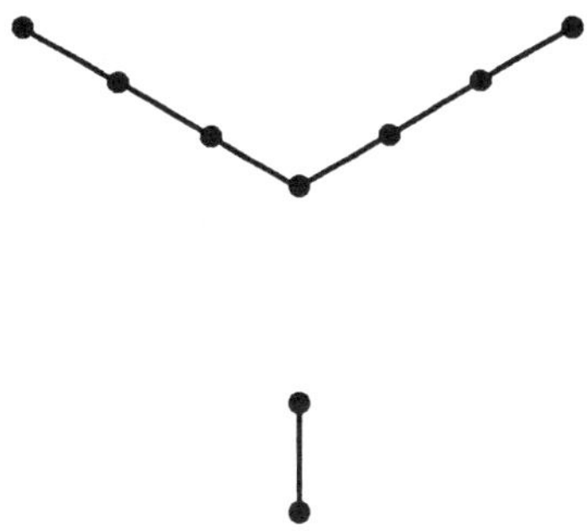

Figure 2.

Now the definition of an $S(a,b)$ group tells us that the elements d', c' generate a group S' conjugate to S, and whose centraliser is therefore another symmetric group of degree 8. We can already see in Figure 2 a large subgroup of this centraliser, namely the group $S_3 \times S_5$ generated by c, d, and a, b'', c'', d''. But it is easy to see that all subgroups $S_3 \times S_5$ in S_8 are conjugate, and so we can suppose that the S_8 centralising the group S' is generated by the elements

$$d, \quad c, \quad b, \quad a, \quad b'', \quad c'', \quad d''$$

playing the respective roles of the transpositions

$$(1,2),\ (2,3),\ (3,4),\ (4,5),\ (5,6),\ (6,7),\ (7,8)$$

as in Figure 3.

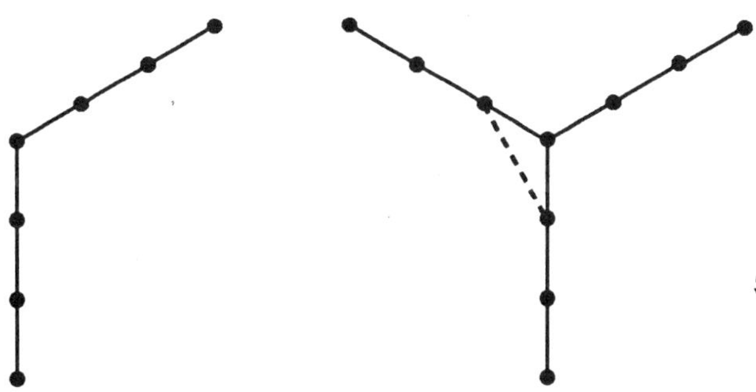

Figure 3. Figure 4.

We now have all the Coxeter relations indicated by the diagram of Figure 4, except that we do not yet know the order of the product bb', for which we have drawn a dotted line in the Figure. We can however show that this order must be either 2 or 3.

To see this, we note that the elements c'', d'' generate a third group S'' conjugate to S, whose centralising S_8 group can be seen to have two distinct subgroups $S_3 \times S_5$, generated by

$$d, c;\ \ a, b', c', d' \qquad \text{and} \qquad d',\ c';\ \ a, b, c, d$$

respectively. So both b and b' must be transpositions in this S_8, and since they are distinct (for one of them commutes with c but not with c', while the other commutes with c' but not with c), the order of their product must be 2 or 3, and we now have all the relations of one of the two Coxeter groups drawn in Figures 5 and 6.

Figure 5.

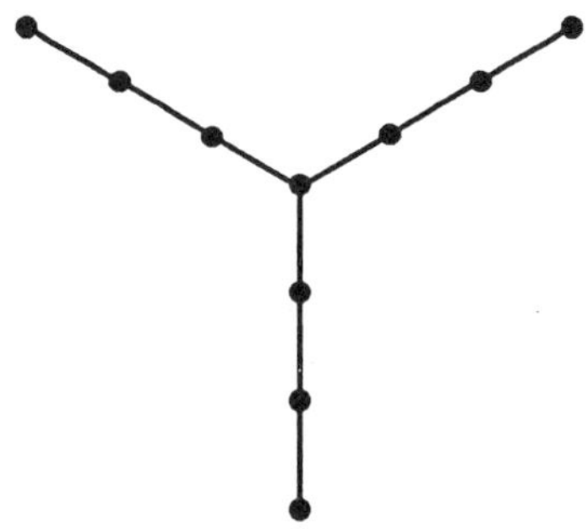

Figure 6.

Now let Y^* be the subgroup of our $S(3,8)$ group Y generated by the elements already mentioned. Then the centraliser of S in Y^* already contains the S_8 which is its centraliser in Y, and so must be this S_8. Again, from Figure 4 we see that the group S'', which is the stabiliser of a certain set of 5 points in this S_8, is already conjugate to S in the centraliser of S', and so is certainly conjugate to S in Y^*. Since Y is a minimal group with these properties, we see that $Y = Y^*$. We summarise:-

THEOREM 1. *Any $S(3,8)$ group Y is a homomorphic image of one of the two Coxeter groups represented by Figures 5 and 6.*

THE TRIVIAL CASE.

The case represented by Figure 5 is easily dealt with. We can replace the generator a by its conjugate by the element bcd. This new element, together with the other generators, satisfies the Coxeter relations of Figure 7, which define S_{11}. (The required relations are easily visible inside the various S_8 subgroups.)

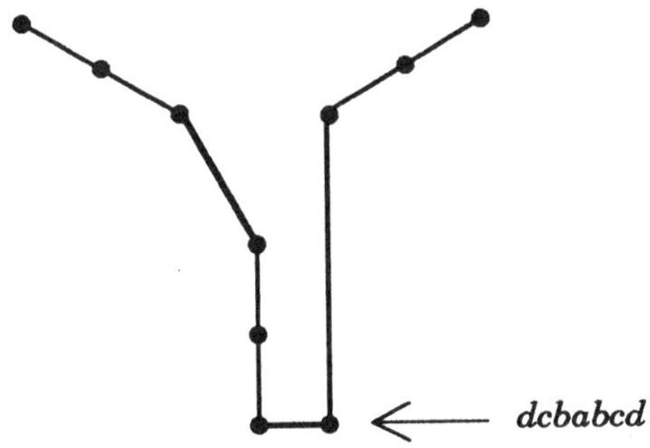

Figure 7.

THE EXTRA RELATIONS.

We shall use the name Y_{pqr} for the group generated by the central generator a of Figure 6, together with the first p of b, c, d, the first q of b', c', d' and the first r of b'', c'', d''. Thus the whole group is Y_{333}. Now the Coxeter relations of the subdiagram Y_{131} define a group W consisting of those symmetries of a coordinate frame of 6 mutually orthogonal vectors and their negatives that involve an even number of sign-changes. To understand this isomorphism, map each node to the reflection $x \to x - (x,r)r$ in the appropriate vector r of Figure 8, and note that the reflection in a vector $0..01\bar{1}0..0$ interchanges the appropriate pair of coordinates, while the reflection in 000011 interchanges the last two coordinates, and changes both their signs.

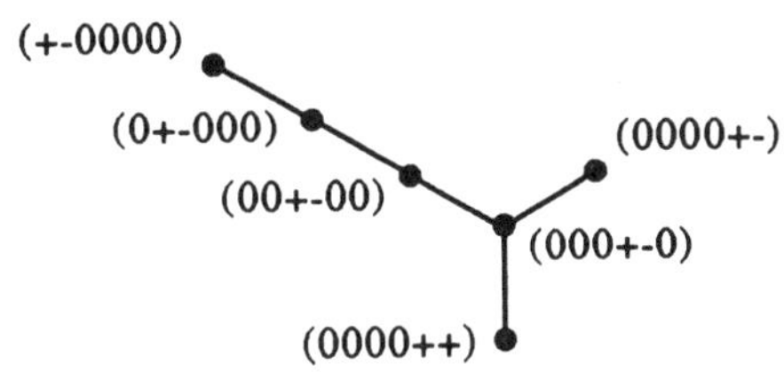

Figure 8. (+ means +1, − means −1)

This group W has a central element z, that negates all 6 basis vectors. In fact z is the 5th power of the product $dcbab'b''$. Since in particular this element centralises c, d, it must lie in the S_8 generated by d',c', b', a, b'', c'', d''. However, it also centralises the three subgroups $\langle d'\rangle$, $\langle b', a, b''\rangle$, $\langle d''\rangle$. But the only elements of S_8 that do this are 1, d', d'', and $d'd''$. We conclude therefore that

$$(dcbab'b'')^5 \text{ must equal one of } 1, d', d'', d'd'' .$$

So we are led to consider the abstract group Y defined by the Coxeter relations of Figure 6, together with the additional relations

$$z = \text{ one of } 1, d', d'', d'd''$$
$$z' = \text{ one of } 1, d, d'', dd''$$
$$z'' = \text{ one of } 1, d, d', dd',$$

where z, z', z'' denote the central elements of the subgroups Y_{311}, Y_{131}, Y_{113}, so that $z = (dcbab'b'')^5$. We shall identify this group as an abstract group. (The argument will show that in fact we must have $z = d'd''$, $z' = dd''$, $z'' = dd'$.)

We now define 10 vectors

$d = 1\bar{1}00\ 0000\ 0000\ 0$, $d' = 0000\ 1\bar{1}00\ 0000\ 0$, $d'' = 0000\ 0000\ 1\bar{1}00\ 0$,

$c = 01\bar{1}0\ 0000\ 0000\ 0$, $c' = 0000\ 01\bar{1}0\ 0000\ 0$, $c'' = 0000\ 0000\ 01\bar{1}0\ 0$,

$b = 001\bar{1}\ 0000\ 0000\ 0$, $b' = 0000\ 001\bar{1}\ 0000\ 0$, $b'' = 0000\ 0000\ 001\bar{1}\ 0$,

$$a = 0001\ 0001\ 0001\ 1,$$

(where we abbreviate $+1$ to $1, -1$ to $\bar{1}$)

in a space in which the quadratic form is the sum of the squares of the first 12 coordinates minus the square of the 13th (so this is a Lorentzian space). It is easy to check that these vectors have norm (squared length) 2, and inner products -1 when joined, 0 when not. This entails that the reflections in these vectors already satisfy the Coxeter relations of Figure 6. We proceed to explore the geometric meaning of the new relations.

Now the central element z of the subgroup Y_{311} acts just by negating the space spanned by d, c, b, a, b', b''. The projection of c' onto the space orthogonal to these must have coordinates of form

$$\text{proj}(c') = pppp\ 01qq\ 00rr\ s\ ,$$

say, where $s = p + q + r$, and we must also have

$$4p = 1 + 2q = 2r = s$$

since all of our generating vectors have the property that the 13th coordinate is the sum of each of the earlier three blocks of four coordinates. We find $p = 1/2$, $q = 1/2$, $r = 1$, $s = 2$. The image of c' under z is therefore

$$2 \cdot \text{proj}(c') - c' = 1111\ 0121\ 0022\ 4\ .$$

This at long last gives us a relation we can use, since c' must represent the same group element as this vector transformed by one of

$$1,\ d',\ d'',\ d'd''\ ,$$

that is to say, as one of the two vectors

$$1111\ 0121\ 0022\ 4 \text{ and } 1111\ 1021\ 0022\ 4\ ,$$

and indeed as the second one, since the first one commutes with

$$0000\ 010\bar{1}\ 0000\ 0\ ,$$

whereas c' does not.

We now transform the "equality"

$$0000\ 01\bar{1}0\ 0000\ 0\ \text{``}=\text{''}\ 1111\ 1021\ 0022\ 4$$

by

$$1000\ 0010\ 0010\ 1\ =\ 1000\ 0010\ 0010\ 1$$

(inner products 1 and -1) to get

$$1000\ 0100\ 0010\ 1\ \text{``}=\text{''}\ 0111\ 1011\ 0012\ 3\ ,$$

and this by

$$0100\ 1000\ 0001\ 1\ =\ 0100\ 1000\ 0001\ 1$$

(inner products -1 and 1) to get

$$1100\ 1100\ 0011\ 2\ \text{``}=\text{''}\ 0011\ 0011\ 0011\ 2\ ,$$

a much more manageable relation. The equality signs within quotes mean that the reflections in the corresponding vectors represent the same element of our group.

The method here is as follows. The reflection of a vector v in a root vector r is given by the formula $v' = v - ir$, where i is the inner product of v and r. The display consists alternately of a pair v, w of vectors linked by the sign "=" to show that they represent the same element of our group, followed by a reflecting vector r (shown twice) together with its inner products with v and w. The next line shows the images v' and w', which must therefore also represent the same group element as each other.

The reflecting vectors we use always have a single 1 in each of the blocks into which we have grouped our coordinates. The reflections in all such vectors are in our group, since these vectors are images of the particular vector 0001 0001 0001 1 under the group $S_4 \times S_4 \times S_4$ defined by the other generators (which group consists merely of the permutations of the coordinates within each block of four).

THE CLASSIFICATION OF $S(3,8)$ GROUPS.

It is now easy to show that this group is finite. We consider the set of elements represented by the vectors obtained from

$1\bar{1}00\ 0000\ 0000\ 0$ (the first set of 18 elements u)

0001 0001 0001 1 (the second set of 64 elements v)

0011 0011 0011 2 (the third set of 54 elements w)

by permuting the coordinates within each block of four, and by permuting the three blocks themselves. The numbers in parentheses are the numbers of group elements rather than vectors – although we get $6 \times 6 \times 6 = 216$ vectors from 0011 0011 0011 2, our relations show that they coincide in sets of 4, and so yield at most $216/4 = 54$ distinct group elements.

Now this set of (at most) 136 elements is a conjugacy class, since it is closed under conjugation by the 10 generators of our group. We can check this as follows. Each of the generators save

$$a = 0001\ 0001\ 0001\ 1$$

merely interchanges two coordinates of some block, so we need only check the closure under a.

If a vector u belongs to the first set of 18 vectors, and has inner product -1 with a, then its image $u + a$ under a is one of the set of 64 vectors in the second line. If instead u had inner product $+1$ with a, we could first replace it by its negative, which represents the same group element. A vector u that has inner product 0 with a is fixed by a.

A vector v from the second line is transformed to one from the first, second, or third line according as it has inner product $+1$, 0, or -1 with a.

We can always replace a vector w from the third set by a vector from that set whose inner product with a is at least 1, by using our relations. If this new inner product is 1, w' transforms back to a vector v from the second line, and if the inner product is 0, w' is fixed.

So our group permutes these 136 elements. Now any group element that permutes them trivially centralises the S_3 subgroup $\langle c'', d'' \rangle$ and so must belong to the S_8 subgroup $\langle d, c, b, a, b', c', d' \rangle$. It must therefore be trivial, since the generators of this latter group are themselves in our set of 136 elements, and S_8 has trivial center. Our group is therefore isomorphic to the group of permutations it induces on this set of 136 elements, which is obviously finite.

This group is readily identified. It would not have affected our calculations if we had read the first 12 coordinates modulo 2. But the group generated by these reflections modulo 2 is the 8-dimensional orthogonal group $GO^-(8,2)$ of minus type over the field of order 2. The corresponding 8-dimensional space is the set of all vectors

$$abcd\ efgh\ ijkl$$

over that field for which $a+b+c+d = e+f+g+h = i+j+k+l$, considered modulo the subspace

$$\{000000000000\ , 000011111111,\ 111100001111,\ 111111110000\}\ .$$

We have therefore proved the theorem:-

THEOREM. *The only $S(3,8)$-groups are the symmetric group S_{11} and the 8-dimensional orthogonal group of minus type over the field of two elements.*

This is a remarkably simple proof for a group characterization theorem.

THE GROUP Y_{552} AND FISCHER'S GROUP Fi_{24}.

An argument exactly analogous to the one we used for $S(3,8)$ groups shows that any $S(5,12)$ group is either isomorphic to S_{17} or is generated by 16 involutions which satisfy the Coxeter relations of the Y_{555} diagram, together with the further relations

$$z = z' = z'' = 1 \ ,$$

where z, z', z'' are the central involutions of the three obvious subgroups Y_{511}, Y_{151}, Y_{115}. (In fact the argument is slightly easier, since for example z centralises a subgroup $S_4 \times S_4 \times S_4$ of S_{12}, and the only element of S_{12} that does this is the identity.)

The Norton-Ivanov theorem now tells us that Y_{555} with these relations added is the Bimonster. In fact, as Soicher has shown, any one of the three extra relations suffices, and indeed we could define the same group by many alternative relations, in particular by the more symmetrical "Spider Relation"

$$(abcab'c'ab''c'')^{10} = 1 \ .$$

I had hoped to prove this theorem using the hyperbolic method, and of course did not succeed in this. However, I did manage to prove (with A.D.Pritchard in [1]) that the subgroup Y_{552} of the resulting group was the group $3Fi_{24}$.

The method is rather elegant. We choose coordinates so that the reflections in the vectors

$$1\bar{1}0000000000\ 00,\ 01\bar{1}000000000\ 00, \ldots ,0000000001\bar{1}\ 00$$

generate the obvious subgroup S_{12} of Y_{552}, while the remaining two generators (on the third arm) are

$$000000111111\ 01$$
$$\text{and}\quad 000000000000\ 1\bar{1}\ ,$$

in a (Lorentzian) space in which the norm of the typical vector

$$abcdefghijkl\ pq$$

is the sum of the squares of the 14 coordinates minus five times the square of $p+q$.

We then derive various consequences of the extra relations. These are neatly described by a table of "alias groups", which tell us that several different vectors represent the same element of our group. For example, we found in the Y_{333} case that the vectors 0011 0011 0011 2 and 1100 1100 0011 2 represent the same element of the group considered there, so that in fact the element represented by the vector 0011 0011 0011 2 is unchanged if we interchange the digits 0 and 1 in any two blocks, leaving the third block unaltered. We say that the alias group of this vector is

$$\{id|id|id,\ (01)|(01)|id,\ (01)|id|(01),\ id|(01)|(01)\}\ .$$

Table of Alias Groups for Fi_{24}:

Typical Vector,v	*Generators for Alias Group of v*	*Number of Images*
$(00000000001\bar{1}\ 00)$	$(1\bar{1})$|id	66
$(000000000000\ 1\bar{1})$	M1|$(1\bar{1})$	1
(000000111111 10)	M2|id	924
(000011112222 11)	M3|id	5775
(000111222333 12)	M4|id, (12)|(12)	30800
(001122334455 14)	M6|id	124740
(0123456789*ab* 1*a*)	M12|id, id|(1*a*),	5040
(001224466788 44)	A6|id	20790
(001234567899 45)	(12)(45)(78)|(12), (1234)(5678)|id, (09)|id	11800
	total	306, 936

(In this table, $\bar{1}$ denotes -1, a denotes 10, b denotes 11.)

In this table, there is a certain uniformity about some of the entries. The group M_n denotes the group of permutations of the set of digits

$$\{0, 1, 2, \ldots, n-1\}$$

generated by the two particular permutations

$$t \longrightarrow n-1-t$$

$$t \longrightarrow \min(2t, 2n-1-2t)\ .$$

For $n = 1, 2, 3, 4, 6, 12$ these groups are S_1, S_2, S_3, A_4, the group PGL(2, 5), (with 0, 1, 2, 3, 4, 5 playing the respective roles of $0, \infty, 1, 2, 3, 4$) and Mathieu's celebrated group on 12 letters (in the "shuffle" numbering).

The alternating group A_6 in the next-to-last entry permutes the set of "numbers" $\{0, 2, 4, 6, 8, \textit{odd}\}$, where "*odd*" denotes either 1 or 7, which can be interchanged without affecting the element represented.

The column headed "number of images" gives the total number of images obtained by letting the group $S_{12} \times S_2$ act by coordinate-permutations. Using the relations implied by the table, it is very easy to show that this set of 306, 936 involutions is a conjugacy class. (One obtains a conjugacy class of thrice this size in the larger group $3Fi_{24}$.)

This gives the easiest way to compute inside Fi_{24} that I know of. The calculations can easily be done by hand. For instance, to transform 000111222333 12 by 010101010101 10 we write one below the other and transform after choosing aliases to maximize the inner product:

thus the vector

$$000111222333\ 12\ ,$$

alias

$$000222111333\ 21$$

transforms by

$$010101010101\ 10$$

(inner product -2) to the desired answer

$$020424131535\ 41\ .$$

It happens that this is already "in standard form", being a permutation of one of the vectors in the Table. This is the usual case, but sometimes we obtain a vector v not yet in standard form. In such a case, we transform v by a permutation of 000000111111 01 chosen to have large inner product with v, so as to obtain a vector w in standard form. We then replace the new vector w by an alias w' of it (chosen to have large inner product with r), and then transform w' back by r to v', an alias for v that is now probably in standard form.

REFERENCES

1. J. H. Conway and A. D. Pritchard, *Hyperbolic reflections for the Bimonster and* $3Fi_{24}$, Groups, Combinatorics and Geometry (eds. M. W. Liebeck and J. Saxl), LMS Lecture Notes 165, 1992, pp. 24–45.
2. A. A. Ivanov, *A Geometric Characterization of the Monster*, Groups, Combinatorics and Geometry (eds. M. W. Liebeck and J. Saxl), LMS Lecture Notes 165, 1992, pp. 46–62.
3. S. P. Norton, *Constructing the Monster.*, Groups, Combinatorics and Geometry (eds. M. W. Liebeck and J. Saxl), LMS Lecture Notes 165, 1992, pp. 63–75.

DEPARTMENT OF MATHEMATICS, PRINCETON UNIVERSITY, PRINCETON, NJ 08540

E-mail address: conway@math.princeton.edu

DIMACS Series in Discrete Mathematics
and Theoretical Computer Science
Volume **11**, 1993

Combinatorial Tools for Computational Group Theory

GENE COOPERMAN AND LARRY FINKELSTEIN

ABSTRACT. A variety of elementary combinatorial techniques for permutation groups are reviewed. It is shown how to apply these techniques to yield faster and/or more space-efficient algorithms for problems including group membership, normal closure, center, base change and Cayley graphs. Emphasis is placed on randomized techniques and new data structures. The paper includes both a survey of recent algorithms with which the authors have been associated, and some new algorithms in the same spirit that have not previously appeared in print. Many of the results include both complexity bounds and pseudo-code, along with comments for faster software implementations.

Contents

1991 *Mathematics Subject Classification.* Primary 20B02, 20B04; Secondary 20P05.
The authors were supported in part by NSF Grants # CCR-8903952 and CCR-9204469.
This paper is in final form and no version of it will be submitted for publication elsewhere.

1. Introduction

The traditional approach in developing algorithms for computing with permutation group has been to study several disjoint problems such as *group membership*, *base change* and *normal closure* as distinct computational problems. While certain themes do appear in more than one context, such as the point stabilizer sequence, or possibly a base change, it has not been clear that progress on one front would necessarily result in progress on a different front. This had been our own viewpoint while searching for improved low-level data structures and algorithmic "subroutines". Each of those "building blocks" was tailored to a particular higher goal, such as group membership [**5, 6, 21**], base change [**18, 22**], a strong generating test [**17**], normal closure [**20**], and spanning lattices for Cayley graphs [**16**].

We no longer view these "building blocks" as algorithmic subroutines, but as solutions to fundamental problems in their own right. These fundamental problems tend to be the bottlenecks in deriving faster algorithms for the higher level problems.

Three examples suffice to illustrate the point. The first is the construction of a suitable data structure for the computation of coset representatives for G/G_x, where G is a permutation group acting on Ω and $x \in \Omega$. The choice of such a data structure is a critical component of most group membership algorithms. The tradeoff typically made in designing such a data structure is the time to recover an arbitrary coset representative versus the number of permutations which must be stored to accomplish this task. Finding the right balance is crucial in reducing the time complexity of group membership while still working within reasonable space constraints and motivated the work described in [**6**].

The second example is the construction of group elements with properties similar to random group elements [**5, 22, 24**]. It has long been known that the availability of truly random group elements leads to a very efficient group membership test [**2, 29, 22**].

The third example is reduction of a generating set to one of smaller size. Construction of a chain of subgroups often leads to combinatorial explosion in the number of generators. Although it has been known for some time how to manage this in the case of permutation groups, the methods developed in [**5**] are sufficiently general to be applied to other domains, such as matrix groups over finite fields. This is useful for many algorithms which test for a certain property by successively constructing generators for a chain of subgroups. This happens, for instance, in testing for solvability, where one must compute the normal closure of the set of commutators formed from the generating set for each successive subgroup in the derived series. Unless care is taken, the number of generators for each subgroup in the derived series will be at least the square of the number in its predecessor.

This paper illustrates this philosophy through a survey of recent results. Some have been previously published and some are new. Those methods that previously appeared in the literature are only sketched here, with pointers to a fuller description in the literature. Greater detail is provided for the "new" results. To fix notation, The permutation group is denoted G, and its degree is n. The new results include:

- an elementary $O(n^4 \log^2 n)$ group membership algorithm with at least $1 - \exp(-n)$ reliability (§2.1).
- an elementary $O(n^3)$ normal closure algorithm (with generators as output) with at least $1 - \exp(-n)$ reliability (§2.2).
- a single, moderately elementary $O(n^2 \log |G| \log n)$ group membership algorithm with at least $1 - 1/n$ reliability (§3.3).

The new results were chosen to emphasize the intuitive simplicity and purely combinatorial nature of the tools involved. The final complexity often represents only a modest improvement or alternative development as compared to our previously published algorithms [**5**, **6**]. The new results may also lend themselves to faster implementations in some situations. We distinguish between *small base groups* (families of groups satisfying $\log |G| = O(\log^c n)$ for some constant c) and *large base groups* (all other families of groups).

We frequently employ the notation $O^{\sim}(f(n)) \equiv O((f(n))(\log^{c'} n))$ for some constant c' (read as "soft oh" of $f(n)$). We also employ the related notation $g(n) = \Omega(f(n))$ and $g(n) = \Theta(f(n))$. $g(n) = \Omega(f(n))$ if and only if $1/g(n) = O(1/f(n))$ and $g(n) = \Theta(f(n))$ if and only if both $g(n) = \Omega(f(n))$ and $g(n) = O(f(n))$. The reliabilities have been included to better estimate the time-reliability tradeoffs. More detailed asymptotic estimates are provided in the body of the paper.

§2 develops the theme of random subproducts, a solution to the second of the three "building block" examples above. As an immediate application, we describe the elementary group membership and normal closure algorithms referred to above. These continue to be the fastest algorithms with exponential reliability, although Babai, Luks and Seress have recently announced a deterministic group membership algorithm with running time $O^{\sim}(|S|n^3)$.

§3 reviews our recent work on group membership algorithms for both large [**5**] and small [**6**] base, which are nearly optimal in a complexity sense made precise in that section. This work is joint with Babai, Luks and Seress. Additional combinatorial techniques, which were required to achieve those results, are developed first.

§3 also describes a near optimal elementary group membership algorithm with further small improvements in complexity. The new algorithm is particularly interesting because it continues a trend towards simplifying the mathematical basis for group membership algorithms. Babai, Luks and Seress had introduced a $O^{\sim}(n^4)$ algorithm [**8**] by using the classification of finite simple groups, and a structure forest analysis in terms of transitivity and primitivity. The $O(n^3 \log^4 n)$ Monte Carlo algorithm introduced in [**5**] requires only the structure analysis. The new approach leads to a $O(n^3 \log^2 n)$ Monte Carlo algorithm and requires neither the classification nor the structure analysis. By introducing a variant of Sims's "sifting" (or "stripping"), the new algorithm returns to some of the simplicity of Sims's original $O(n^6)$ algorithm [**35**].

Next, §4 reviews a sequence of results on base changes which are especially useful for the case of small base groups. §5 reviews joint work with Luks allowing transformations of other problems into group membership, often via a base change. §6 discusses a non-polynomial-time algorithm for computing and storing Cayley graphs and Schreier coset graphs, which we have found important in applications. The more applied tone is continued in §7 with a short discussion of our experience at software implementation of some algorithms presented in this paper.

1.1. Notation and Basic Concepts.

Let G be a permutation group acting on an n-element set Ω with G specified by a generating set S, and let $\alpha = (\alpha_1, \alpha_2, \ldots, \alpha_n)$ be a fixed ordering of the points of Ω. The *point stabilizer sequence* of G relative to α is the chain of subgroups

$$G = G^{(1)} \supseteq G^{(2)} \cdots \supseteq G^{(n)} = \{1\}$$

where $G^{(i)} = G_{\alpha_1, \ldots, \alpha_{i-1}}$, $1 \le i \le n$. S is called a *strong generating set* for G relative to α if

$$\langle S \cap G^{(i)} \rangle = G^{(i)}, \ 1 \le i \le n.$$

The central problem for group membership is the construction of a strong generating set. This allows one to efficiently construct a right transversal $T^{(i)}$ for $G^{(i)}/G^{(i+1)}$ for each i, a set of right coset representatives for $G^{(i)}/G^{(i+1)}$. Construction of a strong generating set S is usually carried out through a variation of Schreier generators. Given a transversal $T^{(i)}$ and generators $S^{(i)}$ for $G^{(i)}$, $G^{(i+1)}$ is generated by the *Schreier generators*

$$\{tg\overline{tg}^{-1} \colon t \in T^{(i)}, g \in S^{(i)}\},$$

where $\overline{tg}$ is the unique element of $T^{(i)}$ such that $tg\overline{tg}^{-1} \in G^{(i+1)}$. With a family of such transversals $\{T^{(i)}\}$ and with $g \in \mathrm{Sym}(\Omega)$, either one can express g as a factored word, $g_{n-1}g_{n-2}\cdots g_1$, for $g_i \in T^{(i)}$, or else $g \notin G$.

§2 relies on the fact that one can efficiently construct a transversal from the strong generators $S^{(i)}$. The time for this construction depends on the data structure for group membership being used. Specific data structures for group membership are defined at the beginning of §3.

The point α_i is called a *base point* relative to α if $|\alpha_i^{G^{(i)}}| \neq 1$. The sequence of all base points $B = (\alpha_{i_1}, \alpha_{i_2}, \ldots, \alpha_{i_b})$, with $i_1 < i_2 < \ldots < i_b$, is called an (ordered) *base* for G relative to α. The significance of a base is that each element g of G is uniquely determined by its base image $(\alpha_{i_1}^g, \alpha_{i_2}^g, \ldots, \alpha_{i_b}^g)$. Note that $b \leq \min(\log|G|, n-1)$. The size of a base may vary with the ordering α, but it is easy to show that two bases relative to two different orderings differ in size by at most a $\log n$ factor.

Finally, since many of the algorithms will be randomized ones, it is useful to review some terminology. A Monte Carlo algorithm is a randomized algorithm whose reliability (probability of success) can be increased at the cost of additional time. A Monte Carlo algorithm is Las Vegas, if it never returns an incorrect answer. (Hence, a Las Vegas algorithm may only return a correct answer or "don't know".) An algorithm is exponentially reliable if the probability of returning an incorrect answer or "don't know" is bounded above by $\exp(-n)$, for n the input parameter. A formal development of these definitions and their consequences is contained in the appendix.

Finally, for convenience, we will often say that a Monte Carlo algorithm runs in $O(f(n))$ time with *with fixed but arbitrarily high probability*. This is shorthand for saying that for any constant p with $0 < p < 1$, there is a constant c dependent on p such that if the algorithm runs for at most $cf(n)$ time, then it will return a correct answer with probability at least p.

2. Random Subproducts

It is known that efficient construction of random group elements would allow efficient randomized algorithms for many group constructions [**2, 22,29**]. Unfortunately in permutation groups, construction of truly random group elements usually depends on first constructing a strong generating set. (Babai has introduced an alternative approach toward approximately random group elements [**4**] which is sufficiently general to work over arbitrary black box groups.) This section introduces *random subproducts*, which can be efficiently constructed without knowing a strong generating set. Random subproducts were first defined for use in orbit calculations by Babai, Luks and Seress [**8**]. Their use in developing randomized algorithms for general group computations was first described in [**5**].

DEFINITION. Given a generating set $S = \{g_1, \ldots, g_r\}$ for a group G, a random subproduct on S is an element w of the form

$$w = g_1^{e_1} g_2^{e_2} \cdots g_r^{e_r},$$

where $e_i = 0$ or 1 with probability $1/2$.

PROPOSITION 2.1. (from [**5, Lemma 2.2**]) *Let S generate a group G and let H be an arbitrary proper subgroup of G. Then a random subproduct w on S is not a member of H, with probability at least* $1/2$.

PROOF. Let $S = \{g_1, \ldots, g_r\}$. There is a largest i for which $g_i \notin H$ (since $H \neq G$). So, w can be decomposed into the form $w = ug_i^{e_i}v$, with $g_{i+1}, \ldots, g_r \in H$. Hence, $v \in H$.

We consider two cases. First, assume $u \in H$. With probability 1/2, $e_i = 1$, and so $w = ug_iv \notin H$. In the second case, assume $u \notin H$. With probability 1/2, $e_i = 0$, and so $w = uv \notin H$. This proves the proposition. □

The parameter L, an upper bound on the length of subgroup chains for a group G, plays an important role in applications of random subproducts. For permutation groups, Cameron, Solomon and Turull [**14**] have shown, using the classification of finite simple groups, a bound $L < 3n/2$ on the length of maximal subgroup chains. (Babai [**3**] had previously demonstrated a bound of $2n-3$ solely using elementary techniques.) For small base groups, $L = O^{\sim}(1)$. For subgroups of the general linear group $GL(n, q)$, there is an upper bound of $L = n^2 \log q$.

As an illustration of the use of random subproducts, consider an arbitrary finitely generated group $G = \langle S \rangle$ with upper bound L on the length of a subgroup chain. By using Proposition 2.1, we are able to show that for $c > 4$, cL random subproducts generate G with probability at least $1 - \exp(-(1 - 4/c)^2 cL/8)$. Hence, a generating set S can be replaced by a generating set of size cL in the time for $O(L|S|)$ group multiplications. A stronger version of this result is presented in §2.3.

From this initial idea of random subproducts, a series of strikingly improved randomized algorithms have been developed, that use little more than this basic idea and some combinatorics. (The parameter n is the permutation degree and S is the generating set. For simplicity, the generating sets are assumed to be of size $O(n)$ for the first three results, and the fourth result shows how to efficiently find such a generating set. The dependence on the size of the generating set and the reliability is given later in the paper.)

- §2.1: Elementary group membership: $O(n^4 \log^2 n)$ (exponential reliability)
- §2.2: Elementary normal closure: $O(n^3)$ (exponential reliability)
- §2.4: Normal closure using random prefixes: $O^{\sim}(n^2)$
- §2.3: Reduction of a generating set S to size $O(n)$: $O(|S| n \log n)$
- §2.3: Testing solvability of matrix groups over finite fields in polynomial time

The first two algorithms are based on simple extensions of random subproducts, *random Schreier subproducts* and *random normal subproducts*. In both cases, an element formed using these extensions has probability at least 1/4 of enlarging the subgroup being incrementally constructed. The resulting chain of subgroups terminates in the target group. Random prefixes are a non-trivial variation of random subproducts.

The methods also apply to arbitrary black box groups. Black box groups were first introduced by Babai and Szemerédi [**10**]. Informally, they can be described as a representation for elements of an abstract group for which group multiplication, group inverse, and recognition of the identity element can all be "effectively" computed. The black box versions are stated in terms of L, a bound on the length of chains of subgroups.

- §2.2: Elementary normal closure of H in G where H and G have generating sets of size s and t respectively: $O((s+t)L)$ multiplications and inverses
- §2.3: Reduction of a generating set S for G to size $O(L)$: $O(|S|\log L)$ multiplications
- §2.4: Normal closure using random prefixes where H and G are given by generating sets of size $O(L)$: $O(L\log^4 L)$ multiplications

2.1. Random Schreier Subproducts: Elementary Group Membership.

For groups $H \subseteq G$, a *transversal* of H in G is a set of coset representatives of H in G. Let G be a finitely generated group with a generating set $S = \{g_1, g_2, \ldots, g_s\}$, let $H \subseteq G$ be a subgroup of finite index, and let $T = \{t_1, \ldots, t_n\}$ be a transversal of H in G. For $g \in G$, let $\overline{g}$ be the unique element $t \in T$ such that $gt^{-1} \in H$. A *random Schreier subproduct* for H with respect to S and T is an instance of a product $(t_1 g\overline{t_1 g}^{-1})^{e_1}(t_2 g\overline{t_2 g}^{-1})^{e_2}\cdots(t_n g\overline{t_n g}^{-1})^{e_n}$ where g is a random subproduct of S, and the e_i are independent random variables uniformly distributed over $\{0,1\}$. Note that a random Schreier subproduct requires at most $|S|-1$ multiplies to compute the random subproduct g and at most $3(|T|-1)$ additional multiplies to complete the computation, hence $O(|S|+|T|)$ multiplies overall. (We are assuming that each $\overline{t_i g}$ can be computed without any multiplies for each i, $1 \le i \le n$.) Random Schreier subproducts were first discovered in 1990, but have only recently appeared in print [**24**].

LEMMA 2.2. *Let $G = \langle S\rangle$ for finite S, and let $H \subseteq G$ be a subgroup of finite index. Let T be a transversal for H in G. Then a random Schreier subproduct h for H with respect to S and T can be computed at the cost of $O(|S|+|T|)$ group multiplications and inverses. Further, for any proper subgroup $K \subset H$,*

$$\mathrm{Prob}(h \notin K) \ge 1/4.$$

PROOF. Let K be an arbitrary proper subgroup of H. Given a generating set $S = \{g_1, \ldots, g_s\}$ for G, there exists a maximal index i and a $\sigma \in T$, such that $\sigma g_i(\overline{\sigma g_i})^{-1} \notin K$. This follows from the fact that $K \subset H$ is a proper subgroup, and the Schreier generators generate all of H. By assumption, $\tau g_j(\overline{\tau g_j})^{-1} \in K$ for all $\tau \in T$ and $j > i$.

Given a random subproduct g on S, we show that with probability at least $1/2$ there exists a $\tau \in T$ such that $\tau g(\overline{\tau g})^{-1} \notin K$. Let u and v be prefixes and suffixes of g uniquely defined by $g = ug_i^{e_i}v$. The calculations below implicitly use the fact that $\overline{\overline{\tau}y} = \overline{\tau y}$ and hence $\overline{\tau}y(\overline{\overline{\tau}y})^{-1} = \overline{\tau}y(\overline{\tau y})^{-1}$. The argument divides into two cases. In the first case, $\tau u(\overline{\tau u})^{-1} \in K$ for all $\tau \in T$. So, with probability $1/2$, $e_i = 1$, which implies $g = ug_iv$ (and so $\sigma u^{-1}g = \sigma g_i v$) and

$$\overline{\sigma u^{-1}}g(\overline{\sigma u^{-1}g})^{-1} = \underbrace{\overline{\sigma u^{-1}}u\sigma^{-1}}_{\in K}\ \underbrace{\sigma g_i(\overline{\sigma g_i})^{-1}}_{\notin K}\ \underbrace{\overline{\sigma g_i}v(\overline{\sigma g_i v})^{-1}}_{\in K} \notin K.$$

In the second case, there is a $\sigma' \in T$ such that $\sigma' u(\overline{\sigma' u})^{-1} \notin K$. So with probability 1/2, $e_i = 0$, which implies $g = uv$ and

$$\sigma' g (\overline{\sigma' g})^{-1} = \underbrace{\sigma' u (\overline{\sigma' u})^{-1}}_{\notin K} \; \underbrace{(\overline{\sigma' u}) v (\overline{\sigma' uv})^{-1}}_{\in K} \notin K.$$

Let h be a random subproduct on $\{\tau g (\overline{\tau g})^{-1} \colon \tau \in T\}$. If $\tau g (\overline{\tau g})^{-1} \notin K$ for some $\tau \in T$, then $h \notin K$ with probability at least 1/2. Since the exponents $\{e'_i\} \subseteq \{0,1\}$ for the random subproduct h are chosen independently of the $\{e_i\}$ for g, there is an overall probability of at least 1/4 that $h \notin K$. □

An important application of Lemma 2.2 occurs in efficiently finding generators of $H \subseteq G$ when only generators for G and a transversal of H in G are known. This is the case in finding generators of a point stabilizer subgroup. The proof of reliability requires Chernoff's bound [**15**].

CHERNOFF'S BOUND. *Let S_t be a random variable equal to the number of successes in t independent Bernoulli trials in which the probability of success is p $(0 < p < 1)$. Let $0 < \epsilon < 1$. Then*

$$\mathrm{Prob}(S_t \leq \lfloor (1-\epsilon)pt \rfloor) \leq e^{-\epsilon^2 pt/2}.$$

THEOREM 2.3. *Let G be a (possibly infinite) group with finite generating set S, let L be an a priori upper bound on the length of subgroup chains in G, and let T be a transversal for $H \subseteq G$. Given a constant $c > 4$, one can construct a set of cL generators for H with probability at least $1 - \exp(-(1-4/c)^2 cL/8)$ using at most $cL(|S| + 3|T| - 4)$ group multiplications and inverses.*

PROOF. We will construct a set $\{h_1, \ldots, h_{cL}\}$ of random Schreier subproducts, which generates H with the stated probability. The construction of each random Schreier subproduct takes at most $|S| + 3|T| - 4$ group operations and so the bound on the number of group operations follows directly. Define the generation of each h_i as a trial, and define the event E_i as the condition that either $h_i \notin \langle \{h_1, \ldots, h_{i-1}\} \rangle$ or $\langle \{h_1, \ldots, h_{i-1}\} \rangle = H$. Then by Lemma 2.2, $\mathrm{Prob}(E_i) \geq 1/4$ for all i. Note that $\mathrm{Prob}(H = \{h_1, \ldots, h_{cL}\})$ is bounded below by the probability that E_i holds for at least L distinct i. If the generation of each h_i is a trial, and the event E_i is a "success", then this probability is in turn bounded below by the probability of at least L successes in cL Bernoulli trials with parameter $p = 1/4$. Setting $t = cL$, $p = 1/4$, and $\epsilon = 1 - 4/c$ in Chernoff's bound, we can estimate the probability as $\mathrm{Prob}(S_t > L) \geq 1 - \exp(-(1-4/c)^2 cL/8)$. □

For permutation groups, G, of degree n, one can apply Theorem 2.3 to finding generators of a point stabilizer group, G_x. For this case, $|T| \leq n$ and $L = 3n/2$ [**14**], yielding the following corollary.

COROLLARY 2.4. *Let G be a permutation group of degree n, generated by S. Given $c > 4$, one can construct a set of cn generators for a point stabilizer subgroup of G with probability at least $1 - \exp(-(1-4/c)^2(3/16)cn)$ using at most $(3/2)cn(|S| + 3n - 4)$ group multiplications and inverses.*

An algorithm to construct the full point stabilizer sequence can be derived through $n-1$ applications of Corollary 2.4. In the following procedure, the subroutine `First-point-moved` is called with a set U of permutations and returns either the first point $j < n$ moved by some element of U or n if each element of U is the identity permutation.

Procedure Construct-SGS
Input: A generating set S for G
Output: A list $(S^{(i_1)} \ldots, S^{(i_m)})$ of generating sets for the point stabilizer sequence of G where $i_1 < i_2 < \ldots < i_m$ and $S^{(i_j)}$ generates $G^{(i_j)}$

Let $\beta_1 \leftarrow$ `First-point-moved`(S)
If $\beta_1 = n$ then return()
Else set $S^{(\beta_1)} \leftarrow S$
Set $j \leftarrow 1$
Loop:
  Let T be a transversal for $(\langle S^{(\beta_j)}\rangle)_{\beta_j}$ in $S^{(\beta_j)}$
  Let U be a set of $18n$ random Schreier subproducts
    with respect to $S^{(\beta_j)}$ and T
  Let $\beta_{j+1} \leftarrow$ `First-Point-Moved`(U)
  If $\beta_{j+1} = n$ then return($S^{(\beta_1)}, \ldots, S^{(\beta_j)}$)
  Else set $S^{(\beta_{j+1})} \leftarrow U$, set $j \leftarrow j+1$, and goto Loop

COROLLARY 2.5. *Given a permutation group $G = \langle S\rangle$ of degree n, procedure* `Construct-SGS` *will return a strong generating set for G in time $O(n^4 + n^2|S|)$ with probability at least* $1 - \exp(-n)$.

PROOF. The proof is by repeated application of Corollary 2.4. The time to compute the first transversal and the $18n$ random Schreier subproducts with respect to $S = S^{(\beta_1)}$ is $O(n^2|S|+n^3)$, and $|S^{(\beta_2)}| = O(n)$. The time for each later iteration is $O(n^3)$. The probability that $\langle S^{(\beta_i)}\rangle = G^{(\beta_i)}$ is at least $1 - \exp(-2n)$ by application of Corollary 2.4 with $c = 18$. The probability that this will hold for all β_i is at least $(1 - \exp(-2n))^n \geq 1 - n\exp(-2n) = 1 - \exp(\log(n) - 2n) \geq 1 - \exp(-n)$. □

REMARK. *The procedure* `Construct-SGS` *can be upgraded to an $O(n^4)$ Las Vegas group membership algorithm by applying the strong generating test presented in* [**17**].

2.2. Random Normal Subproducts: Elementary Normal Closure.

In applications to group membership, one often needs only to find generators (not necessarily strong) of a normal closure. In this case, algorithms that are faster than group membership are available [**5**]. An elementary algorithm ($O(n^3)$) is presented first, followed by a more complicated $O(n^2 \log^4 n)$ algorithm in §2.4. The $O(n^3)$ normal closure algorithm uses the idea of incremental subgroup construction described in §2. However, instead of random subproducts, it uses *random normal subproducts.* Random normal subproducts were discovered in 1990, but have not yet appeared in print. Since the techniques described here apply to black box groups, the discussion will be formulated in terms of a known upper bound L for the length of chains of subgroups.

The problem of constructing strong generators for a normal closure is distinct from finding any generating set, and has long had an unclear relation to the problem of group membership. Our recent work with Luks [**20**], discussed in §5, showed that for permutation groups, normal closure can be efficiently reduced to group membership. In particular, construction of strong generators of a normal closure requires $O^{\sim}(n^3)$ time in general and $O^{\sim}(n)$ time for a small base group.

Let $H \subseteq G$, with $H = \langle U \rangle$ and $G = \langle S \rangle$. A *random normal subproduct with respect to* U *and* S is a conjugate of the form h^g where h is a random subproduct on U and g is a random subproduct on S.

LEMMA 2.6. *Let* $H = \langle U \rangle$ *and* $G = \langle S \rangle$ *be subgroups of a (possibly infinite) group* K *Suppose that* H *is not normalized by* G. *Let* h^g *be a random normal subproduct with respect to* U *and* S. *Then*

$$\text{Prob}(h^g \in \langle H^G \rangle \setminus H) \geq 1/4.$$

PROOF. By hypothesis, the normalizer $N_G(H)$ of H in G is a proper subgroup of G. Hence if g is a random subproduct on S, then $\text{Prob}(g \notin N_G(H)) \geq 1/2$ by Lemma 2.2. So, $H^g \cap H$ is proper in H^g with probability at least $1/2$. Let h be a random subproduct on $U = \{h_1, h_2, \ldots, h_t\}$. The random normal subproduct $h^g = ({h_1}^{e_1} \cdots {h_t}^{e_t})^g = ({h_1}^g)^{e_1} \cdots ({h_t}^g)^{e_t}$, h^g can also be viewed as a random subproduct on $\{{h_1}^g, \ldots, {h_t}^g\}$, a generating set for H^g. Note that the choice of the random $\{e_1, \ldots, e_t\} \subseteq \{0,1\}$ is independent of those chosen for g. Given that $H^g \cap H$ is proper in H^g, it follows again from Lemma 2.2 that $\text{Prob}(h^g \notin H) = \text{Prob}(h^g \notin H^g \cap H) \geq 1/2$. Combining this with the probability $\text{Prob}(H^g \cap H \neq H^g) \geq 1/2$ yields $\text{Prob}(h^g \notin H) \geq 1/4$. □

THEOREM 2.7. *Let* H *and* G *be groups with* $H = \langle U \rangle$ *and* $G = \langle S \rangle$, *and let* L *be an a priori upper bound on the length of a chain of subgroups of* $\langle H^G \rangle$. *Let* $c > 4$ *be a given constant. Then one can construct a set of* cL *generators for* $\langle H^G \rangle$ *with probability at least* $1 - \exp(-(1 - 4/c)^2 cL/8)$ *using at most* $cL(|S| + |U| + (cL + 1)/2)$ *group multiplications and inverses.*

PROOF. Let $U' = U$ initially. Let h^g be a random normal subproduct with respect to U' and S, and then augment U' to $U' \cup \{h^g\}$. Repeat the process cL times. By Lemma 2.6, while $\langle U' \rangle$ is not normalized by G, each new random normal subproduct added to U' has probability at least $1/4$ of enlarging the subgroup $\langle U' \rangle$. The proof now follows as in Theorem 2.3. □

The following corollary uses the fact that if b is the size of an arbitrary base for a permutation group G of degree n and L is the length of a maximal subgroup chain in G, then $L \leq b \log n$. (This can be easily shown by first noting that $L \leq \log(|G|)$ and then observing that $|G| \leq n^b$, hence $log(|G|) \leq b \log(n)$.)

COROLLARY 2.8. *Let H and G be permutation groups of degree n with $H \subseteq G$ and let G have a base of size b. Assume that both H and G are given by generating sets of size at most $b \log n$. Then for a given constant $c > 4$, a generating set of size at most $cb \log n$ for $\langle H^G \rangle$ can be obtained with probability at least $1 - 1/n^{kb}$ for $k = c(1-4/c)^2(\log_2 e)/8$, using $cb^2 \log^2 n$ group multiplications and inverses. In particular, if G is a small base group, then the algorithm runs in time $O^{\sim}(n)$.*

2.3. Reduction of Generators.

Reduction of generators to a set of size $O(L)$ is an important principle when an algorithmic construction yields too many generators. Two examples are presented to illustrate this principle: an elementary group membership algorithm which runs in time $O^{\sim}(n^4)$; and a polynomial time test for solvability and nilpotence of a matrix group defined over a finite field.

The proof of the result on reduction of generators uses a *random subproduct on S of length ℓ*, a random subproduct on a sequence of of length ℓ, with elements randomly drawn (with replacement) from S. The proof is based on ideas first presented in [**5, Theorem 2.3**]. It avoids the explicit computation of one of the constants for purposes of brevity.

THEOREM 2.9. (from [**5, Theorem 2.3**]) *Let $G = \langle S \rangle$ be a finite group. Let L be a known upper bound on the length of all subgroup chains in G. Then for any fixed parameter p such that $0 < p < 1$, with probability at least p one can find a generating set S' with $|S'| = O(L \log(1/(1-p)))$, using $O(|S| \log L \log(1/(1-p)))$ group operations.*

PROOF. Let $|S| = s$. Assume $s > 4L$ or else there is nothing to do. Starting from $S' = \emptyset$, successively place in S', cL random subproducts on S, each of length s/L, for some $c > \lceil 4e \rceil$. We claim that after this initial phase, with probability at least $1 - e^{-cL/(16e)} > 1 - e^{-c/(16e)}$, fewer than L of the elements of S are outside of $\langle S' \rangle$. (e is the base of the natural logarithm.)

The claim can be rephrased in the language of indicator functions. Let $g_1, g_2, \ldots, g_{cL}$ denote the random subproducts on S produced by the procedure. Let ξ_i be the indicator variable taking value $\xi_i = 1$ if $g_i \notin \langle g_1, \ldots, g_{i-1} \rangle$ or if $|S \setminus \langle g_1, \ldots, g_{i-1} \rangle| < L$. Otherwise, set $\xi_i = 0$. Note that by the definition of L, $\sum_{i=1}^{cL} \xi_i > L \Rightarrow |S \setminus \langle g_1, \ldots, g_{cL} \rangle| < L$.

Next, we compute a uniform lower bound for $\text{Prob}(\xi_i = 1)$, independent of i and $\{g_1, \ldots, g_{i-1}\}$. If $|S \setminus \langle g_1, \ldots, g_{i-1} \rangle| \geq L$, then at least L of the s elements of S are not in $\langle g_1, \ldots, g_{i-1} \rangle$. So, a random sequence on S (with replacement) of length s/L contains at least one element of $S \setminus \langle g_1, \ldots, g_{i-1} \rangle$ with probability at least $1 - (1 - L/s)^{s/L} > 1/e$. Since g_i is a random subproduct on S of length s/L, by Proposition 2.1 $g_i \notin \langle g_1, \ldots, g_{i-1} \rangle$ with probability at least $1/(2e)$. Thus, in the case that $|S \setminus \langle g_1, \ldots, g_{i-1} \rangle| \geq L$, $\text{Prob}(\xi_i = 1) = \text{Prob}(g_i \notin \langle g_1, \ldots, g_{i-1} \rangle) > 1/(2e)$. In the other case, if $|S \setminus \langle g_1, \ldots, g_{i-1} \rangle| < L$, then $\text{Prob}(\xi_i = 1) = 1$. Hence, $\text{Prob}(\xi_i = 1) > 1/(2e)$ for all i, independently of $\{g_1, \ldots, g_{i-1}\}$.

This implies $\text{Prob}(\xi_i = 1 \mid \xi_1 = a_1, \dots, \xi_{i-1} = a_{i-1}) > 1/(2e)$ independently of $a_1, \dots, a_{i-1}$. So, $\text{Prob}(\sum_{i=1}^{cL} \xi_i \geq L)$ is bounded below by the probability of at least L successes in cL Bernoulli trials with probability parameter $p' = 1/(2e)$. Therefore one can apply Chernoff's bound (§2.1). Taking $\epsilon = 1 - 1/(cp')$, $t = cL$ and $S_t = \sum_{i=1}^{cL} \zeta_i$, we have

$$\text{Prob}(|S \setminus \langle S' \rangle| < L) \geq \text{Prob}(\sum_{i=1}^{cL} \xi_i > L) \geq 1 - e^{-\epsilon^2 p' t/2}.$$

For $c > 4e$, $\epsilon > 1/2$ and so $\text{Prob}(|S \setminus \langle S' \rangle| < L) \geq 1 - e^{-cL/(16e)} \geq 1 - e^{-c/(16e)}$. Note that as c goes to infinity, the lower bound on the last probability goes to 1, independently of s and L.

After this initial phase, we choose a new constant c' and continue for $\log_2 L$ more rounds. After round $i-1$ assume that $|S - \langle S' \rangle| < L/2^{i-1}$. The set S' will be augmented by $c'L/2^i$ products of random sequences of S of length $2^{i-1}s/L$ during round i. We shall show that with a suitable reliability, $|S - \langle S' \rangle| < L/2^i$ after round i. If $|S - \langle S' \rangle| < L/2^i$ already after round $i-1$, nothing need be proved. Otherwise, $L/2^i \leq |S - \langle S' \rangle| < L/2^{i-1}$, and define $\tilde{n} = 2^{i-1}s/L$, $\tilde{p} = |S - \langle S' \rangle|/s \geq L/(2^{i-1}s)$, and $\tilde{q} = 1 - \tilde{p} > (1 - L/(2^{i-1}s))$. So, a random sequence of S (with replacement) of length $\tilde{n} = 2^{i-1}s/L$ will include exactly one element of $S - \langle S' \rangle$ with probability $\tilde{n}\tilde{p}\tilde{q}^{\tilde{n}-1} > (2^{i-1}s/L)\,(L/(2^i s)) \times (1 - L/(2^{i-1}s))^{(2^{i-1}s/L)-1} > (1/2e)\,(1 - L/(2^{i-1}s))^{-1} > 1/(2e)$. If a product of the elements of the random sequence is added to S', then $|S - \langle S' \rangle|$ is diminished by at least one with probability at least $1/(2e)$. So we form $c'L/2^i$ such random sequences and add the corresponding products to S'. For a sufficiently large constant c', if $|S - \langle S' \rangle| < L/2^{i-1}$ was true after round $i-1$, then $|S - \langle S' \rangle| < L/2^i$ in round i, with probability at least $1 - 2^{-L/2^{i+1}}$.

After round number $\log_2 L$, $|S - \langle S' \rangle| = 0$ and S' generates G. The probability of error on any round subsequent to the initial one is at most $\sum_{i=1}^{\log_2 L} 2^{-L/2^{i+1}} < 1/2$. Finally, it is clear from Chernoff's bound that any fixed probability p of the theorem can be achieved by always taking a multiple of $\log_2(1/(1-p))$ times as many random subproducts and products over random sequences as indicated in the proof. □

REMARK. *The algorithm in the proof of Theorem 2.9 is Monte Carlo, since any probability p of correctness can be achieved at the cost of extra time.*

In application of Theorem 2.9 to the construction of subgroup chains such as the point stabilizer sequence or the commutator series, each subgroup in the chain is usually constructed using a generating set for its predecessor in the chain together with some defining condition. Without taking care to reduce the size of the generating set used for the construction, it is possible to have an exponential blowup in the number of generators as one proceeds down the subgroup chain.

As a first illustration of the importance of reduction of generators, we sketch an elementary $O(n^3 k \log^2 n)$ group membership algorithm [**5**], where k is the maximum orbit length. The algorithm consists of at most n point stabilizer constructions according to the algorithm described in Corollary 2.10 below. In this case, $L < 3n/2$. The algorithm associated with Corollary 2.10 consists of forming Schreier generators (of which there are at most nk), and then applying Theorem 2.9 to find $O(n)$ generators of the point stabilizer subgroup.

COROLLARY 2.10. (from [**5, Corollary 2.8**]) *Let $G = \langle S \rangle$ be a permutation group of degree n. Assume $|S| = O(n)$. Let $x \in \Omega$. Then for any fixed parameter p such that $0 < p < 1$, with probability at least p one can find a set of $O(n \log(1/(1-p)))$ generators of the point stabilizer subgroup G_x in Monte Carlo time $O(n^2 k \log n \log(1/(1-p)))$ where k is the length of the G-orbit of x.*

To achieve the elementary group membership, one forms the $O(n)$ point stabilizer subgroups using Corollary 2.10 with $p = 1-1/(2n)$. Each point stabilizer requires $O(n^2 k \log^2 n)$ time and achieves a reliability of at least $1-1/(2n)$. Thus, one achieves a strong generating set in $O(n^3 k \log^2 n)$ Monte Carlo time with reliability at least $1/2$.

A second illustration of the use of reduction of generators is a Monte Carlo polynomial time test for solvability and nilpotence of a matrix group $G = \langle S \rangle$ with S a set of $n \times n$ matrices over $GF(q)$. The crude bound $L = n^2 \log(q)$ suffices. The algorithm alternates between reducing the size of the generating set and computing generators for the next group in the derived series or descending central series via the normal closure algorithm. The reduction of generators employs probability of success $p = 1 - 1/(2L)$. After at most L iterations, if each of the resulting generators is the identity, then the test is satisfied and the algorithm asserts that the group is solvable or nilpotent respectively. Constants can be chosen so that the probability of error is most some constant p. Since each iteration takes polynomial time in n and $\log q$ and since there are at most L iterations with $L = n^2 \log q$, the combined test is polynomial in n and $\log q$. This gives rise to the following proposition.

PROPOSITION 2.11. (from [**5, Corollary 1.4**].) *A subgroup of $GL(n,q)$ can be tested for solvability and nilpotency by a Monte Carlo algorithm in the time for $O(L^3 \log^2 L)$ multiplications and inverses, with $L = n^2 \log q$.*

The estimate of $L = n^2 \log q$ is asymptotically tight. To see this, note that if $q = p^n$, for p a prime, then a Sylow p-subgroup of $GL(n,q)$ has order $p^{(\log_p q)n(n-1)/2}$. Hence for p fixed, only r varies, and $L = (\log_p q)n(n-1)/2 = \Omega(n^2 \log q)$.

2.4. Random Prefixes: Normal Closure Revisited.

An alternative normal closure algorithm which runs in time $O^{\sim}(n^2)$ for permutation groups of degree n will now be discussed. The key idea is the use of random prefixes to produce new elements for the group under construction.

DEFINITION. A *random prefix* of a sequence of group elements $(g_1, g_2, \cdots, g_k)$ is an instance of a product of the first i elements, for i a random integer from 1 through k.

To see how random prefixes are used in the normal closure algorithm, assume that the maximal length of a chain of subgroups of G is L and that G and H are each given by generating sets of size $O(L)$. For a suitably large m ($m = \Theta(L\log^2 L)$), form lists S_G and T of length m initialized with the generating sets for G and H respectively and padded to the end with the identity element. Let $r = \Theta(\log(m))$ and let $\pi_1, \ldots, \pi_r$ be random permutations of $\{1, \ldots, m\}$ chosen at the beginning of the computation. Apply randomly chosen elements π_i and π_j to S_G and T respectively, let g be a random prefix of $S_G^{\pi_i}$ and h a random prefix of T^{π_j}. Replace one of the original identity elements of T by h^g. Then with fixed but arbitrarily high probability, after $O(mr^2)$ rounds, the list T will contain generators for $\langle H^G \rangle$. In §2.3, we showed how to reduce the generating set T to one of size $O(L)$.

The proof of correctness depends on being able to "spread out" the distance between two arbitrary elements in the list S_G or T using some permutation in S_m. A set of permutations having this property is called a set of ϵ-spreaders. The proof shows that the set of random permutations $\{\pi_1, \ldots, \pi_r\} \subseteq S_m$ is in fact a set of ϵ-spreaders. This relies on the construction of special permutations in S_m, called *ϵ-spreaders.* Given an arbitrary subset of generators, at least one of $\log L$ ϵ-spreaders is guaranteed to permute a list of m generators, so that the sum of distances of each element of the arbitrary subset to its nearest neighbor from the same subset is bounded below by ϵ. Efficient deterministic construction of ϵ-spreaders remains an important open question. An answer would allow a modified normal closure algorithm achieving both exponential reliability and a smaller time complexity.

THEOREM 2.12. (from [**5, Theorem 3.5**]) *Let G and $H \subseteq G$ be finite groups and assume all subgroup chains in G have length at most L. Assume further that G and H are given by generating sets of size $O(L)$. Then one can construct $O(L)$ generators for $\langle H^G \rangle$ with fixed but arbitrarily high probability using $O(L\log^4 L)$ group operations.*

COROLLARY 2.13. (from [**5, Theorem 3.5**]) *Let H and G be permutation groups of degree n with $H \subseteq G$. Assume that both H and G have generating sets of size at most $O(n)$. Then a generating set of size $O(n)$ for $\langle H^G \rangle$ can be obtained with fixed but arbitrarily high probability using $O(n\log^4 n)$ group multiplications and inverses. If G is a small base group ($\log|G| = O^\sim(1)$), then $O^\sim(1)$ such group operations are required.*

3. Nearly Optimal Group Membership

This section reviews our work on "nearly optimal" group membership algorithms. The general group membership algorithm [**5**] operates in $O^\sim(n^3)$ time for generating sets of size $O(n)$. It is nearly optimal in the sense that its time approaches the time for Gaussian elimination. Gaussian elimination for matrices over a finite field is in some sense equivalent to a special case of this algorithm, and Gaussian elimination requires $O(n^3)$ time. The more specialized group membership algorithm for small base groups [**6**] operates in $O^{\sim}(n)$ time for generating sets of size $O^{\sim}(n)$. The time $O^{\sim}(n)$ is the same as the time to read the input, in the sense of "soft O", and so is considered nearly optimal for small base groups. Finally, a novel, elementary group membership algorithm is described, which duplicates the "near optimality" achieved [**BCFLS**] with a small gain in complexity.

The nearly optimal algorithms are all Monte Carlo. However, an exponentially reliable (or Las Vegas or deterministic) algorithm continues to elude us. This is an important open question, which would be solved by either an exponentially reliable or Las Vegas strong generating test operating in $O^{\sim}(n^3)$ time. (Note that a deterministic $O^{\sim}(n^3)$ strong generating test brings no special advantage, since, when combined with a Monte Carlo $O^{\sim}(n^3)$ group membership algorithm, it would yield a new group membership algorithm that is still only Las Vegas.) To date, we have presented a $O(|S|n^3)$ deterministic strong generating test [**17**], and Babai, Luks and Seress have also derived a $O^{\sim}(|S|n^3)$ deterministic strong generating test [**8**].

3.1. Efficient Data Structures for Schreier Trees and Group Membership.

Let $G = \langle S \rangle$ be a permutation group acting on $\Omega = \{\alpha_1, \alpha_2, \ldots, \alpha_n\}$. *Schreier trees* are often used for their space-efficiency in storing coset representatives of G_α in G for $\alpha \in \Omega$.

A *Schreier tree* for $G^{(i)}$ is a directed labeled tree $\mathcal{T}$ with root node α_i, nodes from $\alpha_i^{G^{(i)}}$, and edge labels chosen from a label set of permutations $S \subseteq G^{(i)}$. A label g for the directed edge (β, γ) satisfies $\beta^g = \gamma$. Coset representatives are obtained by taking the product of edge labels along a path from the root to the specified node. A *Schreier vector data structure* for G relative to the ordering α is a sequence $\{\mathcal{T}_i\}_{i=1}^n$ of *Schreier trees*, where $\mathcal{T}_i$ is a Schreier tree relative to $G^{(i)}$. $\mathcal{T}_i$ is *complete* if the set of nodes of T_i coincides with $\alpha_i^{G^{(i)}}$ and the Schreier vector data structure is *complete* if $\mathcal{T}_i$ is complete for $1 \leq i \leq n$. Schreier trees were first discovered by Sims [**35**], where they were called Schreier vectors. We have revised the terminology in keeping with more recent computer science concepts.

The permutation g is said to *sift through* $\{\mathcal{T}_i\}_{i=1}^n$ if we can write g in the form

$$g = g_{n-1} g_{n-2} \cdots g_1,$$

where g_i is a coset representative obtained from $\{\mathcal{T}_i\}$ (using the procedure described above) and $\alpha_i^{g g_1^{-1} \cdots g_{i-1}^{-1}} = \alpha_i^{g_i}$. The number of multiplications and inverses required to sift an element is proportional to the sum of the depths of the Schreier trees. In the case where the Schreier vector data structure is complete, the above factorization has two important implications. The first is the ability to generate random elements of G according to the uniform distribution. This is achieved by creating an element g whose factorization has the above form with the elements g_i chosen at random according to the uniform distribution from amongst the cosets for $G^{(i+1)}$ in $G^{(i)}$ defined by $\mathcal{T}_i$. The second implication is a test for membership in G of an arbitrary permutation. The test consists of attempting to factor an arbitrary permutation through $\{\mathcal{T}_i\}_{i=1}^n$. The factorization will succeed if and only if the element belongs to G. This was Sims's original group membership test [**35**].

Efficient Schreier trees must satisfy two goals. They must use a relatively small number of labels from G. This becomes critical as n approaches 1,000,000. Second, they must have shallow depth, so as to make recovery of a coset representative efficient.

It has been argued that Schreier trees built up by breadth-first search on some generating set are usually shallow. Nevertheless, there are important counter-examples. For example. just consider a Schreier tree for the cyclic group Z_n with one generator, for n large. Another example is $PSL(2,q)$, acting naturally on the points of the projective line, where the last point stabilizer subgroup is cyclic of order $(q-1)/(2,q-1)$. Another natural example is the symmetric group with generating set of size $n-1$, $\{(1\ 2), (2\ 3), \ldots, (n-1\ n)\}$.

3.1.1. *Short Schreier Trees.*

In the special case when random elements are available, one can build *short Schreier trees* with $O(\log n)$ labels and $O(\log n)$ depth. This is always the case, in the context of a base change, and was the basis for a fast randomized base change algorithm [**18, 22**]. The following key observation was discovered independently, but variations of it have appeared from as early as 1965 [**1, 7, 25**].

Let G act transitively on Ω. For fixed $P \subseteq \Omega$ and random $g \in G$,

$$E(|P^g - P|) = |P|\ |\Omega \backslash P|/n.$$

We identify P with the current nodes of a partially built Schreier tree. As long as fewer than half of the elements of Ω are in the tree, each random element has a fixed positive probability of expanding the tree by a constant factor. Thereafter, with fixed positive probability, each random element will decrease by a constant factor the number of nodes of Ω not yet in the tree.

This observation is the basis of the following theorem.

PROPOSITION 3.1. (from [**22, Theorem 3.5**]) *Let $\mathcal{O} = \alpha^G$. Then there exists a constant $c > 0$ (we can take $c = 21$) such that for all $\delta \geq 1$, $\lceil \delta c \log_2 |\mathcal{O}| \rceil$ random group elements suffice to build a short Schreier tree, with probability at least $1 - 1/|\mathcal{O}|^{2\delta}$.*

It is still an open problem to deterministically build short Schreier trees (depth $O(\log n)$) efficiently. This would eliminate the need for a source of random elements, and possibly further improve the time. We have almost accomplished this goal in the important class of *small base* permutation groups, as discussed in the next section.

3.1.2. *Cube Schreier Trees.*

Cubes [**10**] were originally invented by Babai and Szemerédi in the theoretical context of black box groups. A constructive adaptation was provided in [**6**] for building *cube Schreier trees* (Schreier trees of depth $O(\log |G|)$ with $O(\log |G|)$ labels). The construction is deterministic, and also is the basis for a fast deterministic cyclic base change [**19**] discussed in §4.

Given a sequence $R = (g_1, \ldots, g_r)$, define the cube

$$C(R) = \{g_1^{e_1} g_2^{e_2} \cdots \; g_r^{e_r} : e_i \in \{0, 1\}\}.$$

It is easy to see that for $g \in G$

$$g \notin C(R)^{-1}C(R) \iff |C(R \cup \{g\})| = 2|C(R)|.$$

Choosing such a g is made efficient by noting that

$$\alpha_1^g \notin \alpha_1^{C(R)^{-1}C(R)} \Rightarrow g \notin C(R)^{-1}C(R).$$

Thus, while $\alpha_1^{C(R)^{-1}C(R)} \neq \Omega$, we can double the size of the cube $|C(R)|$ by appending a g_{r+1} such that $\alpha_1^{g_{r+1}} \notin \alpha_1^{C(R)^{-1}C(R)}$. After at most $\log_2 |G|$ such steps, $\alpha_1^{C(R)^{-1}C(R)} = \Omega$ and $length(R) \leq \log_2 |G|$. A Schreier tree with depth at most $2 \log_2 |G|$ can then be built using labels $R^{-1} \cup R$ in the obvious way. Any Schreier tree with upper bound $2 \log_2 |G|$ on its depth and the number of labels is a *cube Schreier tree.* If one has a cube Schreier tree for each $G^{(i)}$ for $1 \leq i \leq n-1$, then the family of such trees is a *cube Schreier vector data structure.*

The cube Schreier trees have important implications for implementations. They were shown [**6**] to be one of the keys to a $O^{\sim}(n)$ algorithm for group membership. However, for efficient implementation, the obvious construction can be modified to the pseudo-code below, while preserving the same asymptotic time and space complexity. As compared to a breadth-first search to building Schreier trees, the cube approach below can often be faster in implementations by selectively adding new group elements that are guaranteed to at least double the size of the cube. At the same time, it provides a theoretical guarantee against very deep Schreier trees. This can be illustrated if the reader applies the following code to the case of Z_n with a single generator.

Procedure *Build-Cube-Schreier-Tree*(S, x)
Input: A generating set S for G and point $x \in \Omega$.
Output: A cube Schreier tree $\mathcal{T}$ for x^G with the property that $\texttt{Labels}(\mathcal{T}) = R \cup R^{-1}$, where $C(R)$ is a non-degenerate cube and $\texttt{depth}(\mathcal{T}) \leq 2|R|$.

Initialize $R \leftarrow ()$
Set $\texttt{root}(\mathcal{T}) \leftarrow \{x\}$, $\texttt{Labels}(\mathcal{T}) \leftarrow ()$
  While there exists $g \in S$ such that $\texttt{Nodes}(\mathcal{T})^g \neq \texttt{Nodes}(\mathcal{T})$ do
    Let $y \in \texttt{Nodes}(\mathcal{T})$ such that $y^g \notin \texttt{Nodes}(\mathcal{T})$
    Let $h = \texttt{Coset-rep}(\mathcal{T}, y)$
    Append hg to R

Build a new $\mathcal{T}$ using breadth first search with $R \cup R^{-1}$ to level $2|R|$

Return$\{\mathcal{T}\}$

PROPOSITION 3.2. (from [**19, Proposition 2.3**]) *Procedure Build-Cube-Schreier-Tree is correct and takes time* $O(n \log^2 |G| + n|S| \log |G|)$.

3.1.3. *Local Expansion (Babai).*

Babai's "local expansion lemma" [**4**] is the basis for an efficient randomized strong generating test for permutation groups [**6**], which requires only $O^{\sim}(|S|n)$ time for small base groups with generators S. As with most strong generating tests, when the generating set is not strong, the algorithm exhibits an element that does not sift through a group membership data structure based on the current generating set.

LEMMA 3.3. (from [**4, Local Expansion**]) *Let* $G = \langle S \rangle$, *and let* $T = S \cup S^{-1} \cup \{e\}$. *Let* T^t *be the set of t-term products of* T. *Then any subset* $D \subseteq T^t$ *of a group* G *satisfying* $|D| \leq |G|/2$ *also satisfies*

$$|Dg \backslash D| \,/\, |D| \geq 1/(4t)$$

for at least one generator $g \in S$.

Babai and Szegedy [**9**] recently found an inequality,

$$|DT \backslash D| \,/\, |D| \geq 2/(2t+1),$$

under suitable hypotheses, that is superior in many applications. In addition, by replacing the original hypothesis by the requirement that the diameter of the Cayley graph G with generators S be strictly larger than the diameter of the embedded subgraph D, they achieve the improved inequality [**9**],

$$|DT \backslash D| \,/\, |D| \geq 2/(t+2).$$

The principle of the strong generating test is to identify D with all factored words of G in S according to some group membership data structure such as Schreier trees. Assuming a base of size b, and assuming Schreier trees of depth f, t can be bounded by $t \leq bf = O(\log^c n)$. So if $|D| \leq |G|/2$, there is a $g \in S$ such that for random $d \in D$, $dg \notin D$ with probability at least $1/4t$. Alternatively, if $|D| \leq |G|/2$, for random $g \in S$ and $d \in D$, $dg \notin D$ with probability at least $1/(4t|S|)$. This test is repeated for each point stabilizer subgroup $G^{(i)}$ and strong generating set $G^{(i)} \cap S$.

Random subproducts further improve the strong generating test. Instead of testing if $dg \in D$ for random $d \in D$ and for each $g \in G$, it suffices to test if $dw \in D$ for w a random subproduct of the generating set S. If $|D| \leq |G|/2$, then $dw \notin D$ with probability at least $1/(64t)$ [**6, Lemma 3.4**], which can be improved to $1/(16t)$ using the newer constant for local expansion. This result should be compared with the probability of at least $1/(4t|S|)$ in the previous paragraph.

3.2. Group Membership: Large and Small Base.

From a complexity viewpoint in computational group theory, group membership is probably the problem most studied. Originally, in practical implementations, the formal algorithms were not competitive with a straightforward heuristic algorithms such as random Schreier Sims [**29**]. Using the new ideas described here, and in collaboration with Babai, Luks and Seress, we now have competitive Monte Carlo implementations for both the large and small base case based on new algorithms with provable complexities [**5, 6**]. (The timing in the first theorem is revised from the original.)

THEOREM 3.4. (from [**5, Theorem 4.1**]) *Given $O(n^2)$ generators of a finite permutation group G acting on n points, one can construct a strong generating set for G in $O(n^3 \log^4 n)$ Monte Carlo time. The space requirement of the procedure is $O(n^2 \log^2 n)$.*

THEOREM 3.5. (from [**6, Theorem 6.2**]) *Let $G = \langle S \rangle \leq Sym(\Omega)$ be a group with $|\Omega| = n$, and suppose that b is the maximal size of a non-redundant base. Then there is a Monte Carlo algorithm that returns a strong generating set for the case of transitive G in $O(n \log^3 |G| + b^3 (\log b) (\log^3 |G|) (\log n) + n|S| \log |G|)$ time. For intransitive G, an extension of the algorithm returns a strong generating set in $O(n \log^3 |G| + nb^2 \log^2 |G| \log(b + \log n) + b^3 (\log b) (\log^3 |G|) (\log n) + n|S| \log |G|)$ time. The constructed strong generating set supports membership testing in $O(n \log |G|)$ time and the memory requirement is $O(nb \log |G| + n|S|)$ during the construction.*

We have written straightforward implementations of about one to two thousand lines for both algorithms in our LISP system. Seress and Weisz have done the same in C, and report on their experience in this proceedings [**34**]. At this time, Leon's heuristic implementation [**29**] of randomized Schreier Sims is usually faster, but does not provide a theoretical guarantee of correctness under polynomial time bounds as do the newer implementations. However, at this stage we have not spent significant resources to achieve the fastest possible implementations, and still hold out the hope of combining heuristic speed with a theoretical proof of correctness.

3.3. Elementary Near Optimal Group Membership.

This section provides a general framework for a randomized group membership algorithm. The algorithm is presented in greater detail, since it has not yet appeared in the literature. The underlying principles of this algorithm are all rather elementary in comparison to the algorithm described in [**5**] yet we are still able to achieve a reduction in the time complexity from $O(n^3 \log^4 n)$ to $O(n^2 \log |G| \log n)$ with reliability at least $1 - 1/n$.

In effect, the algorithm of this section provides a theoretically sound mechanism for a fast Sims-type group membership using random elements. It has been observed by several authors, both theoretically [**2**] and heuristically in implementations [**29**], that if one could efficiently construct random elements of an arbitrary group, G, from generators of G, then one could derive a faster group membership algorithm. Typically, such an algorithm finds random generators of $G^{(2)}$ according to some distribution, and then repeats the process. This section continues that tradition, but also invokes the idea of cubes [**10**], as specialized to Schreier trees [**6**] (see §2.7), and maintains several global data structures, $S^{(i)}$, $C^{(i)}$, and $D^{(i)}$.

The global data structure $S^{(i)}$ represents a set of group elements of $G^{(i)}$ to be constructed, such that $\cup_{j=i}^{n-1} S^{(j)}$ generates $G^{(i)}$. $C^{(i)}$ is the cube over $S^{(i)}$. (For $S^{(i)} = \{g_{i,1}, \ldots, g_{i,k}\}$, $C^{(i)} = g_{i,1}^{e_1} \cdots g_{i,k}^{e_k}$ for $e_j \in \{0,1\}$.) The cube $D^{(i)} = C^{(n-1)}C^{(n-2)} \cdots C^{(i)}$ is derived from $C^{(i)}$, and is used purely as a notational convenience to describe the algorithm. Initially, $S^{(i)} = \emptyset$ and $D^{(i)} = C^{(i)} = \{1\}$. One need not store any data structure corresponding to $D^{(i)}$. The points $\{\alpha_i\}$ are the permutation domain. Sets such as $\alpha_i^{C^{(i)}}$ and $\alpha_i^{D^{(i)^{-1}}D^{(i)}}$ are re-computed whenever needed (although caching schemes are possible).

The algorithm is presented in a top-down manner. The overall control structure is that of a Sims-type algorithm: development of the point stabilizer sequence in a top-down fashion. The auxiliary routine *Random-Point-Stabilizer-Element* returns randomized elements of the point stabilizer subgroup, that play the role of Schreier generators. The auxiliary procedure *Deep-Sift* is used for "sifting" or "stripping" these pseudo-Schreier generators. Calculation of the pseudo-Schreier generators requires prior calculation of a transversal for the point stabilizer subgroup. This is the job of *Complete-Fundamental-Orbit.*

The algorithm is described in a somewhat general setting because we believe that the control structure can be modified to work with more specific classes of groups, such as small base groups, by invoking a different version of the auxiliary routine *Random-Point-Stabilizer-Element.* This routine should have the property that while $\langle D^{(i+1)}\rangle \neq G^{(i+1)}$, *Random-Point-Stabilizer-Element(i)* returns an element of $G^{(i+1)} \setminus \langle D^{(i+1)}\rangle$ with probability at least p. The implementation of *Random-Point-Stabilizer-Element* returns a random Schreier subproduct which works in all cases with $p = 1/4$ by Lemma 2.2. If a different version is used, then it may require a different value for p. Additionally, the two constants c and c' must be chosen based on the differing probability p.

Procedure *Strong-Generating-Set*(S, r)
Input: generating set $S \subseteq G$ and $r \geq n$
Parameters: constants $c > 0$, $0 < c' < 1$ chosen appropriately
Output: strong generating set $\{S^{(i)}\}$
Time: $O(nb \log|G| \log r + k'b \log r + n|S| \log|G|)$,
  where k' is the time for a single call to *Random-Point-Stabilizer-Element*
Reliability: at least $1 - 1/r$, provided $r \geq n$
  For $i \leftarrow 1$ to $n-1$ do
    Initialize $S^{(i)} \leftarrow \emptyset$
    Initialize $C^{(i)} \leftarrow \{1\}$

Set $S^{(1)} \leftarrow S$
For $i \leftarrow 1$ to $n-1$ do
  If $S^{(i)} \neq \emptyset$ then [$S^{(i)}$ constructed during previous calls to *Deep-Sift*]
    Complete-Fundamental-Orbit(i)
    [We now have generators for $G^{(i)}$
     and coset representatives for $G^{(i)}/G^{(i+1)}$]
    Repeat
      Initialize $successes \leftarrow failures \leftarrow 0$
      Loop for $c \log r$ iterations [c appropriate constant]
        Set $g \leftarrow$ *Random-Point-Stabilizer-Element*$(i+1)$
        [While $\langle D^{(i+1)} \rangle \neq G^{(i+1)}$, $\text{Prob}(g \in G^{(i+1)} \setminus \langle D^{(i+1)} \rangle) \geq p$.]
        Deep-Sift$(i+1, g)$
        If *Deep-Sift* returned "success" then
          $successes \leftarrow successes + 1$
        Else $failures \leftarrow failures + 1$
    Until $successes < c'(successes + failures)$ [for a constant $c' < p$]
    [Now, $\langle D^{(i+1)} \rangle = G^{(i+1)}$]
Return($\{S^{(i)}\}$)

Procedure *Deep-Sift*(i, g)
Input: $1 \leq i \leq n-1$; $g \in G^{(i)}$
Output: Either a level $i' \geq i$ and $g' \in G^{(i')}$ such that $|\alpha_{i'}^{C^{(i')}\{g',1\}}| = 2|\alpha_{i'}^{C^{(i')}}|$ or else "fail"; "fail" occurs only if $g \in {D^{(i)}}^{-1} D^{(i)}$
Time: $O(n \log |G|)$
  If $|\alpha_i^{C^{(i)}\{g,1\}}| = 2|\alpha_i^{C^{(i)}}|$ then [$g \notin {C^{(i)}}^{-1} C^{(i)}$]
    Set $S^{(i)} \leftarrow S^{(i)} \cup \{g\}$
    Set $C^{(i)} \leftarrow C^{(i)}\{g, 1\}$
    Return(i, g)
  Else if $i = n-1$ then
    Return("fail")
  Else [$\alpha_i^{C^{(i)}g} \cap \alpha_i^{C^{(i)}} \neq \emptyset$]
    Set $u \leftarrow$ arbitrary element of $C^{(i)}$ such that ${\alpha_i}^u \in {\alpha_i}^{C^{(i)}g^{-1}}$
    Set $u' \leftarrow$ arbitrary element of $C^{(i)}$ such that ${\alpha_i}^{ug} = {\alpha_i}^{u'}$
    Return(*Deep-Sift*$(i+1, ugu'^{-1})$)

Verifying the output property of *Deep-Sift* reduces to demonstrating that the property $g \notin {D^{(i)}}^{-1} D^{(i)}$ is preserved in recursive calls. To see this, note that $g \notin {D^{(i)}}^{-1} D^{(i)} \Rightarrow D^{(i)} g \cap D^{(i)} = \emptyset \Rightarrow D^{(i+1)} ug \cap D^{(i+1)} u' = \emptyset \Rightarrow D^{(i+1)} ugu'^{-1} \cap D^{(i+1)} = \emptyset \Rightarrow ugu'^{-1} \notin {D^{(i+1)}}^{-1} D^{(i+1)}$. Let j be the smallest k such that $D^{(k)} = C^{(k)}$. If $g \notin {D^{(i)}}^{-1} D^{(i)}$ then *Deep-Sift* returns (i', g') for which $i \leq i' \leq j$ and $g' \in G^{(i')} \setminus {C^{(i')}}^{-1} C^{(i')}$. This satisfies the output property.

Further, it will be crucial in later arguments to observe that no matter how often *Deep-Sift* is successively called with different arguments, there can never be more than $\log|G|$ successful calls to *Deep-Sift*. This follows from observing that the number of elements added to the cube $C^{(i)}$ can never grow to more than $\lfloor \log n_i \rfloor$ for each i. Further, the bound $\log n_i$ on the length of the cube $C^{(i)}$ shows that $O(n \log n_i)$ time is spent at the ith level, and $O(n \log|G|)$ time can be spent over all recursive calls from *Deep-Sift*.

Procedure *Complete-Fundamental-Orbit*(i)
Input: $1 \leq i \leq n-1$
Output: $\{C^{(i)}\}$ such that $\alpha_i^{D^{(i)^{-1}}D^{(i)}} = \alpha_i^{\langle D^{(i)}\rangle}$
Time: $O(n \log|G| + k)$, where k is the time for all subsequent calls to Deep-Sift
  While $\alpha_i^{D^{(i)^{-1}}D^{(i)}} \neq \alpha_i^{\langle D^{(i)}\rangle}$ do
    [Note that $\langle D^{(i)}\rangle = \langle \cup_{j\geq 1} S^{(j)}\rangle$]
    Let $u \in \langle D^{(i)}\rangle$ be such that $\alpha_i{}^u \notin \alpha_i^{D^{(i)^{-1}}D^{(i)}}$
    Deep-Sift(i, u)
  Return($\{C^{(i)}\}$)

Since the length of $D^{(i)}$ will always be bounded by $\log|G|$, computation of images of α_i can be done in $O(n \log|G|)$ time. At most one such set of computations occurs without a corresponding call to *Deep-Sift*, and the remaining computations are clearly dominated by time for the corresponding call to *Deep-Sift*, $O(n \log|G|)$. Hence the overall time of $O(n \log|G| + k)$ follows.

THEOREM 3.6. *Assume that Random-Point-Stabilizer-Element*($i+1$) *(as described in the code for Strong-Generating-Set) produces a g satisfying* $\text{Prob}(g \in G^{(i+1)} \setminus \langle D^{(i+1)}\rangle) \geq p$ *(for fixed p with $0 < p < 1$) whenever* $\langle D^{(i+1)}\rangle \neq G^{(i+1)}$. *If k' is the time for a single call to Random-Point-Stabilizer-Element, then Strong-Generating-Set completes in time $O(nb \log|G| \log r + k'b \log r + n|S| \log|G|)$ with reliability $O(1 - 1/r)$ for $r \geq n$ for appropriate values of the constants c and c'.*

PROOF. Note that *Strong-Generating-Set* makes at most $((1/c')\log|G| + cb \log r)$ calls to *Deep-Sift*. This is seen by noting that the "Until" condition of the the "Repeat ... Until" block will not trigger while $successes \geq c'(successes + failures)$. Thus, this condition will hold on all except the last iteration of the block. On these earlier iterations, $successes + failures \leq (1/c')successes \leq (1/c')\log|G|$ over the life of the algorithm. This follows since *successes* can be identified with the number of successful calls to *Deep-Sift*, which was previously shown to be bounded by $\log|G|$. Beyond that, when $successes < c'(successes + failures)$, there are exactly b iterations of the "Repeat ... Until" block with $c \log r$ calls to the inner "Loop". Thus, there are at most $((1/c')\log|G| + cb \log r)$ calls to *Deep-Sift*.

The time for each call to *Random-Point-Stabilizer-Element* and *Deep-Sift* is k' and $n \log|G|$, respectively. From the previous paragraph, we know that they are each called at most $((1/c')\log|G| + cb \log r)$ times. So, over the life of the algorithm, the calls to those routines from *Strong-Generating-Set* account for at most $((1/c')\log|G| + cb \log r)(n \log|G| + k') = O(nb \log|G| \log r + k'b \log r)$

time when $r \geq n$.

Next, we must bound the time for calls to *Complete-Fundamental-Orbit.* Since there are at most $O(\log|G| + b\log r)$ calls to *Deep-Sift* over the life of the algorithm, the time bound is $O(nb\log|G| + nb\log|G|\log r + k'b\log r)$, which is dominated by the previous times for $r \geq n$.

Finally, we prove the stated reliability. At level i, as long as $G^{(i+1)} \neq \langle D^{(i+1)}\rangle$ throughout the execution of the "Repeat... Until" block, we would like the probability that $successes \geq c'(successes + failures)$ to be sufficiently high. This will ensure that the termination criteria will not be invoked prematurely and the algorithm will not proceed incorrectly to level $i+1$. By the above analysis, there are at most $(1/c')\log|G|/(c\log r) = \log|G|/(cc'\log r)$ blocks for which $successes \geq c'(successes + failures)$. In order to have at least $1 - 1/r$ overall reliability, the probability of exiting a single block with the termination criterion $successes < c'(successes + failures)$ while $G^{(i+1)} \neq \langle D^{(i+1)}\rangle$ should be at most $1/(r\log|G|/(cc'\log r))$.

We invoke Chernoff's bound (see §2.1) with $t = successes + failures = c\log r$ and $S_t = successes$. Set $c' = (1-\epsilon)p$. (The parameter p is determined by the particular procedure *Random-Point-Stabilizer-Element.*) To ensure an overall probability of error of at most $1/r$ among all $\log|G|/(cc'\log r)$ blocks, or an error of at most $1/(r\log|G|/(cc'\log r))$ for each block, we invoke Chernoff's bound to satisfy

$$\begin{aligned}\operatorname{Prob}(S_t < c't) &\leq \operatorname{Prob}(S_t \leq \lfloor(1-\epsilon)pt\rfloor)\\ &\leq e^{-\epsilon^2 pc(\log r)/2} \leq \frac{1}{r}\cdot\frac{1}{\log|G|/(cc'\log r)}, \qquad \text{for } c' = (1-\epsilon)p.\end{aligned}$$

If $r \geq n$ (which implies $r \geq b \geq \log|G|/\log r$), this will be true if $\epsilon^2 pc(\log r)/2 \geq \ln(r^2/(cc'))$ for some value of c, since $\ln(r^2/(cc')) \geq \ln(r\log|G|/(cc'\log r))$. The inequalities $r\log r \geq \log|G|$ and $|G| \geq 2$ imply that $r > 1.5$, and it is clear that some value of c will satisfy the previous inequality for $r > 1.5$. This will then yield the reliability of $1 - 1/r$. □

COROLLARY 3.7. *If the procedure Random-Point-Stabilizer-Element returns a random Schreier subproduct, then for* $r = n \geq 10$ *and parameter values* $c' = 0.05$ *and* $c = 20$, *the algorithm Strong-Generating-Set performs in* $O(n^2\log|G|\log n + n|S|\log|G|)$ *time with reliability at least* $1 - 1/n$.

PROOF. Note that a random Schreier subproduct has probability at least $p = 1/4$ of lying in $G^{(i+1)} \setminus \langle D^{(i+1)}\rangle$. The time $O(n^2)$ for a random Schreier subproduct will enter into the time $k' = O(n\log|G| + n^2)$ for *Deep-Sift*. However, the term $k'b\log r$ will still be bounded by $n^2\log|G|\log n$. The rest is easily seen by setting $\epsilon = 0.8$, noting that $c' = (1-\epsilon)p$, and verifying the Chernoff inequality of the previous proof for $c = 20$. □

4. Base Change

Let G be a subgroup of S_n and let S be a strong generating set for the point stabilizer sequence of G relative to an ordering $\alpha = \alpha_1, \ldots, \alpha_n$ of $\{1, 2, \ldots, n\}$. A *change of base* is the construction of a strong generating set S' for G relative to a new ordering α'. Base change is a crucial algorithm in many important group computations and plays an especially important role in many of the backtracking algorithms currently used to solve problems for which efficient algorithms are not currently known. A *cyclic base change* occurs when α' is obtained from α through a *right cyclic shift*. In this case, α and α' satisfy the relationship

$$\alpha = \alpha_1, \ldots, \alpha_{r-1}, \alpha_r \ldots, \alpha_{s-1}, \alpha_s, \alpha_{s+1}, \ldots, \alpha_n$$

$$\alpha' = \alpha_1, \ldots, \alpha_{r-1}, \alpha_s, \alpha_r, \ldots, \alpha_{s-1}, \alpha_{s+1}, \ldots, \alpha_n$$

It is this special case of base change that is usually required in many backtracking algorithms [**12, 13, 30**].

We quickly review existing algorithms for performing a change of base and then describe new deterministic and randomized results for the special case of cyclic base change.

The first algorithm for performing a change of base was given by Sims [**35**]. This algorithm employs a clever trick and uses a Schreier vector data structure for representing the point stabilizer sequence. Sims's trick was generalized in [**11**] to show that a deterministic cyclic base change could be performed in time $O(n^2)$ using $O(n^2)$ space. This improved the running time of Sims's algorithm from $O(n^5)$ to $O(n^3)$. This speedup was due in part to the use of a new data structure for representing the point stabilizer sequence, first described by Jerrum [**28**] and referred to as a labeled branching,

The labeled branching data structure requires $\Theta(n^2)$ storage for transitive G and so is impractical when n is large. In the context of a base change, one already has a strong generating set S for G relative to α and a Schreier vector data structure $\{\mathcal{T}_i\}_{i=1}^n$ which can be used to compute random elements of G. Ideally, one would like to use $\{\mathcal{T}_i\}_{i=1}^n$ to compute random elements in time $O(n \log |G|)$. If the sum of the depths of $\mathcal{T}_i$ is $O(\log |G|)$, then this is the case and $\{\mathcal{T}_i\}_{i=1}^n$ is said to be a *short Schreier vector* data structure. Under this assumption, the randomized base change algorithm described in [**18**] takes time $O(n \log^2(|G|))$ and builds a short Schreier vector data structure relative to α'. The following result ensures that with some additional preprocessing time, one can always assume that $\{\mathcal{T}_i\}_{i=1}^n$ is short.

PROPOSITION 4.1. (from [**19, Theorem C**]) *Given a strong generating set S for G relative to α, one can compute with reliability $1 - 1/n$ a short Schreier vector data structure in time $O(n \log^2 |G| + n|S| + n \log n)$.*

The proof of Proposition 4.1 uses the methods of §3.1.1 and §3.1.2 in a novel way to achieve the time bound. For insight into how this is accomplished, assume that $\{\mathcal{T}_j\}_{j=i+1}^n$ is a short Schreier vector data structure for $G^{(i+1)}$ (relative to

α) which we want to extend to $G^{(i)}$. Let $n_i = |{\alpha_i}^{G^{(i)}}|$. The first step is to build a sequence R of elements of $G^{(i)}$ such that $|G^{(i+1)}C(R)| = |G^{(i+1)}|2^{|R|}$. If R is a maximal length sequence with this property, then $|R| \le \log n_i$ and

$${\alpha_i}^{G^{(i)}} = {\alpha_i}^{C(R)^{-1}G^{(i+1)}C(R)}.$$

Since each element of $G^{(i+1)}$ can be written as a word in the labels of $\mathcal{T}_j$, $j < i$, of length at most $c\log(|G^{(i+1)}|)$, it then follows that each coset representative of $G^{(i)}$ can be computed as a word of length at most $c\log|G^{(i)}|$. In particular, a random element for $G^{(i)}$ can be computed using at most $c\log^{|G^{(i)}|}$ multiplies. We can now use Proposition 3.1 to build a short Schreier tree $\mathcal{T}_i$ for ${\alpha_i}^{G^{(i)}}$ using at most $c\log n_i$ random elements. The timing for the ith iteration can be shown to be $O(n\log|G^{(i)}|\log n_i)$ (assuming that $|S| \le \log|G|$) and this leads to the final result.

THEOREM 4.2. (from [**18, Theorem 5.3**]) *Given a strong generating set S for G relative to an ordering α, one can with reliability $1 - 1/n$ perform a base change relative to an arbitrary ordering α' in $O(n\log^2|G| + n|S| + n\log n)$ time using $O(n\log|G|)$ space. Furthermore the algorithm returns a short Schreier vector data structure with respect to the new ordering.*

The next series of results concerns the important case of a cyclic base change. A *cube Schreier vector data structure* is one for which each Schreier tree can be represented by a cube $C^{(i)}$ of length at most $\log|G^{(i)}|$ such that $\alpha_i^{G^{(i)}} = \alpha_i^{C^{(i)^{-1}}C^{(i)}}$.

THEOREM 4.3. (from [**18, Theorem A**]) *Given a strong generating set S for G, a deterministic cyclic base change algorithm can be described that requires $O(n\log^2|G| + n|S| + n\log n)$ time and $O(n\log|G|)$ space. Furthermore, the algorithm returns a cube Schreier vector data structure for G relative to the new ordering.*

The key to the proof of Theorem 4.3 is the technique described in [**11**] in conjunction with the following result on the construction of cube Schreier trees.

PROPOSITION 4.4. (from [**19, Proposition 2.3**]) *Let S be a strong generating set for G relative to some ordering α. Then one can compute, in time $O(n\log^2|G| + n|S| + n\log n)$, both a sequence R of group elements of G such that $C(R)$ is non-degenerate and a complete cube Schreier vector data structure $\{\mathcal{T}_i\}_{i=1}^n$ with the following property. For each $\mathcal{T}_i$,* $\texttt{Labels}(\mathcal{T}_i) \subseteq R_i \cup R_i^{-1}$ *where $R_i = G^{(i)} \cap R$ is a prefix of R of length at most $\log|G^{(i)}|$, for $1 \le i \le n$. In particular, $\{\mathcal{T}_i\}_{i=1}^n$ requires $O(n\log|G|)$ space.*

The proof of Proposition 4.4 follows easily from the ideas used in the procedure *Build-Cube-Schreier-Tree*. It is interesting to note that within the same time bound, one can build a sequence R with $C(R)$ non-degenerate, such that every orbit of $G^{(i)}$, $1 \le i \le n$, can be represented by a cube Schreier tree with labels chosen from a suitable prefix of R.

Using randomized methods one can give a substantial theoretical improvement to Theorem 4.3 under the hypothesis of availability of fast generation of random group elements. This is satisfied if a *short Schreier vector data structure* is available.

THEOREM 4.5. (from [**19, Theorem B**]) *Assume random elements can be computed in time* $O(n \log |G|)$. *Let* b *be the size of a base relative to some ordering* α *and let* α' *be an ordering obtained from* α *by a right cyclic shift. Then a randomized cyclic base change algorithm can be described which has probability at least* $1 - 2/n$ *of using* $O(nb \log^2 n)$ *time and* $O(nb \log n)$ *space. Furthermore the algorithm returns a short Schreier vector data structure with respect to the new ordering.*

The proof of Proposition 4.5 depends on the following observation: If H is a subgroup of a finite group G, U a complete set of right coset representatives for H in G and $S \subseteq G$ a set of mutually independent uniformly random elements of G, then $T = \{g\bar{g}^{-1} : g \in S\}$ is a set of mutually independent and uniformly random elements of H. (For $g \in G$, $\bar{g}$ is the unique element of U so that $H\bar{g} = Hg$.) This observation is applied in the context of sifting down a subgroup chain of a permutation group G. The subgroup chain is chosen consistently with the problem of performing a right cyclic shift. Initially, we have a set of $c \log n$ mutually independent elements of G for some constant c. As we move down the subgroup chain, each element is multiplied by a suitable coset representative to produce a set of $c \log n$ mutually independent random elements for the next subgroup in the chain. These random elements are sufficient to build a short Schreier tree for that next subgroup using Proposition 3.1.

5. Reduction of Other Constructions to Group Membership

Another important construction [**20**] shows how to reduce certain group constructions to the point stabilizer construction (equivalent to group membership). In some situations, a partial strong generating set will already be known, and efficiency will be further improved through a base change algorithm. By carefully choosing generators of an initial group, and constructing a point stabilizer sequence with respect to a specified ordering of the points, one can then change the ordering and extract other groups. This method turns out to be surprisingly powerful, yielding efficient algorithms for normal closure, center, commutator subgroups and intersection of two groups, when one group is contained in the normalizer of the other. Further, all constructions yield a strong generating set for the constructed group in essentially the time to carry out the point stabilizer sequence, or less.

Let G and H be distinct permutation groups acting on an n-element set A. Let $\tilde{A} = A_1 \dot{\cup} A_2$, where A_1 and A_2 are copies of A. Let D be the subgroup of $Sym(\tilde{A})$ consisting of all elements of the form $\{(g, g) : g \in G\}$. Let $K = \langle D \cup (1 \times H) \rangle \subseteq G \times G$. Since each element of K can be written uniquely in the form (g_1, g_2)

for suitably chosen elements $g_1 \in G$, and $g_2 \in \langle H, G\rangle$, we can define projection maps $f_1\colon K \to G$, and $f_2\colon K \to \langle H, G\rangle$ according to the rules $f_1((g_1, g_2)) = g_1$ and $f_2((g_1, g_2)) = g_2$. Finally, let $K_1 = f_1(K_{A_2})$ and $K_2 = f_2(K_{A_1})$ where K_{A_1} and K_{A_2} are the pointwise stabilizer subgroups in K of A_1 and A_2 respectively. To fix notation, for elements x, y of a group we refer to $x^{-1}yx$ as the *conjugate* of y by x and denote it by y^x.

THEOREM 5.1. (from [**20, Theorem 2.1**]) *Under the above conditions,* $K_1 = \langle H^G\rangle \cap G$ *and* $K_2 = \langle H^G\rangle$.

PROOF. Observe that for sufficiently large k, an arbitrary element of K can always be expressed in the form $\bar{g}_0\bar{h}_1\bar{g}_1\bar{h}_2\bar{g}_2\cdots\bar{h}_k\bar{g}_k$, where $\bar{g}_1, \cdots, \bar{g}_k \in D$ and $\bar{h}_1, \cdots, \bar{h}_k \in 1 \times H$. Letting $\bar{g}_i = (g_i, g_i)$ and $\bar{h}_i = (1, h_i)$ (1 is the identity), it follows that

$$\begin{aligned} K &= \{(g_0g_1\cdots g_k,\, g_0h_1g_1h_2g_2\cdots h_kg_k) \mid g_i \in G, h_i \in H\} \\ &= \{(g_0g_1\cdots g_k,\, g_0g_1\cdots g_kh_1^{g_1\cdots g_k}h_2^{g_2\cdots g_k}\cdots h_k^{g_k}) \mid g_i \in G, h_i \in H\}. \end{aligned}$$

The last equation is similar to the one used to prove Schreier's lemma [**26, Lemma 7.2.2**]. In light of this description of K, we can "solve for g_0" in order to find the following characterizations of K_1 and K_2.

$$\begin{aligned} K_1 &= \{g_0g_1\cdots g_k \mid g_0g_1\cdots g_kh_1^{g_1\cdots g_k}h_2^{g_2\cdots g_k}\cdots h_k^{g_k} = 1, g_i \in G, h_i \in H\} \\ &= \{(h_1^{g_1\cdots g_k}h_2^{g_2\cdots g_k}\cdots h_k^{g_k})^{-1} \mid (h_1^{g_1\cdots g_k}h_2^{g_2\cdots g_k}\cdots h_k^{g_k})^{-1} = g_0g_1\cdots g_k \in G, \\ &\qquad\qquad g_i \in G, h_i \in H\} \\ &= \langle H^G\rangle \cap G \end{aligned}$$

$$\begin{aligned} K_2 &= \{g_0g_1g_2\cdots g_kh_1^{g_1\cdots g_k}h_2^{g_2\cdots g_k}\cdots h_k^{g_k} \mid g_0g_1\cdots g_k = 1, g_i \in G, h_i \in H\} \\ &= \{h_1^{g_1\cdots g_k}h_2^{g_2\cdots g_k}\cdots h_k^{g_k} \mid g_1, \ldots, g_k \in G, h_1, \ldots, h_k \in H \text{ are arbitrary}\} \\ &= \langle H^G\rangle \quad \square \end{aligned}$$

One of the most interesting deductions from this method is that construction of strong generators for a normal closure is no harder than construction of strong generators for a group of degree at most twice the original degree. Since group membership is in $O^{\sim}(n^3)$ time for large base groups, the time for normal closure is approximately a factor of 8 times slower than group membership. For small base groups, group membership is in $O^{\sim}(n)$ time, and the time will be approximately a factor of 2 slower.

The utility of the theory has been enhanced by our recent results. Consider the situation where G normalizes H. Under this assumption, it follows that K can be factored as $K = D(1 \times H)$. If in addition, strong generating sets are known for G and H respectively, then this factorization can be used to directly compute a strong generating set for K [**20, Lemma 2.2**]. This reduces the computation of $G \cap H$ to that of a base change. Strong generating sets for G and H can be computed in time $O(n^2\log(|G||H|)\log n + n(|S_G| + |S_H|)\log(|G||H|))$

(Corollary 3.7) with reliability at least $1-1/n$. Note that $|K| \leq |G||H|$. Applying Theorem 4.2 allows one to perform a randomized base change on K in time $O(n\log^2(|G||H|) + n(|S_G| + |S_H|) + n\log n)$, which then yields the following enhancement of [**20, Theorem 1(iii)**].

THEOREM 5.2. *Suppose G and H act on n points and G normalizes H. Let G and H be generated by S_G and S_H respectively. Then a strong generating set for $G \cap H$ can be computed in time $O(n^2\log(|G||H|)\log n + n(|S_G|+|S_H|)\log(|G||H|))$ with reliability at least $1 - 2/n$.*

In a journal version of [**20**], the authors together with Luks will in fact extend the method described in Theorem 5.1 to show that additional group constructions may be reduced to the point stabilizer sequence. One that is particularly interesting and indicates the flavor of the method is the construction of the commutator subgroup. Using the notation earlier, let $G \subseteq Sym(A)$ and and let $\tilde{A} = A_1\dot{\cup}A_2\dot{\cup}A_3$, where A_1, A_2 and A_3 are copies of A. Let $D_1 \subseteq Sym(A_1\dot{\cup}A_2)$ consist of all elements of the form $\{(g,g): g \in G\}$ and let D_2 be the analogous subgroup of $Sym(A_2\dot{\cup}A_3)$. Let $K = \langle (D_1 \times 1) \cup (1 \times D_2)\rangle$. Then the projection K_3 of $K_{A_1\dot{\cup}A_2}$ onto A_3 is the commutator subgroup of G.

6. Compact Data Structures for Cayley and Schreier Coset Graphs

Given a generating set Φ for a group G, the *minimum word problem* is to express an arbitrary element of G as a word in Φ of minimal length in $\Phi \cup \Phi^{-1}$. This problem is known to be *P-space complete* [**27**]. Nevertheless, one practical solution is to calculate a minimum depth spanning tree of the Cayley graph for the group relative to Φ. The *Cayley graph* of G with respect to generators Φ is a labeled graph $\widetilde{G}$ such that the nodes of $\widetilde{G}$ correspond to the group elements of G, and an edge (g,h) of $\widetilde{G}$ (for $g,h \in G$) is labeled by a generator $\phi \in \Phi \cup \Phi^{-1}$ such that $g\phi = h$.

We have developed two distinct techniques [**16**] to construct spanning trees. One is more space-efficient. It requires $\log_2 5$ bits per node ($|G|(\log_2 5)/8$ bytes) plus a much smaller amount of additional space for temporary data structures. However, as part of a space-time tradeoff, it operates in time $O(|G|(d + |\Phi|b^2))$ instead of the $O(|G||\Phi|)$ time of a traditional breadth-first implementation. (Here b is the base size of G represented as a permutation group of degree n, and d is the diameter of the Cayley graph.) The second technique is more CPU-efficient than the traditional breadth-first implementation, while retaining comparable space usage. While we have analyzed the formal worst-case complexity in terms of some more obscure parameters, the *typically* observed time in computer experiments is more relevant. It was $\Theta(|G|(b + d))$. Computer experiments have convinced us that for most computers and groups, space is more of a bottleneck than time, and so the first technique is the more valuable one.

Because the space-efficient data structure requires $\log_2 5$ bits per node of intermediate storage and $\log_2 3$ bits of final storage, we have referred to this as a

2-bit data structure, where $\log_2 4 = 2$ bits was jokingly considered as an average of $\log_2 3$ and $\log_2 5$ bits. An alternative formulation exists that uses only 2 bits per node for intermediate storage, but at the cost of some additional time.

The 2-bit data structure was used **[23]** to enumerate the group of order 3,674,160 for Rubik's 2-cube using only 1 megabyte of RAM on a SUN-3. Where $|G|$ is still larger, even the space-efficient method runs out of space, and approximate solutions (short words) need to be obtained by using a chain of subgroups and developing spanning trees for the respective *Schreier coset graphs* [**16**]. Finally, the techniques can often be generalized to other abstract groups containing an effective group multiplication and an effectively computable dense mapping between the group elements and the integers.

The 2-bit method can best be understood by comparison with a standard breadth-first search algorithm for building a spanning tree for a Cayley graph. Such an algorithm would require an array of length $|G|$. It might use hashing into an array to determine if a node labeled by a permutation has previously been seen. Pointers to neighbors, or at least to the parent, could then be stored in each entry.

There are two important principles for designing the 2-bit data structure. First, we assume a perfect, invertible hashing function, f. This is an encoding function $f : G \to \{0, 1, \ldots, |G| - 1\}$ along with its inverse. It is well known in computational group theory [**16**] that such a function is efficiently computable. This provides an implicit index into the array, so that the key (the permutation) need not be stored for each array slot.

Second, each array entry need not store pointers to some of the adjacent nodes. Instead, it suffices to store one of five values in each entry. (Hence, the requirement for $\log_2 5$ bits per node.) The numbers 0 to 2 indicate that the distance is known from the current node to the root modulo 3. (This is equivalent to the depth, if we view the data structure as representing a tree.) The number 3 represents a (frontier) node which is adjacent to a node with label from 0 to 2. The number 4 is for all other (unexplored) nodes.

Note that this construction suffices to find a shortest path from a node g to h. When considering the path as a sequence of labels drawn from the generating set Φ, this is the same as a path from the node $h^{-1}g$ to the identity. So, the problem reduces to finding a shortest path from a node to the root. Shortest paths are determined by repeatedly determining parent nodes until the root node is reached. Formally, the process reduces to calling *parent*(), initially on $f(h^{-1}g)$, and then calling it repeatedly on the result until the root node is reached. The function *parent* and associated data structures are defined as follows.

Let D be an array with $|G|$ entries (indexed from 0 to $|G| - 1$) capable of storing 5 distinct values. For $i \in [0, |G| - 1]$, let

$$\textit{node-neighbor}(i) = \{j \in [0, |G| - 1]\colon\ f^{-1}(i)\, g = f^{-1}(j) \text{ for some } g \in \Phi\}.$$

For i such that $D[i] \in \{0, 1, 2\}$, define $\textit{parent}(i) = \{j \in \textit{node-neighbor}(i)\colon\ D[j] =$

$D[i] - 1 \mod 3\}$. Note that for all nodes except the root node, *parent*(i) must be non-empty, since $D[i]$ can be interpreted as the depth of a spanning lattice. *parent*(i) may have more than one element, and so the data structure should be interpreted as a spanning lattice rather than a spanning tree.

Procedure *Build-Compact-Data*
Input: A generating set Φ *for* G
Output: The array D giving the 2-bit data structure.
Global Variables: A labeled branching data structure for G relative to α (or any other data structure, such as a Schreier vector, for computing f and f^{-1} in *node-neighbor*).

```
Initialize D[0] ← 0 and all other entries in D ← 4 [unexplored]
For each k ∈ node-neighbor(0), set D[k] ← 3 [frontier]
Set level ← 0
[Loop through nodes which are at the current level.]
LOOP:
    Set level ← level + 1
    For i ← 0 to |G| − 1 do
        If D[i] = 3 then
            If ∃j ∈ node-neighbor(i) with D[j] = level − 1 mod 3 then
                Set D[i] ← level mod 3
                For each k ∈ node-neighbor(i) with D[k] = 4 do
                    Set D[k] ← 3 [frontier]
    If any new nodes were set to 3 in current loop, go to LOOP
    Else return D
```

THEOREM 6.1. (from [**16, Theorem 4.3**]) *Let T be the time for each call to the encoding function f, and T' the time for each call to f^{-1}. Then the array D can be constructed in time $O(|G|(d + |\Phi|T + T'))$ using space of $\log_2 5$ bits per node plus space and time required for building group membership data structures associated with f.*

The *Schreier coset graph* of G with respect to a subgroup H and generating set Φ is a labeled graph $\widetilde{G}$ such that the nodes of $\widetilde{G}$ correspond to the set of right cosets G/H of G, and an edge $(Hg, H\bar{g})$ of $\widetilde{G/H}$ (for $Hg, H\bar{g} \in G/H$) is labeled by a generator $\phi \in \Phi \cup \Phi^{-1}$ such that $Hg\phi = H\bar{g}$. The 2-bit data structure for storing Cayley graphs extends to Schreier coset graphs relative to a subgroup H as long as there is an efficiently computable $1-1$ function $f_H : G/H \to \{0, 1, \ldots, |G|/|H| - 1\}$. In the case of permutation groups, we developed such a function [**16**]. Previously, Richardson had developed a technique [**31**] for finding lexicographically smallest elements of a left coset suitable for hashing of cosets. This function f_H acts as a perfect hash function, and requires the newer cyclic base changes of §4 for efficiency. Since the inverse of f_H is not necessarily efficiently computable, in many interesting examples a better alternative is the use of an intermediate subgroup K with $H \subset K \subset G$ such that $|G/K|$ and

$|K/H|$ are sufficiently small to support an encoding function based on factoring.

7. Software Implementations

Most of the tools and algorithms described here have been implemented in our LISP system, comprising approximately 25,000 lines, built up with limited manpower over about five years. Primarily, this manpower consisted of ourselves and our Ph.D. student, Namita Sarawagi, all working on a very part-time basis. The system was also used for small projects done by other students. LISP was chosen because the interactive environment coupled with an integrated debugger is convenient for rapid prototyping, while not sacrificing speed in compiled mode. In a small number of time-critical routines (primarily permutation multiply and permutation inverse), C routines are linked in to achieve close to optimal speed. A second important feature is a wide variety of data types, and flexible coercion between those data types. This has been especially important in experimenting with new data structures.

Having our own software system available was especially important in the beginning, in exploring the effect of using different low-level data structures. This facility has given us valuable insights into why certain data structures sometimes behave better than others in practice. Conversely, better theoretical understanding of algorithms has often led us to improved implementations.

Thus, we are often able to closely correlate our experimental times with theoretically derived asymptotic complexities. This has been especially true in the case of randomized algorithms, where the worst case examples tend to be "blurred" by the randomness, and the probabilistic worst case analysis tends to be close to the average case. For example, this was our experience with a randomized base change algorithm **[18]**, where, after scaling by a constant factor, experimental times agreed with theoretically predicted times within a factor of 1.5. This comparison was carried out over 10 permutation groups with orders ranging from 10^7 through 10^{55}, degrees ranging from 48 to 106,080, and base sizes ranging from 4 to 52.

Appendix. Randomized Algorithms: Terminology

Although the terminology below and its motivation are widely understood in theoretical computer science, some of it is less widely known among those in computational group theory. Since it is difficult to find a standard exposition, this brief review is included.

A *randomized algorithm* is one which, for a fixed input, may require varying amounts of time and return different outputs upon different invocations. In the absence of statements to the contrary, there is no guarantee that a randomized algorithm will return a correct answer.

DEFINITION M_1. *A randomized algorithm is a Monte Carlo algorithm if there is a function $f(n)$ such that for any probability $p < 1$, there is a reliability* *parameter t such that if the algorithm is allowed to run for time $tf(n)$ on an input whose size is parametrized by n, then it computes the correct answer for that input with reliability (probability of success) at least p. In such a situation, one says that the algorithm runs in $f(n)$ Monte Carlo time.*

In descriptions of a Monte Carlo algorithm, the reliability parameter, t, is sometimes omitted as an explicit input if the algorithm is to be run with a fixed reliability. A Monte Carlo algorithm is *Las Vegas* if it never returns an incorrect answer. Hence, a Las Vegas algorithm may only return a correct answer or "don't know". We say that an algorithm *runs in $O(g(n))$ Monte Carlo (Las Vegas) time*, if it is a Monte Carlo (Las Vegas) algorithm running in $f(n)$ time with $f(n) = O(g(n))$.

DEFINITION M_2. *(exponential reliability) A Monte Carlo algorithm has exponential reliability if there is a function $f(n)$ and a reliability parameter t such that if the algorithm is allowed to run for time $tf(n)$ on an input whose size is parametrized by n, then it computes the correct answer for that input with reliability (probability of success) at least $1 - e^{-n}$. In such a situation, one says that the algorithm runs in $f(n)$ Monte Carlo time with exponential reliability.*

The next theorem motivates the importance of exponential reliability over multiple invocations of an algorithm.

THEOREM A.1. *Suppose algorithm A runs in $O(f(n))$ Monte Carlo (Las Vegas) time with exponential reliability. If algorithm A' consists of invoking A for $g(n) \geq 3$ times on an input of size at most $n + \ln(g(n))$, then a variation of algorithm A' runs in $O(f(n+\ln(g(n)))\, g(n))$ Monte Carlo (Las Vegas) time with exponential reliability. If f and g are both polynomial functions, then the variation of algorithm A' runs in $O(f(n)g(n))$ polynomial time.*

Proof. Note that for a fixed input size n, an exponentially reliable algorithm can achieve higher reliability by simulating a larger input size (larger n). Hence, a variation of algorithm A' can simulate an input to A of size $n + \ln(g(n))$. The variation then runs in the stated time. The probability of error of each invocation of A is at most $e^{-(n+\ln(g(n)))}$, and so the overall probability of error is at most $g(n)e^{-(n+\ln(g(n)))} = e^{-n}$, as stated in the theorem. □

In many applications, the reliability does not increase exponentially with the input size, but it does increase exponentially with the amount of time allotted to it (the reliability parameter t). The next theorem addresses that. It assumes that each invocation receives an input of the same size, and is allowed to run for the same time. It is clear how to extend this to more general settings.

THEOREM A.2. *Let algorithm A be a Monte Carlo algorithm such that if it is allowed to run for $ctf(n)$ time for some constant c and some reliability parameter t, then its reliability will be exponential in t (at least $1-e^{-t}$). If algorithm A' consists of invoking A for $g(n) \geq 3$ times and allowing A to run for $2ct\ln(g(n))f(n)$*

time per invocation, then algorithm A' runs in $O(tf(n)g(n)\ln(g(n)))$ Monte Carlo (Las Vegas) time with reliability exponential in t.

Proof. Let $t = 2t'\ln(g(n))$ be the reliability parameter for A (for t' an arbitrary parameter). If algorithm A' is an invocation of algorithm A for $g(n) \geq \lceil e \rceil$ times with each invocation running for $ctf(n)$ time, and if algorithm A' is allowed to run for $g(n)ctf(n)$ time, then the probability of an error in A' over the $g(n)$ trials of A is at most $g(n)e^{-t} = g(n)/(g(n))^{2t'} \leq 1/(g(n))^{t'} \leq 1/e^{t'}$. So, A' has reliability at least $1 - e^{-t'}$, and A' is a $O(f(n)g(n)\ln(g(n)))$ Monte Carlo algorithm with reliability parameter t'. □

Note that a $O(tf(n))$ Monte Carlo algorithm that is exponentially reliable in the reliability parameter t can also be considered as a $O(nf(n))$ Monte Carlo algorithm that is exponentially reliable (in the input n). However, the previous theorem provides a finer estimate for such algorithms.

Acknowledgements

The authors would like to acknowledge many important discussions with László Babai, Gene Luks, Namita Sarawagi, and Ákos Seress. The authors would also like to acknowledge William Kantor for his careful reading and many helpful comments in earlier versions of this manuscript.

References

[1] D. Aldous, *On the Markov chain simulation method for uniform combinatorial distributions and simulated annealing*, Prob. Eng. and Informational Sciences **1** (1987), 33–46.

[2] L. Babai, *Monte-Carlo algorithms in graph isomorphism testing*, Université de Montréal Tech. Rep. D.M.S. 79-10, Dep. Math. et Stat., 1979.

[3] ———, *On the length of chains of subgroups in the symmetric group*, Comm. in Algebra **14** (1986), 1729–1736.

[4] ———, *Bounded round interactive proofs in finite groups*, SIAM J. Discr. Math. (to appear).

[5] L. Babai, G. Cooperman, L. Finkelstein, E.M. Luks, and Á. Seress, *Fast Monte Carlo algorithms for permutation groups*, Proc. 23rd ACM STOC, 1991, pp. 90–100.

[6] L. Babai, G. Cooperman, L. Finkelstein, and Á. Seress, *Nearly linear time algorithms for permutation groups with a small base*, Proc. of the 1991 International Symposium on Symbolic and Algebraic Computation (ISSAC '91), Bonn, 1991, pp. 200–209.

[7] L. Babai and P. Erdős, *Representation of group elements as short products*, Annals of Discrete Mathematics **12** (1982), 27–30.

[8] L. Babai, E. Luks, and Á. Seress, *Fast management of permutation groups*, Proc. 29th IEEE FOCS, 1988, pp. 272–282.

[9] L. Babai and M. Szegedy, *Local expansion in symmetrical graphs*, Combinatorics, Probability and Computing **1** (1992) (to appear).

[10] L. Babai and E. Szemerédi, On the complexity of matrix group problems I, Proc. 25th IEEE FOCS, 1984, pp. 229–240.

[11] C.A. Brown, L. Finkelstein, and P.W. Purdom, *A new base change algorithm for permutation groups*, SIAM J. Computing **18** (1989), 1037–1047.

[12] G. Butler, *Computing in permutation and matrix groups* II: *Backtrack Algorithm*, Math. Comp. **39** (1982), 671–680.

[13] G. Butler and C. Lam, *Isomorphism testing of combinatorial objects*, J. of Symbolic Computation **1** (1985), 363–381.

[14] P.J. Cameron, R. Solomon and A. Turull, *Chains of subgroups in symmetric groups*, J. of Algebra **127** (1989), 340–352.

[15] H. Chernoff, *A measure of asymptotic efficiency for tests of a hypothesis based on the sum of observations*, Annals of Math. Statistics **23** (1952), 493–507.

[16] G. Cooperman and L. Finkelstein, *New methods for using Cayley graphs in interconnection networks*, Discrete Applied Mathematics **37/38** (1992), 95–118 (to appear).

[17] ———, *A strong generating test and short presentations for permutation groups*, J. Symbolic Computation **12** (1991), 475–497.

[18] ———, *A random base change algorithm for permutation groups*, J. Symbolic Computation (to appear).

[19] ———, *A fast cyclic base change for permutation groups*, Proceedings of the International Symposium on Symbolic and Algebraic Computation (ISSAC 92), 1992.

[20] G. Cooperman, L. Finkelstein and E. Luks, *Reduction of group constructions to point stabilizers*, Proceedings of the International Symposium on Symbolic and Algebraic Computation (ISSAC 89), ACM Press, 1989, pp. 351–356.

[21] G. Cooperman, L. Finkelstein and P.W. Purdom, *Fast group membership using a strong generating test for permutation groups*, Computers and Mathematics (E. Kaltofen and S.M. Watt, eds.), Springer-Verlag, 1989, pp. 27–36.

[22] G. Cooperman, L. Finkelstein and N. Sarawagi, *A random base change algorithm for permutation groups*, Proc. of 1990 International Symposium on Symbolic and Algebraic Computation, (ISSAC 90), ACM Press and Addison-Wesley, 1990, pp. 161–168.

[23] ———, *Applications of Cayley graphs*, Applied Algebra, Algebraic Algorithms and Error-Correcting Codes (AAECC-8) (S. Sakata, ed.), Lecture Notes in Computer Science **508**, Springer-Verlag, 1991, pp. 367–378.

[24] G. Cooperman, L. Finkelstein and B. York, *Parallel implementations of group membership and the method of random subproducts*, Proc. of 1992 Dartmouth Institute for Advanced Graduate Studies in Parallel Computation Symposium (DAGS'92) (also Northeastern University Technical Report NU-CCS-90-17) (D. Johnson, F. Makedon, and P. Metaxas, eds.), 1992, pp. 94–100.

[25] P. Erdős and A. Rényi, *Probabilistic methods in group theory*, J. d'Analyse Math. **14** (1965), 127–138.

[26] M. Hall, Jr., *The Theory of Groups*, Macmillan, 1959.

[27] M. Jerrum, *The complexity of finding minimal length generator sequences*, Theoretical Computer Science **36** (1985), 265–289.

[28] ———, *A compact representation for permutation groups*, J. Algorithms **7** (1986), 60–78.

[29] J. Leon, *On an algorithm for finding a base and strong generating set for a group given by a set of generating permutations*, Math. Comp **35** (1980), 941–974.

[30] ———, *Computing automorphism groups of combinatorial objects*, Computational Group Theory (M. D. Atkinson, ed.), Academic Press, 1984, pp. 321–337.

[31] D. Richardson, *GROUP: A Computer System for Group-theoretical Calculations*, M.Sc. Thesis, Univerity of Sydney, 1973.

[32] N. Sarawagi, G. Cooperman and L. Finkelstein, *Group membership for groups with primitive orbits*, this volume.

[33] ———, *Computational Group Theory and Applications to Search*, Ph.D. Thesis, Northeastern University.

[34] Á. Seress and I. Weisz, *PERM: A Program Computing Strong Generating Sets*, this volume.

[35] C.C. Sims, *Computation with permutation groups*, Proc. Second Symposium on Symbolic and Algebraic Manipulation (S.R. Petrick, ed.), ACM Press, 1971, pp. 23–28.

College of Computer Science, Northeastern University, Boston, MA 02115

E-mail address: gene@ccs.northeastern.edu and laf@ccs.northeastern.edu

DIMACS Series in Discrete Mathematics
and Theoretical Computer Science
Volume **11**, 1993

Efficient Computation of Isotypic Projections for the Symmetric Group

PERSI DIACONIS AND DANIEL ROCKMORE

March 1, 1992

ABSTRACT. Spectral analysis on the symmetric group S_n calls for computing projections of functions defined on S_n and its homogeneous spaces, onto invariant subspaces. In particular, for the analysis of partially ranked data, the appropriate homogeneous spaces are given as quotients by Young subgroups. Here the naive character theoretic approach to computing projections requires $O(n \cdot n!)$ operations.

In this paper two types of polynomial time algorithms (quadratic in the size of the homogeneous space) are presented for partially ranked data. The first approach makes use of a more careful organization of the character theoretic computation and is applicable to arbitrary finite groups and their homogeneous spaces. The second approach makes use of the techniques of the combinatorial Radon transform.

1. Introduction

Let G be a finite group acting transitively on a set X. Often X is called a **homogeneous space** for G. Let $L(X)$ denote the vector space of complex-valued functions on X. Then $L(X)$ naturally admits a (permutation) representation ρ of G defined by

$$(\rho(s)(f))(x) = f(s^{-1}x) \tag{1.1}$$

for all $s \in G$ and all $x \in X$. As a representation space for G, $L(X)$ has a natural decomposition as a direct sum of isotypic subspaces

---

1991 *Mathematics Subject Classification.* Primary 20C40, 65U05; Secondary 05-04, 05E10.

The first author was supported in part by NSF Grant DMS 86-00235. The second author was supported in part by an NSF Mathematical Sciences Postdoctoral Fellowship. He also thanks Laci Babai and the Department of Computer Science at the University of Chicago for their hospitality while much of this was written.

This paper is in final form and no version will be submitted elsewhere.

$$L(X) = \oplus_{i=1}^{m} V_i.$$

The general problem may be stated as given a vector $f \in L(X)$ how may we quickly compute the projection of f onto any one (or some subset) of the V_i?

We were led to considering this calculation in the course of analyzing data on certain homogeneous spaces for the symmetric group. To explain, let S_n denote the symmetric group on n letters. Let λ be a partition of n. Thus,

$$\lambda = (\lambda_1, ..., \lambda_k)$$

where $\lambda_i \geq \lambda_{i+1} > 0$ and $\lambda_1 + ... + \lambda_k = n$. The corresponding Young subgroup $S_\lambda \leq S_n$ permutes the first λ_1 coordinates among themselves, the next λ_2 coordinates among themselves, and so on. Let X^λ denote the quotient space S_n/S_λ. The associated permutation representation of S_n on $M^\lambda = L(X^\lambda) = \{f : X^\lambda \longrightarrow \mathbb{R}\}$ is then of dimension $n!/\prod_i \lambda_i!$.

As motivation, consider the California Lotto game. This involves a choice of a random unordered 6-element subset of $\{1, 2, \ldots, 49\}$. Gamblers try to guess at the 6-set. The collection of 6-sets can be represented as the set of cosets, $S_{49}/S_6 \times S_{43}$. Indeed, S_{49} acts transitively on the 6-sets and the subgroup fixing $\{1, 2, 3, 4, 5, 6\}$ is $S_6 \times S_{43}$. Here, $\mid X^{(43,6)} \mid = 13,983,816$. About 10^6 people play the game every week. This leads to a function $f : X^{(43,6)} \longrightarrow \mathbb{R}$, $f(s)$ being the number of people choosing the subset s. The state takes about half of the total bet and splits the rest among the people choosing the winning 6-set.

Gamblers are interested in the function $f(s)$ - if one can pick a rarely chosen 6-set, one avoids having to split the prize. Chernoff [**8**] shows how this can actually lead to bets with positive expectation for the Massachusetts Lottery. The state is also interested in $f(s)$ - if there are no winners, the lottery is "rolled over" and betting activity increases. To make accurate fiscal projections, the number of roll-overs must be predicted and the data in $f(s)$ is clearly relevant.

One analysis of $f(s)$ begins by looking at the popularity of individual numbers. This is measured by averages like

$$\overline{f}(i) = \sum_{s \ni i} f(s).$$

Following this one might look at the popularity of pairs, triples and so on. The functions on 6-sets decompose naturally into distinct invariant irreducible subspaces (denoted $S^{(49-j,j)}$),

$$M^{(43,6)} = S^{49} \oplus S^{(48,1)} \oplus ... \oplus S^{(43,6)}.$$

The invariant subspace $S^{(49-j,j)}$ naturally measures the "pure" contribution of the popularity of the various j-sets. The analysis described above amounts to computing the projection of f onto these invariant subspaces. This approach is called **spectral analysis** by Diaconis ([**10**],[**11**]) who develops the subject in some detail.

Spectral analysis can be applied to partially ranked data of "shape λ". This involves n items (e.g. a list of 40 movies). Respondents are asked to choose from a list of n items their favorite λ_1 items (but not rank within) then their next favorite λ_2 items, and so on. For example, ranking the top 10 movies (in order) out of 40 leads to data of the shape (30,1,1,...,1) (10 ones). Sample surveys and the "q-sort" data collected by psychologists leads to other shapes. In each case spectral analysis offers a systematic method of analysis ([**10**], [**5**]).

Projections onto isotypic subspaces is a mainstay of the practical implementation of the spectral analysis approach to data. This generalizes to some degree the classical sums of squares decomposition of the standard ANOVA approach [**13**]. Computation of the isotypic projections is followed by computation of inner products of the projections with a sequence of easily interpretable vectors. Computation of the inner products is well understood. This paper focusses on the computation of the isotypic projections.

In this paper two algorithms are given for computation of the isotypic projections. The first method employs a more careful organization of the character theoretic approach and in fact, applies in the general case of decomposing the permutation representation of an arbitrary finite group G and homogeneous space X. This is developed fully, with special attention paid to the case in which $L(X)$ is multiplicity-free - as is the case for k-sets of an n-set. The results are developed as a series of reductions, each yielding greater computational improvements. Efficiency however, is gained at the cost of stronger assumptions on explicit knowledge of the representation theory of the associated symmetry group. Each subsection here contains an idea that is worth isolating. The strongest general result is contained in Theorem 2.4 with a slightly stronger result (Theorem 2.6) holding for distance transitive graphs.

The second algorithm uses ideas related to the **combinatorial Radon transform.** Use of an appropriate Radon transform (and its inverse) yield the decomposition of $M^{(n-k,k)}$. The work in this section is rather preliminary. Possible generalizations of this approach are discussed.

2. A character theoretic approach

2.1. Generalities. Let G be a finite group acting transitively on the finite set $X = \{x_1, ..., x_n\}$. Then X is called a **homogeneous space** for G. For any $x \in X$ let $Stab(x)$ denote the subgroup of elements of G which fix x. Let $L(X)$ denote the vector space of complex-valued functions on X. Then $L(X)$ naturally defines a representation ρ of G. This is precisely the permutation representation of G on the quotient space $G/Stab(x_i)$ for any i. (Equivalently this is the representation obtained by inducing the trivial representation from $Stab(x_i)$ to G.) As a vector space this is naturally equivalent to the complex vector space generated by the points $\{x_1, \ldots, x_n\}$. This associates a function $f \in L(X)$ to the point $f(x_1)x_1 + \ldots + f(x_n)x_n$.

Following Serre [**22**], the **isotypic decomposition** of $L(X)$ may be explained as follows. Let $\rho_1, \ldots, \rho_h$ be a complete set of inequivalent irreducible representations of G and let χ_i denote the character of ρ_i. As a representation space for G, $L(X)$ has a direct sum decomposition into irreducible G-invariant subspaces. That is,

$$L(X) = \oplus_{i=1}^{m} U_i. \tag{2.1}$$

where

$$\rho(s)U_i = U_i \qquad (U_i \neq 0)$$

and no nontrivial subspace of U_i has this property. For each i, $1 \leq i \leq m$, $\rho(s)$ restricted to U_i gives an irreducible representation of G. For each j, $1 \leq j \leq h$, let V_j be the subspace of $L(X)$ given by the direct sum over all U_i which define representations equivalent to ρ_j. Thus, the decomposition (2.1) may be rewritten as

$$L(X) = \oplus_{i=1}^{h} V_i$$

where V_i is the ith **isotypic** subspace of ρ. (Note that some of these V_i may be 0.)

2.2. Computing projections via characters. The following classical theorem (see e.g. [22], Theorem 8) defines the projection matrices.

THEOREM 2.1. *With the above notation the projection matrix p_i of V onto V_i along $\oplus_{j \neq i} V_j$ is given by the formula*

$$p_i = \frac{dim(\rho_i)}{|G|} \sum_{s \in G} \chi_i(s)^* \rho(s).$$

Theorem 2.1 is completely general. In our particular case, where ρ is the permutation representation on $G/Stab(x_1)$ more can be said.

THEOREM 2.2. *Let X be a homogeneous space for G and let ρ be the associated permutation representation of G in $L(X)$. Let $\chi_1, \ldots, \chi_h$ be the irreducible characters of G and $\rho_1, \ldots, \rho_h$ be a corresponding set of irreducible representations.*

If $f = f^{(1)} + \cdots + f^{(h)}$ is the isotypic deomposition of $f \in L(X)$ (some $f^{(i)}$ may be 0) then for all $i \leq h$ and all $x \in X$:

$$f^{(i)}(x) = \frac{dim(\rho_i)}{|G|} \sum_{s \in G} \chi_i(s) f(sx). \tag{2.2}$$

PROOF. The definition of the permutation representation (1.1) and the definition of the projection p_i (Theorem 2.1) immediately yield the sequence of equalities,

$$\begin{aligned} f^{(i)}(x) = p_i f(x) &= \frac{dim(\rho_i)}{|G|} \sum_{s \in G} \chi_i^*(s)(\rho(s)f)(x) \\ &= \frac{dim(\rho_i)}{|G|} \sum_{s \in G} \chi_i^*(s) f(s^{-1}x). \\ &= \frac{dim(\rho_i)}{|G|} \sum_{s \in G} \chi_i(s^{-1}) f(s^{-1}x), \end{aligned}$$

with the last equation following from the fact that $\chi_i^*(s) = \chi_i(s^{-1})$. □

Theorem 2.2 gives the following simple, albeit improvable, upper bound on the time needed to compute one projection.

COROLLARY 2.1. *The projections of f onto any isotypic subspace may be computed in at most* $|X||G|$ *operations.*

Remark: The computational model used here counts a single complex multiplication and addition as one "operation". As such, this does not take into account the inner group multiplication (i.e. sx for $s \in G$, $x \in X$) needed to compute the point for which the function value is retrieved.

As a first step in reducing the number of calculations, consider the summation appearing in (2.2),

$$\sum_{s \in G} \chi_i(s) f(sx). \tag{2.3}$$

The idea is to decompose (2.3) into subsums which will occur repeatedly as the values $f^{(i)}(x)$ are computed over all $x \in X$. These subsums need be computed once and then stored to be recalled when needed. To do this, a closer investigation of the structure of the quotient space $G/Stab(x_i)$ is useful.

Fix $\{s_1, ..., s_n\} \subseteq G$ such that

$$(i)\ s_1 = identity$$

$$(ii)\ s_i x_1 = x_i.$$

Also, let $H = Stab(x_1)$.

PROPOSITION 2.1. *With all the notation as above,*
(i) $Stab(x_i) = s_i H s_i^{-1}$.
(ii) Coset representatives for $G/Stab(x_i)$ *are given by the set*

$$\{s_i s_j s_i^{-1} \mid 1 \leq j \leq n\}.$$

Let

$$\chi_i^{(j)} = \sum_{s\in H} \chi_i(s_j s). \tag{2.4}$$

As will emerge, these $\chi_i^{(j)}$ are "generalized spherical functions". The following proposition shows that to compute projections, only certain sums involving the $\chi_i^{(j)}$ are needed.

PROPOSITION 2.2. *With notation as in Theorem 2.2 and (2.4),*

$$f^{(i)}(x_k) = \frac{dim(\rho_i)}{|\, G\, |} \sum_{j=1}^{n} f(s_k x_j)\chi_i^{(j)}. \tag{2.5}$$

PROOF. Consider first the computation of $f^{(i)}(x_1)$. Ignoring the outer scalar factor, the summation admits the following decomposition.

$$\begin{aligned}
\sum_{s\in G} \chi_i(s) f(s x_1) &= \sum_{j=1}^{n}\sum_{s\in H} f(s_j s x_1)\chi_i(s_j s) \\
&= \sum_{j=1}^{n} f(s_j x_1) \sum_{s\in H} \chi_i(s_j s) \\
&= \sum_{j=1}^{n} f(x_j)\chi_i^{(j)}
\end{aligned}$$

The subsums $\{\chi_i^{(j)}\}$ in fact determine the ith projection. To calculate $f^{(i)}(x_k)$ where $k \neq 1$, (2.2) gives

$$f^{(i)}(x_k) = \frac{dim(\rho_i)}{|\, G\, |} \sum_{s\in G} f(s x_k)\chi_i(s).$$

Again, considering only the summation,

$$\begin{aligned}
\sum_{s\in G} f(s x_k)\chi_i(s) &= \sum_{j=1}^{n} \sum_{t\in Stab(x_k)} f(s_k s_j s_k^{-1} t x_k)\chi_i(s_k s_j s_k^{-1} t) \\
&= \sum_{j=1}^{n}\sum_{s\in H} f(s_k s_j s_k^{-1} s_k s s_k^{-1} x_k)\chi_i(s_k s_j s_k^{-1} s_k s s_k^{-1}) \\
&= \sum_{j=1}^{n}\sum_{s\in H} f(s_k s_j s x_1)\chi_i(s_k s_j s s_k^{-1}) \\
&= \sum_{j=1}^{n} f(s_k x_j) \sum_{s\in H} \chi_i(s_k s_j s s_k^{-1}) \\
&= \sum_{j=1}^{n} f(s_k x_j) \sum_{s\in H} \chi_i(s_j s)
\end{aligned}$$

where the first equality follows from Proposition 2 and the last line follows from the fact that χ_i is a class function. So, finally

$$f^{(i)}(x_k) = \frac{dim(\rho_i)}{\mid G \mid}\sum_{j=1}^{n} f(s_k x_j)\chi_i^{(j)}.$$

□

Thus, the following initial reduction is obtained.

LEMMA 2.1. *For any fixed i, the projections onto the isotypic V_i may be computed in at most $O(\mid G \mid + \mid X \mid^2)$ operations.*

PROOF. We need at most $O(\mid G \mid)$ operations to calculate the subsums $\chi_i^{(j)}$ for all j. Then, for each x_k, $\mid X \mid$ additional multiplications to calculate $f^{(i)}(x_k)$. □

Note: The subsums $\chi_i^{(j)}$ need only be computed once - ever - and then stored. In situations in which X remains constant and f changes frequently and must be analyzed repeatedly, the above idea may prove useful. Of course, this does have the drawback that if $\mid G \mid$ is large (e.g. 49!) then these startup costs may prove prohibitive.

To reduce this $O(\mid G \mid)$ term, the subsums $\chi_i^{(j)}$ must be analyzed more closely. Since χ_i is a class function for G, for any t in the normalizer of H in G (denoted $N_G(H)$),

$$\begin{aligned}
\chi_i^{(j)} &= \sum_{s\in H} \chi_i(ts_j st^{-1}) \\
&= \sum_{s\in H} \chi_i(ts_j t^{-1}tst^{-1}) \\
&= \sum_{s\in H} \chi_i(ts_j t^{-1}s).
\end{aligned}$$

Thus,

LEMMA 2.2. *If s_j and s_k are conjugate by elements of the normalizer of H in G then*

$$\chi_i^{(j)} = \chi_i^{(k)}.$$

Pause for a moment to consider this reduction for a particular class of examples of interest, the homogeneous spaces for S_n given by its action on the collections of k-sets of an n-set where $k \leq \frac{n}{2}$. The canonical n-set is $\{1, ..., n\}$ and S_n acts in the usual way on the collection of subsets of size k. Let this homogeneous space be denoted as $X^{(n-k,k)}$. This may be realized by the obvious action of S_n on the space of binary n-tuples with k ones. (Here the k-set $\{i_1, ..., i_k\}$ with $i_j < i_{j+1}$ is represented by the n-tuple with 1's in positions i_j and 0's elsewhere.) The stabilizer of the point $\{1, ..., n-k\}$ is the subgroup $S_{n-k} \times S_k$ where S_{n-k} is the subgroup acting only on the elements $\{1, ..., n-k\}$

and S_k the subgroup which permutes $\{n-k+1,...,n\}$. Let $M^{(n-k,k)}$ denote the the vector space $L(X^{(n-k,k)})$. The isotypic decomposition of these spaces is well-known. In section 4 this will be given in more detail, but for now it will suffice to state without proof,

THEOREM 2.3. *The isotypic decomposition of* $M^{(n-k,k)}$ *where* $k \leq \frac{n}{2}$ *is given by*

$$M^{(n-k,k)} = S^{(n)} \oplus S^{(n-1,1)} \oplus ... \oplus S^{(n-k,k)}$$

where $S^{(n-j,j)}$ *denotes an irreducible representation of* S_n *of dimension* $\binom{n}{j} - \binom{n}{j-1}$.

To compute the projections onto the $S^{(n-j,j)}$, the coset representatives of $S_n/S_{n-k} \times S_k$ must first be determined.

PROPOSITION 2.3. *Coset representatives for* $S_n/S_{n-k} \times S_k$ *are given by the identity and all products of transpositions of the form*

$$(i_1 j_1)...(i_r j_r)$$

where $1 \leq r \leq k$ *and both*

$$n-k+1 \leq j_1 < ... < j_r \leq n$$

and

$$1 \leq i_1 < ... < i_r \leq n-k.$$

PROOF. Consider the action of S_n on the binary n-tuples with k ones and in particular consider the action on the vector

$$(\underbrace{0,0,...,0}_{n-k\ 0's},\underbrace{1,1,...,1}_{k\ 1's}).$$

By inspection, the above permutations yield all other such n-tuples. □

It is easy to see that for $k < n/2$, the normalizer of $S_{n-k} \times S_k$ in S_n is just $S_{n-k} \times S_k$. For $k = n/2$ it is isomorphic to a wreath product, $S_2\ wr\ S_{n/2}$, containing $S_{n/2} \times S_{n/2}$. The following lemma determines the number of orbits for the coset representatives under the action of $S_{n-k} \times S_k$ by conjugation.

LEMMA 2.3. *The coset representatives given for* $S_n/S_{n-k} \times S_k$ *in Lemma 2.2 are divided into* $k+1$ *orbits with respect to the action of* $S_{n-k} \times S_k$ *by conjugation.*

PROOF. By inspection one can see that under the action of $S_{n-k} \times S_k$ the coset representatives, as given in proposition 3, break up into equivalence classes determined by the number of transpositions needed to write them. These coset representatives can be taken to be the set

$$\{1, (1,n), (1,n-1)(2,n), ..., (1,n-k+1)\cdots(n-k,n)\}.$$

□

Examples: (These have been grouped by conjugacy class with respect to the action of the stabilizing subgroup.)

(1) Coset Representatives for $S_6/S_5 \times S_1$:

$$\{1\},\ \{(16), (26), (36), (46), (56)\}$$

(2) Coset Representatives for $S_6/S_4 \times S_2$:

$$\{1\}, \{(16), (26), (36), (46), (15), (25), (35), (45)\},$$
$$\{(15)(26), (15)(36), (15)(46), (25)(36), (25)(46), (35)(46)\}$$

(3) Coset Representatives for $S_6/S_3 \times S_3$:

$$\{1\}, \{(16), (26), (36), (15), (25), (35), (14), (24), (34)\},$$
$$\{(14)(25), (14)(35), (14)(26), (14)(36),$$
$$(24)(35), (24)(36), (15)(26), (15)(36), (25)(36)\},$$
$$\{(14)(25)(36)\}$$

This reduction gives the following improvements. These will be further refined later.

COROLLARY 2.2. *For fixed $j \leq k$ at most*

$$O((k+1)(n-k)!k! + \binom{n}{k}^2)$$

operations are need to calculate the projection of any $f \in M^{(n-k,k)}$ onto $S^{(n-j,j)}$.

PROOF. At most $(n-k)!k!$ operations are needed to calculate each of the $k+1$ different $\chi_i^{(j)}$. Then for each of the $\binom{n}{k}$ points x at most another $\binom{n}{k}$ will be required to calculate $f^{(i)}(x)$. □

COROLLARY 2.3. *At most*

$$O(k((k+1)(n-k)!k! + \binom{n}{k}^2))$$

operations are needed to calculate the projections onto every $S^{(n-j,j)}$.

The reductions in this section have in fact been implemented and used. See [**5**] for such an example.

2.3. Computing projections via matrix representations. As noted previously, assuming that this is a computation which is to be performed frequently, the character subsum calculation can be viewed as a "start-up" cost. The actual "interesting" computation takes only $O(\binom{n}{k}^2)$ operations. This compares very favorably with the naive bound of $(n!)^2$.

However, the associated start-up cost for the character theory approach of section 2.2 may still prove to be prohibitive. For instance, in the example cited in the introduction, in which we wished to compute projections from $M^{(43,6)}$, this would require performing at least 43! operations. Clearly further reduction is necessary. This may be achieved if certain explicit matrix representations are well-understood. This is now explained.

The sum $\chi_i^{(j)}$ is a sum of traces of a certain collection of matrices. Let ρ_i be an explicit matrix representation with character χ_i. Then,

$$\begin{aligned} \chi_i^{(j)} &= \sum_{s\in H} \chi_i(s_j s) \\ &= \sum_{s\in H} trace(\rho_i(s_j s)) \\ &= trace(\rho_i(s_j)\sum_{s\in H}\rho_i(s)). \end{aligned}$$

Consider the inner sum $\sum_{s\in H}\rho_i(s)$. As this is the sum of the values of a matrix representation over an entire subgroup (the representation ρ_i restricted to the subgroup H) this is "known". To explain, recall that for any irreducible unitary representation η of a group K,

$$\sum_{t\in K}\eta(t) = \begin{cases} \mid K\mid & \text{if } \eta = 1_K \\ d_\eta \times d_\eta \quad 0 \text{ matrix otherwise} & \end{cases} \tag{2.6}$$

where 1_K denotes the trivial representation of K ([**22**], section 2.2). Let $I_H(\rho_i)$ be the number of times that the trivial representation occurs in ρ_i restricted to H. Then recall,

$$I_H(\rho_i) = \langle \rho_i \downarrow H, 1_H\rangle = \langle \rho_i, 1_H \uparrow G\rangle. \tag{2.7}$$

In (2.7) restriction of a representation is denoted by $\downarrow$, while $\uparrow$ denotes induction. The notation $\langle\eta,\psi\rangle$ denotes the usual inner product of two representations. Then (2.7) follows from Frobenius reciprocity ([**22**], Theorem 13). Note that $1_H \uparrow G$ is just the representation ρ.

Say that a matrix representation ψ of G is H-**adapted** if for any $t \in H$,

$$\psi(t) = \begin{pmatrix} \eta_1(t) & 0 & \cdots & 0 \\ 0 & \eta_2(t) & \cdots & 0 \\ \vdots & \vdots & \ddots & 0 \\ 0 & 0 & \cdots & \eta_l(t) \end{pmatrix} \tag{2.8}$$

where each η_i is an irreducible unitary representation of H.

Then application of (2.6) shows that

$$\chi_i^{(j)} = trace(\rho_i(s_j) \begin{pmatrix} \mid H \mid & 0 & \dots & 0 \\ 0 & \mid H \mid & \dots & 0 \\ \vdots & \vdots & \ddots & \vdots \\ 0 & 0 & \dots & 0 \end{pmatrix})$$

where the inner matrix has $I_H(\rho_i)$ occurrences of $\mid H \mid$ on the diagonal, 0 elsewhere. Thus,

$$\chi_i^{(j)} = \mid H \mid \sum_{k=1}^{I_H(\rho_i)} (\rho_i(s_j))_{kk}. \tag{2.9}$$

Finally, note that if s_l and s_j are in the same $H\backslash G/H$ double coset, then assuming that the representation ρ_i is H-adapted,

$$(\rho_i(s_j))_{k,k} = (\rho_i(s_l))_{k,k}.$$

Hence, a further reduction in the start-up cost is obtained.

THEOREM 2.4. *Let ρ_i be an irreducible representation of G corresponding to the isotypic subspace V_i such that ρ_i is H-adapted. Then assuming the precomputation of the set of matrix elements*

$$\{(\rho_i(s_j))_{k,k} \mid 1 \le j \le m, 1 \le k \le I_H(\rho_i)\},$$

the projection onto V_i may be computed in at most

$$O(\mid X \mid^2 + \mid H\backslash G/H \mid \cdot I_H(\rho_i))$$

operations.

Remark: Theorem 4 depends upon the knowledge of explicit matrix representations for those representations occuring in the permutation representation on $L(X)$. Furthermore, assuming that the representations are known, they must be given in bases that are H-adapted. Assuming the existence of some realization of the representations as matrix representations the latter problem of constructing an appropriate change of basis has been solved (cf. [**4**], section 2 or [**21**], Appendix).

For the symmetric group S_n, Young's orthogonal form provides explicit formulas for a complete set of unitary irreducible matrix representations (cf. [**18**], section 25) evaluated at the set of pairwise adjacent transpositions. Additionally, for the permutation representations arising from quotients by Young subgroups, the decomposition into irreducibles is given by Young's Rule (cf. [**18**], section 14).

For an arbitrary group, explicit matrix representations may not be known. Babai and Ronyai [**2**] treat the problem of constructing a complete set by decomposing the group algebra.

It is worth pointing out that more generally, it may also prove fruitful to view the problem of computing projections in the context of matrix algebras. Babai, Friedl and Stricker [**1**], extending work of Friedl and Ronyai [**16**] show that deterministic polynomial time (in the dimension) algorithms exist for decomposing semisimple matrix algebras over number fields (closed under conjugate-transpose) into their simple components. There has been no investigation of the practicality of these algorithms.

2.4. The multiplicity-free case. A particular case of interest, which includes $M^{(n-k,k)}$, is that in which $L(X)$ is **multiplicity-free**. In this case $I_H(\eta)$ is 1 or 0 for each representation η of G. So,

$$L(X) = V_1 \oplus \ldots \oplus V_l$$

where the V_i are not only isotypics, but also inequivalent irreducible subspaces. Recall that (G, H) is said to be a **Gelfand pair** if the permutation representation of G associated to G/H is multiplicity-free. Assume that ρ_i is an irreducible representation which occurs in $L(X)$ in this case and furthermore assume that it is given as an H-adapted matrix representation of G. Then for any $t \in H$,

$$\rho_i(t) = \begin{pmatrix} 1 & 0 \\ 0 & B(t) \end{pmatrix}$$

where $B(t)$ is block diagonal

$$B(t) = \begin{pmatrix} \eta_1(t) & 0 & \cdots & 0 \\ 0 & \eta_2(t) & \cdots & 0 \\ \vdots & \vdots & \ddots & \vdots \\ 0 & 0 & \cdots & \eta_m(t) \end{pmatrix}$$

and the η_j are non-trivial unitary matrix representations of H (depending on ρ_i). Then for all $s \in G$ the H-bi-invariant function that picks out the 1,1 entry of the matrix $\rho_i(s)$ is called the ith **spherical function** of the Gelfand pair (G, H). Let these be denoted σ_i, then

$$\chi_i^{(j)} = \mid H \mid \sigma_i(s_j). \tag{2.10}$$

Thus, computation of the projection onto the ith isotypic (denoted $f^{(i)}$) is given as a **spherical transform**. That is, recall

$$f^{(i)}(x_k) = \frac{\chi_i(1)}{|G|} \cdot \sum_{j=1}^{n} f(s_k x_j)\chi_i^{(j)} \tag{2.11}$$

$$= \frac{\chi_i(1)}{|G|} \cdot |H| \cdot \sum_{j=1}^{n} f(s_k x_j)\sigma_i(s_j) \tag{2.12}$$

where (2.12) follows from (2.10). To reduce this computation further we take advantage of the H-bi-invariance of the spherical functions. For this, consider the action of the subgroup H on X. By applying Frobenius reciprocity (or equivalently, using "Mackey theory"), it can be shown that the number of orbits is equal to the number of nontrivial isotypic subspaces. Thus the partitioning into orbits can be written as

$$X = \Omega_1 \coprod \cdots \coprod \Omega_l$$

so that l is also the number of H, H double cosets.

This partitioning allows for a rewriting of (2.12) by

$$f^{(i)}(x_k) = \frac{\chi_i(1)}{|G|} \cdot |H| \cdot \sum_{r=1}^{l} \sum_{x \in \Omega_r} f(s_k x)\sigma_i(\Omega_r) \tag{2.13}$$

$$= \frac{\chi_i(1)}{|G|} \cdot |H| \cdot \sum_{r=1}^{l} \sigma_i(\Omega_r) \sum_{x \in \Omega_r} f(s_k x) \tag{2.14}$$

$$= \frac{\chi_i(1)}{|G|} \cdot |H| \cdot \sum_{r=1}^{l} \sigma_i(\Omega_r) f_k(\Omega_r) \tag{2.15}$$

where $f_k(\Omega_r) = \sum_{x \in \Omega_r} f(s_k x)$. Finally, write

$$\widehat{f_k}(i) = |H| \cdot \sum_{r=1}^{l} f_k(\Omega_r)\sigma_i(\Omega_r) \tag{2.16}$$

and call (2.16) the ith **spherical transform** of f_k.

Note that at most $|X|$ additions are needed to compute any given f_k, so at most $|X|^2$ are required to compute all f_k. At most $|H\backslash G/H|$ operations are required to compute any single spherical transform, so at most $|H\backslash G/H| \cdot |X|$ operations are needed for each of the spherical transforms necessary for a single projection. This is summarized by the Theorem 2.5 below.

THEOREM 2.5. *Let X be a homogeneous space for G that yields a multiplicity-free representation of G. Then, assuming that the spherical functions for the associated Gelfand pair are known, the projection onto any one of the isotypics may be calculated as a sequence of spherical transforms (eq. 2.16). This requires at most $|X|^2$ additions, followed by at most an additional $|H\backslash G/H| \cdot |X|$*

operations. To compute the projections onto every isotypic would require an additional $| H\backslash G/H |^2 \cdot | X |$ *operations.*

COROLLARY 2.4. *All isotypic projections for* $M^{(n-k,k)}$ *may be computed using* $\binom{n}{k}^2$ *additions and* $\binom{n}{k}(k+1)^2$ *operations.*

2.5. Distance transitive graphs. As a final reduction in this section, the interesting case of multiplicty-free representations arising from distance-transitive graphs is considered. In this case, the set X is to be considered as an undirected graph with the usual metric given by edge-distance between points. It is assumed that G acts as isometries on the graph, so that $d(sx, sy) = (x, y)$ for all $s \in G$ and $x, y \in X$. The graph is said to be **distance transitive** if given any two pairs of points (x, y), (x', y') such that $d(x, y) = d(x', y')$, there exists $s \in G$ such that $(sx, sy) = (x', y')$.

Let Ω_k denote the set of points of distance k from the basepoint x_1. Then the algebra of functions constant on each Ω_k is isomorphic to the algebra of H-bi-invariant functions on G ($H = Stab(x_1)$), which has basis equal to the set of associated spherical functions. See Brouwer, Cohen and Neumaier [**7**] or Stanton [**23**] for excellent surveys of the subject of distance transitive graphs and their spherical functions as well as for pointers to other relevant literature.

For the spherical functions which arise from distance transitive graphs, spherical transforms may be computed more efficiently than by direct computation. Recent work of Driscoll, Healy and Rockmore shows that for such graphs, the **discrete spherical transform** defined as the set of spherical transforms $\{\widehat{f}(i)\}_i$, can be computed in at most $O(n \log^2 n)$ operations versus n^2 operations required for direct computation. Their algorithm makes use of the fact that for distance transitive graphs, the spherical functions satisfy a three-term recurrence. See [**15**] for details.

Such graphs may be also studied from the point of view of association schemes. Bannai and Ito [**3**] give a full treatment of the subject.

The set of k-sets of an n-set naturally form a graph where two k-sets are joined if and only if they differ by a single element. This graph is in fact distance transitive and is sometimes referred to as the Johnson scheme in the association scheme literature. The associated spherical functions can be written as Hahn polynomials so that the isotypic projections can be computed as Hahn polynomial transforms. In [**15**] this example is worked out in full detail (section 3.3). Using this result gives a final reduction for computing projections for $M^{(n-k,k)}$.

THEOREM 2.6. *All isotypic projections for* $M^{(n-k,k)}$ *may be computed using at most* $\binom{n}{k}^2$ *additions followed by* $O(\binom{n}{k} k \log^2 k)$ *operations.*

3. Radon Transform Approach

In the particular case of k-sets of an n-set, special techniques are available. We believe that these techniques extend to other standard examples (see section 3.3).

3.1. k-sets of an n-set. For $k \leq n/2$, let X denote the $\binom{n}{k}$ k-element subsets of $\{1, \ldots, n\}$. Abbreviate $L(X) = M^{(n-k,k)}$ as M^k in this section. For $0 < j \leq k \leq n/2$, define the **Radon transform** $R : M^j \longrightarrow M^k$ by summing over subsets:

$$Rf(t) = \sum_{s \subset t} f(s)$$

for $| s |= j$ and $| t |= k$.

This type of combinatorial Radon transform was introduced by Bolker [**6**]. Kung [**20**] gives an extensive survey. Diaconis and Graham [**12**] discuss algorithmic aspects.

Each of these authors prove that this particular version of the transform is one-to-one when $0 < j \leq k \leq n/2$. Graham, Li and Li [**17**] give an explicit left inverse. To describe this, choose a basis of delta functions for M^k:

$$\delta_t(t') = \begin{cases} 1 & \text{if } t = t', \\ 0 & \text{otherwise.} \end{cases}$$

Choose a similar basis for M^j. Then the Radon transform can be regarded as an $\binom{n}{k}$ by $\binom{n}{j}$ matrix, R, with entries

$$R_{ts} = \begin{cases} 1 & \text{if } s \subset t, \\ 0 & \text{otherwise} \end{cases} \tag{3.1}$$

for all $| s |= j$ and $| t |= k$.

The left inverse R^- is an $\binom{n}{j}$ by $\binom{n}{k}$ matrix satisfying $R^- R = I$, the $\binom{n}{j}$ by $\binom{n}{j}$ identity matrix. Graham, Li and Li [**17**] give R^- explicitly as

$$R^-_{s,t} = \frac{(-1)^{k-j}(k-j)}{(-1)^{|t-s|} \mid t-s \mid} \frac{1}{\binom{n-j}{|t-s|}} \tag{3.2}$$

with $| t - s |$ denoting the cardinality of the set of elements in t but not in s.

It follows from the definitions that both R and R^- are S_n-homomorphisms. To state the first lemma, regard M^j and M^k as inner product spaces by defining

$$\langle \delta_s \mid \delta_{s'} \rangle = \begin{cases} 1 & \text{if } s = s', \\ 0 & \text{otherwise} \end{cases}$$

LEMMA 3.1. *For R and R^- defined above let $\pi = RR^-$. Then π is a projection with range isomorphic to M^j.*

PROOF. Clearly $\pi^2 = \pi$. By multiplying out, the general entry of π can be written down explicitly. If t and t' are k-sets, the t, t' entry of π only depends on the cardinality of $| t \cap t' |= z$ (say). So $0 \leq z \leq k$ and

$$\pi_{tt'} = \frac{k-j}{n-j} \sum_{u=0}^{j} (-1)^{j-u} \frac{\binom{z}{u}\binom{k-z}{j-u}}{\binom{n-z-1}{n-u-1}}. \tag{3.3}$$

This clearly shows that $\pi_{tt'} = \pi_{t't}$. So, π is a projection. Since π commutes with the action of S_n, the range of π is an invariant subspace. Let δ_s be a basis for M^j as s runs through j-sets. The $\bar{\delta}_t = R\delta_t$ are linearly independent and in the range of π which is thus isomorphic to M^j. □

The decomposition of M^k into irreducibles is well-known to be multiplicity-free. Using the notation of James [**18**] this decomposition is given by

$$M^k \cong S^{(n)} \oplus S^{(n-1,1)} \oplus S^{(n-2,2)} \oplus \cdots \oplus S^{(n-k,k)} \tag{3.4}$$

$$\cong M^{k-1} \oplus S^{(n-k,k)}. \tag{3.5}$$

See Stanton [**23**] for a direct proof. Bolker [**6**] gives a proof using Radon transform arguments. Using induction Lemma 1 yields

COROLLARY 3.1. *Let $\pi^{(j)}$ be the projection from M^k described in Lemma 1. Then a projection onto the irreducible subspace $S^{(n-j,j)} \subset M^k$ can be computed as $\pi^{(j)} - \pi^{(j-1)}$, $1 \le j \le k$, with $\pi^{(0)}$ equal to the projection onto $S^{(n)}$ given by $1/\binom{n}{k} \cdot J$. (J denotes the $\binom{n}{k}$ by $\binom{n}{k}$ matrix of all ones.)*

The rows and columns of the projections matrices $\pi^{(j)}$ are indexed by the k-sets, with any given entry only depending on the cardinality of the intersection of the associated k-sets (see eqn. (3.3)). Computation of this intersection requires at most on the order of $k \log k$ operations. Consequently, computation of this matrix will require at most $O(\binom{n}{k}^2 k \log k)$ operations. Thus to obtain all projection matrices requires at most $O(\binom{n}{k}^2 k^2 \log k)$ operations. To compare, the startup cost required of the spherical function approach is $\binom{n}{k}^2$ additions. Either approach is feasible.

Having built the projection matrices, the isotypic projections may be computed as a sequence of k matrix-vector multiplications.

THEOREM 3.1. *To compute the projections of an arbitrary element of M^k onto each of the $k+1$ isotypic subspaces via Radon transform methods requires a startup cost of at most $O(k^2 \log k \binom{n}{k}^2)$ operations and then an additional $k\binom{n}{k}^2$ operations.*

3.2. Further directions. The techniques of section 3.1 seem to present many possiblities for further investigation. We close here by indicating a few of these.

1. $GL_n(\mathbb{F}_q)/GL_{n-k}(\mathbb{F}_q) \times GL_k(\mathbb{F}_q)$. The quotient $S_n/S_{n-k} \times S_k$ has a "q-analogue" given by $GL_n(\mathbb{F}_q)/GL_{n-k}(\mathbb{F}_q) \times GL_k(\mathbb{F}_q)$. The associated permutation representation of $GL_n(\mathbb{F}_q)$ is given by the action of $GL_n(\mathbb{F}_q)$ on k-dimensional subspaces of $\mathbb{F}_q^n$. As for k-sets of an n-set, this is a multiplicity-free representation (and even a distance transitive graph with associated spherical functions given by the q-Hahn polynomials). The techniques of the Radon transform are again available. If $X^k(q)$ denotes the set of k-dimensional subspaces of F_q^n, then for $j < k \leq n/2$ define $R : L(X^j(q)) \longrightarrow L(X^k(q))$ by

$$Rf(y) = \sum_{x \subset y} f(x).$$

2. General Young subgroups and parabolic subgroups. Let $\lambda \vdash n$ and let S_λ denote the corresponding Young subgroup of S_n and G_λ the corresponding parabolic subgroup of $GL_n(\mathbb{F}_q)$. Work of James suggests that the techniques of the Radon transform could be applicable in this situation for computing projections onto not only isotypic subspaces, but even onto a complete set of irreducible subspaces. For S_n see ([**18**], section 17) and for $GL_n(\mathbb{F}_q)$ see ([**19**], section 15).

3. Connections between Radon transform and character theory. There must be a clear connection between the material of section 2 and the that of section 3.1. This has not yet been investigated.

4. We would be remiss if we did not point out the closely related area of noncommutative fast Fourier transforms. See Clausen [**9**] and Diaconis and Rockmore [**14**] and the references cited there. These ideas speed computation of the discrete Fourier transform by successive restriction through a tower of subgroups. Such restrictions have not been invoked in any of the above approaches. This idea may be of use for the computation considered here and provide even further speedups.

References

1. L. Babai, K. Friedl, and M. Stricker, *Decomposition of *-closed algebras in polynomial time*, Technical Report, University of Chicago, Department of Computer Science, 1992.
2. L. Babai and L. Ronyai, *Computing irreducible representations of finite groups*, Math. Comp. **55** (1990), 705–722.
3. E. Bannai and T. Ito, *Algebraic Combinatorics I. Association Schemes.* Benjamin/Cummings Pub. Co., Inc., Menlo Park, CA 1984.
4. U. Baum, *Existence and efficient construction of fast Fourier transforms on supersolvable groups*, Ph.D. Thesis, University of Bonn, 1991.
5. L. Beckett and P. Diaconis, *Spectral analysis for discrete longitudinal data*, Technical Report, Department of Mathematics, Harvard University, 1991.
6. E. Bolker, *The finite Radon transform*, Contemporary Math **63** (1987), 27–50.
7. A. E. Brouwer, A. M. Cohen and A. C. Neumaier, *Distance Regular Graphs*, Springer-Verlag, Berlin, 1990.
8. H. Chernoff, *An analysis of the Massachusetts numbers game*, Math. Intelligencer, **3** (1981), 166–172.
9. M. Clausen, *Fast generalized Fourier transforms*, Theoret. Comp. Sci. **67** (1989), 55–63.
10. P. Diaconis, *Spectral analysis for ranked data*, Ann. of Stat., **17** (1989), 349-379.

11. P. Diaconis, *Group Representations in Probability and Statistics*, Inst. of Math. Stat., Hayward, CA, 1989.
12. P. Diaconis and R. Graham, *The Radon transform on* $\mathbb{Z}_2^k$, Pacific J. of Math. **118** (1985), 323–345.
13. P. Diaconis and D. Rockmore, *Representation theoy and ANOVA for designed experiments*, Technical Report, Harvard University, Department of Mathematics, 1991.
14. P. Diaconis and D. Rockmore, *Efficient computation of the Fourier transform on finite groups*, Jour. of AMS **3** (1990), 297–332.
15. J. Driscoll, D. Healy, and D. Rockmore, *Fast spherical transforms on distance transitive graphs*, Technical Report, Dartmouth College, Department of Mathematics and Computer Science, 1991.
16. K. Friedl and L. Ronyai, *Polynomial time solutions of some problems in computational algebra*, Proc. 17th ACM STOC, 1983, 153–162.
17. R. L. Graham, R. Li, and W. Li, *On the structure of t-designs*, SIAM J. Alg. and Disc. Meth. **1** (1980), 8–14.
18. G. D. James, *The Representation Theory of the Symmetric Group*, Lecture Notes in Math., vol. 682, Springer-Verlag, Berlin, 1978.
19. G. D. James, *Representations of General Linear Groups*, London Math. Soc. Lecture Notes, vol. 94, Cambridge Univ. Press, Great Britain, 1984.
20. J. P. Kung, *Radon transforms in combinatorics and lattice theory*, Contemporary Math. **57** (1986), 36–74.
21. D. Rockmore, *Fast Fourier analysis for finite groups*, Ph.D. Thesis, Harvard University, 1989.
22. J. P. Serre, *Linear Representations of Finite Groups*, Springer-Verlag, NY , 1986.
23. D. Stanton, *Orthogonal polynomials and Chevalley groups*, Special Functions: Group Theoretical Aspects and Applications. (ed. R. Askey, et. al.), 1984, 87–128.

DEPT. OF MATHEMATICS, HARVARD UNIVERSITY, CAMBRIDGE, MASSACHUSETTS 02138

DEPT. OF MATHEMATICS AND COMPUTER SCIENCE, DARTMOUTH COLLEGE, HANOVER, NEW HAMPSHIRE 03755

DIMACS Series in Discrete Mathematics
and Theoretical Computer Science
Volume 11, 1993

Constructing Representations of Finite Groups

JOHN D. DIXON

ABSTRACT. This paper considers the problem of constructing a matrix representation for a finite group affording a given (ordinary) irreducible character. The method described has several desirable properties such as representing the matrices over a small extension of Q, and providing a quadratic invariant for the representation. Some examples of the computations are included.

A great deal is known about practically efficient methods of computing the irreducible characters of a finite group, and some of these methods have been built into systems such as CAYLEY and GAP. Parker's MEATAXE is a wonderfully effective tool for computing with modular representations of groups over small finite fields. On the other hand, the situation with regard to computing ordinary representations of finite groups is less satisfactory. Although there has been recent work on this problem (see, for example, [**1,2,3,5,6**]), the methods described do not seem to have been implemented for significant examples, and indeed in most cases the authors explicitly state their intention to deal with the questions of theoretical complexity rather than practical computation.

One of the difficulties in considering this question is that, unlike the case of characters, there is no natural canonical representation in a set of equivalent representations of a given group, so certain arbitrary choices have to be made. Some desirable criteria for choosing a satisfactory representation are:

(i) the entries of the representing matrices should be expressed exactly in terms of a few simple irrationalities;
(ii) the number field over which the representation is expressed should be of as small a degree as possible over $\mathbb{Q}$;
(iii) the representation should consist of unitary matrices.

1991 *Mathematics Subject Classification*. Primary 20C40.
The author wishes to acknowledge support in part by an operating grant from the Natural Sciences and Engineering Research Council of Canada.
This paper is in final form and no version of it will be submitted for publication elsewhere.

The criteria (ii) and (iii) tend to conflict. A possible compromise is to drop the condition (iii) and instead ask for information about the (essentially unique) quadratic invariant for the representation; for a unitary representation this invariant is $x_1^2 + \ldots + x_n^2$ (see Note 2 below).

The object of the present paper is to describe a rather *ad hoc* method of computing an irreducible representation of a group when the corresponding character is supplied. We shall make no attempt to analyze the method theoretically, but some examples of the kinds of representations which have been handled are given at the end.

Let G be a finite group of order g and let χ be an absolutely irreducible character for G of degree $d = \chi(1)$ which is afforded over some field $F \subseteq \mathbb{C}$. Then

$$e(= e_\chi) := (d/g) \sum_{x \in G} \chi(x^{-1})x$$

is a central idempotent in FG which corresponds to a representation ρ of FG with kernel $(1-e)FG$ and image $\mathrm{Mat}(d, F)$ (see Satz V.5.15 of [**8**]). The restriction of ρ to eFG gives an isomorphism onto $\mathrm{Mat}(d, F)$. If we consider the action of G on eFG by right multiplication we see that eFG is an FG-module of dimension d^2 which affords the character $d\chi$. Our object is to find a module which affords χ by constructing a submodule of eFG of dimension d in the form $aeFG$. It follows from the isomorphism between eFG and $\rho(FG)$ that any element $a \in FG$ with the property that $\rho(a)$ has rank 1 defines such a submodule. To find a suitable element a, we can look for an element $b \in FG$ for which $\rho(b)$ has an eigenvalue $\mu \in F$ with an eigenspace of dimension 1. If $\rho(b)$ has minimal polynomial $m(X)$, and if a is any nonzero scalar multiple of $g(b)$ where $g(X) := m(X)/(X - \mu)$, then $\rho(a)$ has rank 1. Simple arguments show that "random" choices of b from the group algebra are likely to have the required property provided we are willing to extend the underlying field.

The construction of a suitable "rank-1" element is considerably simplified when we can find a subgroup of G such that the restriction of χ to this subgroup contains a constituent of degree 1 and multiplicity 1. G. Glauberman has given examples which show that, even for solvable groups, there are examples where no such subgroups exist; see [**9**]. However, it appears that, at least for small degrees, it is frequently quite easy to find such a subgroup, and indeed this has been true for all of the examples we have wanted to deal with. In the description below we therefore restrict ourselves exclusively to this case.

(R1) Choose a subgroup $H \leq G$ as small as possible such that the restriction χ_H has at least one constituent θ with degree 1 and multiplicity 1. Put $h :=\mid H \mid$.

(R2) Define

$$a := \sum_{z \in H} \theta(z^{-1})z \in FH$$

and verify that $a^2 = ha$. The trace tr $\rho(a) = h < \theta, \chi_H >= h$ because

θ occurs as a constituent with multiplicity 1 in χ_H; thus $\rho(h^{-1}a)$ is an idempotent with a single eigenvalue equal to 1 and all other eigenvalues 0. In particular, $\rho(a)$ has rank 1.

(R3) Set $f := (g/d)ae$ where e is the central idempotent defined above. Then

$$f = \sum_{t \in G} \alpha(t)t \text{ where } \alpha(t) := \sum_{z \in H} \theta(z^{-1})\chi(zt^{-1}) \text{ for all } t \in G.$$

A simple computation shows that $\alpha(t^{-1}) = \overline{\alpha(t)}$ and so the $g \times g$ matrix $A := [\alpha(yx^{-1})]_{x,y \in G}$ is hermitian.

(R4) It follows from the introductory paragraph that the module fFG affords the character χ for G. Since $fx = \sum_{t \in G} \alpha(tx^{-1})t$, the xth row of A gives the components of fx in FG, and so rank $A = \dim fFG = d$. If $x_1, \ldots, x_d$ index a set of linearly independent rows of A then, because A is hermitian of rank d, a simple argument using linear algebra shows that the $d \times d$ submatrix $[\alpha(x_j x_i)]$ is nonsingular. Define $A(x)$ to be the $d \times d$ matrix $[\alpha(x_j x^{-1} x_i)]$ for each $x \in G$, and note that $A(1)$ is a nonsingular hermitian matrix by the choice of $x_1, \ldots, x_d$. Then the lemma below shows that the mapping

$$x \mapsto A(x)A(1)^{-1} \tag{1}$$

defines a representation of degree d for G over F which affords the character χ as desired. Although no good algorithm seems to be known to choose a set of independent rows, it appears in practice that a random selection of rows quickly yields a set with the desired properties.

LEMMA. *If the indices $x_1, \ldots, x_d$ correspond to a set of linearly independent rows for the matrix A, then the mapping (1) defines a representation of G of degree d affording χ.*

Remark I am indebted to Christian Wieners (Köln) for pointing out to me that a version of this result already appears in Frobenius' paper [**7**].

PROOF. First recall that fFG is an FG-module of dimension d which affords the character χ, and that $fx_1, \ldots, fx_d$ are linearly independent elements of FG by the choice of $x_1, \ldots, x_d$. Therefore $fx_1, \ldots, fx_d$ is an F-basis for fFG, and so the mapping $x \mapsto [\xi_{ij}]$ defined by

$$(fx_i)x = \sum_{j=1}^{d} \xi_{ij}(fx_j)$$

is a representation of G affording χ. This latter relation is equivalent to

$$\alpha(tx^{-1}x_i) = \sum_{j=1}^{d} \xi_{ij}\alpha(tx_j) \text{ for all } t \in G \text{ and } i = 1, \ldots, d.$$

In particular, $A(x) = [\xi_{ij}]A(1)$, and so $[\xi_{ij}] = A(x)A(1)^{-1}$. Thus (1) defines the required representation.

Notes 1. If $v \in H$ then a simple calculation shows that $\alpha(tv) = \theta(v)^{-1}\alpha(t)$ for all $t \in G$. Thus the rows of A corresponding to x and $v^{-1}x$ are linearly dependent. In particular, this shows that the x_i chosen in (R4) must come from distinct right H-cosets.

2. Define $R(x) := A(x)A(1)^{-1}$, so R is a representation of G affording χ. Then it follows easily that the hermitian conjugate $A(x)^* = A(x^{-1})$ and hence that $A(1) = R(x)A(1)R(x)^*$ for all $x \in G$. On the other hand, the positive definite matrix $C := \sum_{t\in G} R(t)R(t)^*$ also has the property that $C = R(x)CR(x)^*$ for all $x \in G$. It is a classical result that there is essentially only one quadratic invariant for each representation of G. (Briefly, the representation $R \otimes R^*$ of G acting on $\mathrm{Mat}(n, F)$ via

$$U \mapsto R(x)UR(x)^*$$

has the character $\chi\chi^*$. Since $< \chi\chi^*, 1_G >=< \chi, \chi >= 1$, the fixedpoint space of this representation has dimension 1. This subspace contains both $A(1)$ and C and so $A(1)$ is a scalar multiple of C.) Thus the quadratic form corresponding to $A(1)$ is the (essentially unique, positive definite) invariant quadratic form for R. In particular, we can factor $A(1)$ in the form DD^*, and then R is equivalent to the unitary representation $x \mapsto D^{-1}R(x)D$.

EXAMPLE. As a simple example, let G be the alternating group A_5 and let χ be the irreducible character of degree 5 for G. Then we have

Class	1	2	3	5_1	5_2
χ	5	1	-1	0	0

where the classes are denoted by the orders of their elements.

In this case when we restrict χ to the cyclic group $H =< (1\,2\,3) >$ the trivial character 1_H occurs with multiplicity 1. Thus, we define

$$\alpha(t) = \sum_{z\in H} \chi(zt^{-1}) \text{ for all } t \in G.$$

After choosing (essentially randomly) a series of seven elements of G we discover that for

$$x_i = I, (1\,2)(3\,4), (1\,3\,4), (1\,4\,2\,3\,5) \text{ and } (1\,4\,3\,5\,2) \text{ for } 1 \le i \le 5$$

the corresponding 5×5 submatrix of $[\alpha(yx^{-1})]$ is nonsingular. Now the representation $x \mapsto A(x)A(1)^{-1}$ gives:

$$(1\,2\,3) \longmapsto \begin{bmatrix} 1 & 0 & 0 & 0 & 0 \\ 0 & 0 & 1 & 0 & 0 \\ -1 & -1 & -1 & 0 & 0 \\ 0 & 0 & 0 & 0 & 1 \\ 1 & 0 & 0 & -1 & -1 \end{bmatrix}$$

and

$$(1\,2\,3\,4\,5) \longmapsto \begin{bmatrix} 0 & 0 & 1 & 0 & 0 \\ -1 & 0 & -1 & 0 & 1 \\ 0 & 0 & 0 & 1 & 0 \\ 0 & 0 & 0 & 0 & 1 \\ -1 & -1 & 0 & 1 & 0 \end{bmatrix}.$$

A similar calculation finds an irreducible representation which affords an irrational character of degree 10 for the Mathieu group M_{11} using the fact that the restriction of this character to a noncyclic subgroup of order 6 contains the trivial character exactly once. Details are given in the appendix below.

Representations of degrees up to 14 for a number of other simple groups have been calculated using the character tables from the ATLAS [**4**]. With Holger Gollan in Essen I have begun a project of constructing a complete list of explicit representations of finite primitive complex linear groups of small degrees exploring the use of this technique as well as others. In the cases considered so far there appears to be little difficulty in finding a suitable subgroup where the restriction of the given irreducible character has a constituent with degree and multiplicity 1.

Appendix

The Mathieu group M_{11} can be generated by the three permutations

$$x = (1\,2\,3\,4\,5\,6\,7\,8\,9\,10\,11), y = (1\,5\,9\,6)(2\,8\,3\,4)$$

and

$$z = (1\,7\,9\,6\,8\,3\,5\,4)(2\,10).$$

It has three irreducible characters of degree 10. One of which is:

Class	1	2	3	4	5	6	8_1	8_2	11_1	11_2
χ	10	-2	1	0	0	1	w	$-w$	-1	-1

where $w^2 = -2$ and we assume that z lies in the class 8_1. Thus χ is determined by the order of the element except in the case where the order is 8. In the case that $u \in M_{11}$ has order 8 we can compute $\chi(u)$ as follows. The centralizer C of z in S_{11} equals $< (1\,7\,9\,6\,8\,3\,5\,4) > \times < (2\,10) >$ and the ATLAS [**4**] shows that $C_0 = C \cap M_{11} \leq A_{11}$ has order 8; thus $C = C_0 \cup (2\,10)C_0$. To determine whether u is conjugate to z in M_{11} it is sufficient to choose any $v \in S_{11}$ such that $v^{-1}uv = z$. Then u is conjugate to z in M_{11} if and only if either v or $v(2\,10)$ lies in M_{11}.

It is evident from the character table for χ that a dihedral subgroup of order 6 in M_{11} has the property that the restriction of χ to this subgroup must contain the trivial character exactly once. Therefore we choose

$$H := < (1\,4\,5)(2\,3\,7)(6\,8\,9), (2\,8)(3\,6)(4\,5)(7\,9) > .$$

Calculations now show that the elements x_i $(1 \leq i \leq 10)$:

$(2\ 3\ 11\ 8\ 5\ 7\ 6\ 10)(4\ 9)$ $(2\ 6\ 5\ 11)(3\ 10\ 7\ 8)$
$(2\ 8\ 6\ 3\ 5\ 10\ 11\ 7)(4\ 9)$ $(1\ 7\ 2\ 6\ 11)(3\ 9\ 5\ 8\ 4)$
$(1\ 3\ 9\ 8\ 6)(4\ 7\ 11\ 10\ 5)$ $(1\ 11)(2\ 8\ 9\ 3\ 4\ 6\ 10\ 5)$
$(1\ 6)(2\ 7\ 10\ 3\ 4\ 11\ 5\ 9)$ $(1\ 10\ 3\ 9\ 6\ 5\ 7\ 11)(4\ 8)$
$(1\ 8\ 2\ 11\ 3\ 9\ 7\ 5\ 4\ 10\ 6)$ $(3\ 9\ 4\ 6)(5\ 8\ 7\ 11\ )$

correspond to linearly independent rows for the matrix A (in this case a total of 13 elements had to be tested.) In terms of these elements we obtain the following matrices and a representation affording χ.

$A(1) =$

$$\begin{bmatrix}
6 & -2 & w & 2-w & -2+2w & -1-2w & 2-w & w & -2 & -1+w \\
-2 & 6 & -2 & -1+2w & -1-w & 2-w & -2+2w & -1-2w & 2-w & 0 \\
-w & -2 & 6 & 0 & 1+2w & -1+2w & -1-w & 2-w & -2+2w & -1+w \\
2+w & -1-2w & 0 & 6 & 2w & -w & 2+w & -2 & -w & w \\
-2-2w & -1+w & 1-2w & -2w & 6 & w & -1 & 0 & 2+w & 3 \\
-1+2w & 2+w & -1-2w & w & -w & 6 & 2w & -w & 2+w & w \\
2+w & -2-2w & -1+w & 2-w & -1 & -2w & 6 & w & -1 & 1 \\
-w & -1+2w & 2+w & -2 & 0 & w & -w & 6 & 2w & 0 \\
-2 & 2+w & -2-2w & w & 2-w & 2-w & -1 & -2w & 6 & 2-w \\
-1-w & 0 & -1-w & -w & 3 & -w & 1 & 0 & 2+w & 6
\end{bmatrix}$$

$A(x) =$

$$\begin{bmatrix}
-2+w & 2 & -1-w & -2 & -1-2w & 2-w & 1 & -2w & 2 & 1-w \\
-w & -1+w & -2-w & 0 & 1 & 0 & -1 & 2w & 1 & w \\
2w & 2 & w & 2w & -2 & 2 & -2+w & -w & 1-w & -2-w \\
-2 & 2 & -1 & 0 & 2-w & 1-w & 1+w & 2-w & 3 & 2-w \\
2 & 1-w & w & 2+w & -1+w & -2-w & 0 & -2-w & -2w & -2-w \\
-2w & w & 1-w & -w & 3 & 0 & -1 & 0 & 1 & 1+w \\
-3 & 2-w & w & 0 & -w & -1 & 0 & -w & -w & 2 \\
-2 & 2 & 1-w & -2+2w & 0 & 1+w & -3 & -2-w & 1+w & 0 \\
1-w & -w & -w & 2 & 2+w & -1-w & w & -2+w & 0 & 2 \\
w & 0 & -w & 2+w & 0 & 0 & 0 & -2-w & -w & 0
\end{bmatrix}.$$

$A(y) =$

$$\begin{bmatrix}
1 & 0 & 2+w & 0 & -2+w & -1+w & -2 & w & -4 & -2+w \\
-4 & 2 & -1 & -1+w & 1-2w & w & -1+w & 1-w & 2-w & 1-2w \\
2-w & -1 & 0 & -1-2w & 0 & -w & 2 & 1+w & -1+w & 0 \\
2+2w & 1-w & -1+2w & 2w & -6 & -w & 1 & 0 & -2-w & -3 \\
2-w & 0 & 0 & 1 & 2w & -w & 2 & 1-w & 1+w & w \\
-2-w & 1 & w & -1+w & -w & -2 & -w & 1-w & 2-w & -2w \\
1+w & -w & 2 & 2+w & -2+w & 0 & -w & -2-w & -1 & -3 \\
2-w & -1 & 1-w & -w & 2 & -w & 1 & w & w & 1 \\
-1 & 0 & -1-w & w & -w & -1 & 0 & -w & 2-w & -1 \\
1+w & -2-w & 0 & 2 & w & -1-w & 2-w & 0 & 1 & -w
\end{bmatrix}.$$

$A(z) =$

$$\begin{bmatrix}
3 & 0 & -2+w & -w & -2 & -2-2w & -2w & -2 & -2w & -1 \\
w & 0 & 0 & w & -1 & 1+w & 1+w & 0 & -2+w & 1 \\
2 & 0 & 2w & 0 & -2+w & -w & 0 & 2+2w & -2 & -2+w \\
2 & 3 & -1 & 1+w & -2+w & 1-2w & 0 & -1-2w & 2 & 0 \\
-2w & -w & 3 & 0 & 1+2w & -2+w & -1+w & 0 & -2+w & -1+w \\
1+2w & -w & -w & 2 & 0 & 2-w & 2+2w & -w & 0 & 1+w \\
2 & -1-w & -2-w & 2-w & 1 & -2w & 2 & -2+w & 2 & 2 \\
w & -1 & -1 & -1-w & -w & 1 & w & -1+2w & -2-w & w \\
-w & -1+w & 2-w & -2-w & 1 & w & 1 & 2-w & -1+2w & 2 \\
-w & -2w & 2+w & 0 & 0 & -2 & 2+w & 0 & -2 & 0
\end{bmatrix}.$$

$R(x) =$

$$\frac{1}{54}\begin{bmatrix} -73 & -4 & 51 & -9 & -86 & -23 & 20 & -27 & 38 & 1 \\ -2 & 28 & -6 & -18 & -10 & 26 & 22 & 54 & 4 & 2 \\ -36 & -144 & 0 & 27 & -18 & 36 & -90 & -81 & 18 & -18 \\ -44 & 130 & 84 & -18 & -4 & -130 & 106 & 0 & 88 & -10 \\ 88 & 172 & 48 & -45 & 44 & -64 & 112 & 27 & 40 & 2 \\ -34 & -10 & 168 & 18 & -80 & -44 & 50 & -108 & 122 & 16 \\ -48 & 78 & 72 & 27 & -42 & -78 & 42 & -27 & 42 & 30 \\ 30 & -42 & -18 & 0 & 24 & 42 & -60 & -54 & -6 & 6 \\ 55 & 148 & 57 & -18 & 14 & -13 & 70 & 54 & -2 & 35 \\ 13 & 196 & 93 & -18 & 20 & -115 & 154 & 0 & 28 & 23 \end{bmatrix}$$

$$+\frac{w}{54}\begin{bmatrix} 7 & 37 & -6 & -45 & 8 & -37 & 31 & 0 & 13 & -7 \\ -22 & -16 & -66 & 18 & 34 & 16 & -28 & 54 & -64 & 4 \\ -18 & 36 & 108 & 0 & -36 & -63 & 63 & -54 & 63 & -9 \\ 83 & 80 & -21 & -36 & 46 & 1 & 32 & 54 & -4 & 7 \\ 50 & 2 & -66 & 18 & 52 & 79 & -37 & 54 & -73 & -5 \\ -23 & 106 & 39 & -18 & -34 & -25 & 64 & 54 & -8 & 23 \\ 39 & 48 & 9 & -27 & -12 & 6 & 3 & 27 & 3 & 24 \\ -21 & -30 & 45 & 27 & -42 & 3 & -12 & -27 & 42 & 3 \\ 11 & -19 & -102 & 18 & 46 & 46 & -40 & 81 & -76 & -20 \\ 89 & 77 & -84 & 18 & 58 & 58 & -34 & 81 & -70 & 10 \end{bmatrix}$$

$R(y) =$

$$\frac{1}{54}\begin{bmatrix} -47 & 118 & 129 & 9 & -82 & -91 & 58 & -27 & 40 & 11 \\ 82 & 40 & -132 & -18 & 86 & 122 & -38 & 162 & -110 & 26 \\ -64 & -130 & 24 & 18 & -86 & 22 & -52 & -54 & 20 & -26 \\ 0 & 0 & 0 & 0 & -54 & 0 & 0 & 0 & 0 & 0 \\ 34 & 118 & -6 & -72 & 26 & -64 & 112 & 54 & 40 & -16 \\ -49 & 92 & 177 & -9 & -92 & -119 & 134 & -81 & 152 & 13 \\ -24 & 12 & 36 & 54 & -12 & -12 & -6 & -54 & -6 & -30 \\ 116 & 158 & 78 & -36 & 22 & 4 & 74 & 54 & 38 & 28 \\ 88 & 172 & -6 & -72 & 44 & -64 & 112 & 108 & 40 & 2 \\ 30 & 174 & -18 & -54 & 60 & -66 & 102 & 108 & -6 & -12 \end{bmatrix}$$

$$+\frac{w}{54}\begin{bmatrix} 50 & 83 & -39 & -36 & 34 & -2 & 17 & 81 & -19 & 4 \\ -43 & -154 & -75 & 45 & 10 & 73 & -94 & -27 & -76 & -29 \\ -56 & 28 & 48 & -45 & -28 & -82 & 76 & -27 & 58 & -16 \\ 0 & 0 & 0 & 0 & 0 & 0 & 0 & 0 & 0 & 0 \\ 50 & 2 & -66 & -9 & 88 & 52 & -10 & 27 & -46 & 4 \\ 28 & 121 & 3 & -18 & 14 & 14 & 43 & 81 & -29 & 8 \\ 6 & 24 & 18 & 0 & -24 & -24 & -12 & 0 & 42 & -6 \\ 7 & -44 & -87 & 9 & 8 & 71 & -50 & 81 & -68 & -7 \\ 50 & -52 & -120 & 18 & 52 & 106 & -64 & 54 & -100 & 22 \\ 87 & -30 & -117 & 0 & 102 & 57 & -66 & 54 & -66 & -15 \end{bmatrix}$$

and $R(z) =$

$$\frac{1}{54}\begin{bmatrix} 07 & 34 & -15 & -18 & 2 & -7 & -8 & 0 & 28 & 5 \\ 66 & -6 & -126 & 0 & 96 & 60 & -24 & 54 & -132 & 24 \\ -79 & -28 & 33 & 18 & -44 & 1 & -22 & 0 & -4 & -29 \\ -34 & 44 & 114 & 18 & -98 & -98 & 50 & -108 & 122 & -2 \\ -48 & 24 & 72 & -27 & -42 & -78 & 42 & -27 & 42 & -24 \\ 51 & 42 & -63 & -27 & 66 & -15 & 6 & 81 & -48 & 3 \\ 24 & -12 & 72 & 27 & -60 & -42 & 6 & -81 & 114 & 12 \\ -98 & -32 & 138 & -18 & -94 & -76 & 52 & -108 & 88 & 8 \\ 6 & 78 & 72 & -54 & -42 & -78 & 96 & 0 & 42 & 30 \\ -5 & -38 & 93 & 9 & -52 & -43 & 28 & -81 & 64 & 5 \end{bmatrix}$$

$$+\frac{w}{54}\begin{bmatrix} 8 & -31 & -3 & -9 & -14 & 31 & -34 & 0 & -16 & 19 \\ -3 & -66 & -9 & 54 & -6 & 39 & -48 & -54 & 6 & -15 \\ -5 & 43 & 12 & -18 & 20 & -70 & 28 & 27 & 10 & -4 \\ 31 & 106 & 39 & -45 & 2 & -25 & 64 & 27 & 46 & 5 \\ 12 & 48 & -18 & -27 & 42 & -21 & 57 & 27 & 3 & -3 \\ 48 & -51 & -45 & 0 & 24 & 24 & -15 & -27 & -15 & 6 \\ -6 & 30 & 36 & -27 & -48 & -3 & 39 & 27 & 39 & 15 \\ 29 & 134 & 33 & -36 & 10 & -53 & 86 & 54 & -4 & 25 \\ 39 & 48 & -45 & -27 & 42 & 33 & 30 & 27 & -24 & -3 \\ -28 & 41 & 51 & -9 & -50 & -14 & 65 & 0 & 29 & 10 \end{bmatrix}$$

REFERENCES

1. L. Babai and K. Friedl, *Approximate representation theory of finite groups*, Proc. Foundations of Computer Science (1991), (to appear).
2. L. Babai and L. Rónyai, *Computing irreducible representations of finite groups*, Math. Comput. **55** (1990), 705–722.
3. C. Brott and J. Neubüser, *A programme for the calculation of characters and representations of finite groups*, Conference on Computational Problems in Abstract Algebra (J. Leech, eds.), Pergamon Press, Oxford, 1970.
4. J.H Conway et al, *Atlas of Finite Groups*, Claredon Press, Oxford, 1985.
5. J.D. Dixon, *Computing irreducible representations of groups*, Math. Comput. **24** (1970), 707-712.
6. W. Eberly, *Computations for algebras and group representations*, Ph.D. thesis, Dept. Computer Science, Univ. Toronto, 1989.
7. G. Frobenius, *Über die Darstellungen der endlichen Gruppen durch lineare Substitutionen II*, Berlin Sitzber. (1899), 482-500.
8. B. Huppert, *Endliche Gruppen I*, Springer-Verlag, Berlin, 1967.
9. G.J Janusz, *Primitive idempotents in group algebras*, Proc. Amer. Math. Soc. **17** (1966), 520-523.

DEPARTMENT OF MATHEMATICS AND STATISTICS, CARLETON UNIVERSITY, OTTAWA, ONTARIO, CANADA K2G0E2. JDIXON@CARLETON.CA

DIMACS Series in Discrete Mathematics
and Theoretical Computer Science
Volume 11, 1993

A Graphics System for Displaying Finite Quotients of Finitely Presented Groups

DEREK F. HOLT AND SARAH REES

ABSTRACT. We describe a graphics package that the authors are currently developing for plotting the finite quotient groups of a given finitely presented group. For the graphics display, an X-Windows server is required, which may be coloured or black and white.

1. Introduction

In this paper, we describe an interactive graphics package that the authors are currently developing for plotting the finite quotient groups of a given finitely presented group G. For the graphics display, an X-windows server is required, which may be coloured or black and white. The user must first make a file containing the given presentation of G; thereafter, all computations are done interactively, by using a mouse to select menu options. A finite quotient $Q = G/M$ of G is represented in the display by a vertex, which can be thought of as being labelled by the normal subgroup M. Given two such vertices M and N such that $|G/N|$ has more prime factors (counted with repetition) than $|G/M|$, N will always appear lower than M. If M is joined to N either by an edge, or by a sequence of edges passing through intermediate vertices, then the inclusion $M \supset N$ holds. Initially, there will be a single vertex G corresponding to the trivial quotient; further vertices are plotted as the corresponding quotients are calculated. The system can output a postscript file for producing a good hard copy of the lattice that has been plotted.

We did not develop the system with any particular classes of groups in mind. However, the order of the quotients that we can plot is highly dependent on the particular example, and on the structure of the quotient itself. Quotients that are extensions of nilpotent groups by groups of small order can often be

1991 *Mathematics Subject Classification*. Primary 20-04, 20F05 .
This paper is in final form and no version of it will be submitted for publication elsewhere.

computed (using the p-quotient algorithm) up to a very large order (such as p^{1000} for a prime p, where can be large). On the other hand, in some examples, such as free groups, and groups with a large number of generators, it might be impractical to plot all quotients even up to index 20 or 30, since there could be far too many of them. The advantage of the interactive approach is that we are able to explore the quotient structure in those directions in which we are interested, and we are never obliged to plot more of them than we want to. In order to have some sort of specific goal in mind, we decided to provide at least potential access to all quotients up to order 10000. To justify this, we prove the following easy result.

PROPOSITION. *Let G be a group of order at most* 10000. *Then G has a normal soluble subgroup N, such that $S \subseteq G/N \subseteq Aut(S)$, where S is either trivial, or a nonabelian simple group, or isomorphic to $A_5 \times A_5$.*

PROOF. Let N be the maximal normal soluble subgroup of G. Then either $N = G$, or G/N has a minimal normal subgroup M/N which is a direct product of isomorphic finite nonabelian simple groups. Since $|G| \leq 10000$, M/N must be either simple or isomorphic to $A_5 \times A_5$. In either case, M/N is self-centralizing in G/N, and so $G/N \subseteq \mathrm{Aut}(M/N)$.

It is clear that this proposition can easily be made valid up to much larger orders by making obvious additions to the list of possibilities for S. Guided by this result, we have provided our system with two principal facilities. The first is to find the epimorphisms from G onto the relevant subgroups of $\mathrm{Aut}(S)$, for all appropriate groups S. This is done interactively by the user, by selecting the group S from a list on a menu. The second facility can be described as follows. Given a finite quotient G/M of G of sufficiently small order (at most 5000 will do in this case), find those quotients G/N of G with $M \supset N$ and M/N an elementary abelian p-group for some prime p. If we can do this, then it is clear from the proposition that we can, in principle, find all quotients of G up to the required order. We achieve this as follows. First, we use an abelianized Reidemeister-Schreier process (as originally programmed by Havas in [**1**]) to compute the largest elementary abelian p-quotient $M/M^{(p)}$ of M for a particular prime p (which the user can select from the menu). Then $M/M^{(p)}$ can be regarded as a $K(G/M)$-module, where K is the finite field of order p, and the normal subgroups N that we are seeking correspond precisely to the $K(G/M)$-submodules $N/M^{(p)}$ of $M/M^{(p)}$. In fact there was already in existence an efficient collection of programs for computing the submodule lattice of a module over a prime field, based on the well-known 'Meataxe' algorithm of Richard Parker (see [**10**]). This had been written at Aachen, and we are grateful to the Aachen group for allowing us to incorporate these programs into our system, and thereby to complete our requirements (see [**8**]).

In addition to these basic facilities, we have provided a number of other options, which experience has suggested are useful in examples which arise in practice. For example, when G/M is not too large, we can compute the complete set of abelian invariants of $M/[M,M]$, and thereby determine which prime numbers p are relevant. This will also sometimes reveal that M and hence G is infinite. We have also provided a version of the Canberra p-quotient algorithm (until recently, known as the NQA; see, for example, [**3**]) to compute the quotients of the lower exponent-p-central series of M. The basic lattice operations intersection and join on the vertices (thought of as normal subgroups of G) are also available; indeed, these can be used to ascertain whether or not one vertex is contained in another when this is not already clear. Although we are not providing sufficient facilities to completely identify a given quotient up to isomorphism, this is possible in many cases. As an easy aid to this, it is possible to count the numbers of elements of each order in a quotient. This is particularly useful if we are trying to distinguish one group from another. The simplest example is the use of this technique to distinguish between the dihedral and quaternion groups of order 8.

Of course there is a considerable overlap between these facilities and those provided by the system SPAS developed in Aachen, and for some particular computations that stretch resources to their limits, the best results will be obtainable by using SPAS followed by another program, such as the integral matrix diagonalization algorithm, as implemented by Havas and Stirling (see [**4**]). Although we hope to improve some of our particular implementations in the future (or incorporate existing implementations), our principal aim has been not so much to push back barriers of effective computability as to provide a convenient and user-friendly interface to this particular problem, which will hopefully enable the nonexpert user to gain a rapid overview of the finite quotient structure of a finitely presented group. We have also been guided by the aim of providing the theoretical capability of finding all quotients of G up to a given order. Unlike in SPAS, we do not make use of the low-index subgroup algorithm (see, for example, Section 6 of [**9**]) to find quotients that have permutation representations of small degree. However, we may well decide in the future to incorporate this, since it can sometimes lead to more rapid determination of particular quotients.

The design of our system is that we have a control program **quotpic** which manages the graphics display and remembers details of the part of the lattice that has been computed so far. This program calls up other programs to perform the individual computations required. These other programs input and output from and to files, which are read by **quotpic**. The user can of course run these programs individually if that should be more convenient, or copy and edit the files for use by other programs, but for most of the time, even for an expert user, it will be more convenient to work from within **quotpic**.

An essential part of the design of the system is that a quotient G/M of sufficiently small order will be stored in a file as the regular permutation representation (or equivalently, the Cayley graph) of G/M. It is for such subgroups M that it is possible to carry out further computations on M in order to find larger quotients G/N with $M \supset N$. All of the programs that construct such G/N are capable of computing the regular permutation representation of G/N

provided that the order is sufficiently small. (This is set by default to 10000, but it can easily be increased up to at least 100000.) For quotients G/N that are too large, this is not possible. We have therefore distinguished between three types of vertices, the green (or circular, for those with a black and white screen), the yellow (or triangular) and the red (or octagonal). The green vertices are those for which a regular permutation representation is stored. For the yellow (or triangular) vertices, a faithful but not regular permutation representation of the quotient is stored, and for the red no permutation representation is stored. Fewer functions are available for yellow and red vertices than for green vertices.

The remainder of this paper is organized as follows. In the second section, there is a more detailed description of the control program **quotpic**. In the third section, we deal with the individual programs that are called up by **quotpic**. Several of these were written in connection with an earlier project of the authors to attempt to decide whether two finitely presented groups are isomorphic or not (see [**7**]), and so they have already been described and do not require so much detail here. In the fourth section, we describe some results and present some illustrations. Finally, in the fifth we outline our plans for the future development of the program.

We are grateful to Michael Ringe in Aachen for providing us with the code for the submodule lattice package, and general assistance in its use, and to Steve Rumsby at Warwick for help with programming the X-windows material. Helpful suggestions that have influenced the design of the system have been received from various people, but particularly from George Havas, Joachim Neubüser, Mike Newman and Eamonn O'Brien.

2. The X-Window program *quotpic*

The program **quotpic** starts by opening a window on the screen, which contains a green circular vertex corresponding to the trivial quotient of G. A click with the left mouse button opens a menu of available functions, and the user can then start to construct the lattice from the top down.

As we have already mentioned, **quotpic** constructs vertices of three different types, depending on the amount of information that is held by the program on the corresponding quotient. Different functions are available at the different types of vertices. Whatever the type of the vertex, the menu listing the functions with which the lattice can be built downward from that vertex is always opened by a click with the left mouse button near to the vertex. A further click in the menu selects a function, and causes a parameter menu to open if the function requires input parameters. For almost all functions a time parameter limits the amount of time that the program will spend on the function (if time runs out, the program will simply terminate the calculation and wait for the user to make another choice). Many functions require a prime parameter to be set; sometimes the user needs to select an exponent, an exponent-p class (i.e. length of the lower exponent-p-central series, see [**3**]) or a permutation group. All such parameters are selected from the parameter menu.

Clicking on the right hand mouse button causes a menu to be opened which allows the selection of general 'housekeeping' tasks. Vertices and edges can be deleted, the lattice can be resized or refreshed, hard copies can be taken of the lattice in its current state (a postscript file from which a hard copy can be printed is always created on exit from the program, but sometimes it is also useful to have a hard copy of intermediate stages of the calculation), the global parameter which controls the maximal degree for which the program would attempt to construct a regular permutation representation can be reset, and the program can be terminated.

Clicking on the middle mouse button allows the repositioning of vertices. The program automatically plots vertices in such a way that quotients of the same order are plotted along a horizontal line. In fact, if we define the *level* of a quotient to be the number of primes dividing its order (counted with repetitions), then, by default, all quotients with the same level are plotted on the same horizontal line. Quotients of the same level but differing orders may be separated vertically using the mouse, but vertices of the same orders may not; thus, if a vertex is moved vertically, all vertices of the same order as it move with it. It is not possible to move a vertex to a position above vertices of a smaller level than itself.

Information about the results of a function call is given in a special output window which opens up as necessary at the bottom of the main program window. The user will be informed if the function was unable to complete for some reason, such as lack of available time, or memory (lack of memory does not cause **quotpic** to exit), or if the result of the calculation was trivial. Also the user will be given information in this window such as how many quotients of the group have been found isomorphic to a particular group in the program's list, or how many intermediate quotients there are between two previously specified ones (see (e) below), and will then be invited to choose how many of these to plot. Small message windows issue running instructions to the user (e.g. to wait, or select a vertex or edge with the mouse).

The full range of functions is available at those (green, circular) vertices corresponding to quotients G/M for which the program holds a regular permutation representation; that is, functions to

(a) calculate the abelian invariants of $M/[M, M]$,
(b) extend downwards by various abelian sections,
(c) calculate the sequence of orders of elements in G/M,
(d) calculate a presentation of M (which is then output to a file),
(e) calculate (and plot as required) all normal subgroups between M and some normal $N \subset M$, where M/N is elementary abelian (and either maximal elementary abelian or a section in a lower exponent-p-central series),
(f) calculate intersections and joins of M with a second normal subgroup

(in general, but not always one for which some, not necessarily regular, permutation representation is stored).

If $G = M$ we are at the top of the lattice, and then have the facility, already mentioned, to find all quotients of G isomorphic to a specified permutation group in our list, and then plot as many of the corresponding new vertices as are required by the user. Of these facilities, the functions to construct intersections and joins are also available at the (yellow, triangular) vertices for which the program holds non-regular permutation representations, and in exceptional situations even at the (red, octagonal) vertices for which no permutation information is held. Provided the order of a quotient is less than the maximum allowed degree for a regular permutation representation, the user has the choice of constructing a regular representation for a yellow vertex, thus turning it into a green vertex, equipped with the full choice of functions. Edges are also divided into categories, according to their origins, and the different categories of edges are plotted with different colours on a colour screen. It is actually the category of the edge joining vertices G/M and G/N which determines whether or not the intermediate normal subgroups can be found; thus this facility is potentially available at any G/M, whatever the type of the vertex.

As a new quotient G/N is built from an existing one G/M, provided the order of G/N is less than 1000, a check is carried out to see whether or not the quotient has already been constructed in some other way and, if it has, the appropriate edges are drawn to the existing vertex, and no new one is constructed. For higher indices however this check is not carried out (because it can be somewhat slow), and it is up to the user to manually check for equality between appropriate quotients using the intersection function. A message window indicates to the user that the automatic equality check has not been carried out.

The program does not automatically search for all possible inclusions between normal subgroups as it constructs them. This would also be a lengthy business. The user can verify a suspected inclusion using the intersection function.

3. The programs called by *quotpic*

As was mentioned in the Introduction, the first major facility required is to compute the epimorphisms from our finitely presented group G onto groups H satisfying $S \subseteq H \subseteq \mathrm{Aut}(S)$, where S is a member of a given list of groups. (Since we are interested in the quotients of G, we regard two such epimorphisms as being equivalent if they have the same kernel or, equivalently, if one is equal to the other followed by an automorphism of H.) This is done using our program **permim**. Since **permim** has already been described in some detail in [**5**] and [**7**] we shall not repeat this here. The relevant groups H are stored as permutation groups in a library of files, which contain bases and strong generating sets for the groups and also representatives of their conjugacy classes. The user has to select the group S, and then all epimorphisms onto all groups H between S and $\mathrm{Aut}(S)$ are sought. Since this number can be very large in some examples, the user is informed at the end of the calculation how many of them there are onto each of these intermediate groups, and must then decide how many of them, if any, to plot on the screen. As with all of the programs that are called by **quotpic**, it is also possible (indeed mandatory) for the user to impose an upper cpu-time limit on the particular process, and it will abort with a message if this limit is exceeded. It is a fairly easy matter for the authors to add additional groups S to the list if required, but this normally requires doing some computations (of conjugacy classes, for example) outside of the system (usually in CAYLEY), and then converting the output to our file format.

The second major facility is provided by the programs that use the Reidemeister-Schreier algorithm for computing presentations of a subgroup H of G (see [**1**] for a description of the first computer implementation of this process). In order to describe these, we have to repeat a small amount of the material in [**7**]. Let P be a transitive permutation group on the finite set $\Omega = \{1, 2, \ldots, t\}$, and let $\phi : G \longrightarrow P$ be an epimorphism, where $G = \langle g_1, g_2, \ldots, g_m | r_1, r_2, \ldots, r_s \rangle$. We shall regard the generators g_i as acting on Ω via ϕ. Let $Q = G_1$ be the stabilizer of the point 1 in P, and let $H = \phi^{-1}(Q)$. Currently, in the applications to **quotpic**, H is always a normal subgroup of G and so P is a regular permutation group, but this is not a necessary assumption for most of what follows (and in future developments to **quotpic** we shall probably relax this restriction). For proofs see, for example, Section 4 of [**9**], bearing in mind that P can be regarded as a complete coset table for G over the subgroup H. For each i in $\Omega - \{1\}$, we can regard one particular equation $j^{g_k} = i$ as being the definition of i, where j is already defined, and 1 is regarded as being defined initially. This associates an element $g(i)$ of G to each point i in Ω, with the property that $1^{g(i)} = i$, where $g(1) = 1$ and $g(i) = g(j)g_k$ with $j^{g_k} = i$ a definition. The transversal $\{g(i) | 1 \leq i \leq t\}$ is therefore a Schreier transversal as defined in [**9**]. To each equation $j^{g_k} = i$ that is not a definition, we can define an element $h(j,k)$ of H by $g(j)g_k = h(j,k)g(i)$. The $h(j,k)$ then generate H. We obtain the relators for H as follows. For each relator r_k of G and each point i in Ω, we can use the equations $g(j)g_k = h(j,k)g(i)$ to obtain an equation $g(i)r_k = w(i,k)g(i)$, where $w(i,k)$ is a word in the $h(j,k)$. Then the $w(i,k)$ form a set of defining relators for H with respect to the generators $h(j,k)$.

There are now several different ways in which we can proceed, depending on what precisely is required. If we want to compute the abelian quotient invariants of H, we apply the standard integer diagonalization algorithm to the presentation, although it is important to check for integer overflow, since this becomes a serious danger when the index of H in G grows beyond a few hundred. At some stage, we hope to install a better implementation that either uses arbitrarily large integers or uses modular techniques as described in [**4**]. To avoid this danger, it is possible for the user to choose an integer $m > 0$ and compute the largest abelian quotient $H/H^{(m)}$ of H of exponent dividing m. The case in which m is a prime p has been implemented separately, since it can be done much more efficiently than the general case. This calculation is feasible for subgroups H of much larger index - in fact, it has completed successfully for indices as high as 50000. There is also an option to compute the largest such quotient $H/H^{(p)}[G, H]$ that is centralized by the conjugation action of G.

Sometimes we simply want a presentation of the kernel. (We need that for instance if we are going to apply the p-quotient algorithm.) This is technically more complicated, because we can no longer assume throughout that the generators $h(j, k)$ of H commute with each other. Instead, we carry out a restricted set of Tietze transformations to remove redundant relators and simplify a relator if some cyclic conjugate of it has more than half of its length in common with another relator. We only eliminate generators, however, when they are either trivial or actually equal to another generator. The reasons for this are discussed in the next paragraph.

Assume now that H is normal in G and that we have used one of the above methods to compute a quotient H/K of H. Notice that, in each of the cases described above, K is characteristic in H and hence normal in G. Furthermore, we have a definition of H/K (such as a power-commutator presentation) in which we can compute efficiently, together with an explicit epimorphism ψ from H to H/K, giving the images of the generators $h(j, k)$ of H. The elements of G/K have the form $\overline{h}g(j)$, for $\overline{h}$ in H/K and $j = 1, 2, \ldots, t$. Since $\overline{h}g(j)g_k = \overline{h}\psi(h(j, k))g(i)$, this enables us to construct immediately the epimorphism from G to the regular permutation representation of G/K. For the applications to **quotpic** it is essential to be able to do this (at least when G/K is reasonably small), which is the principal reason why we have kept the original generators $h(j, k)$ of H. As is described in **[9]**, it is often possible to simplify the presentation of H considerably by applying further Tietze transformations that eliminate redundant generators, or indeed to use one of the versions of the Reidemeister-Schreier process that uses fewer generators of H to begin with. This would make it considerably more complicated for us to compute the required permutation representation of G/K. In the future, however, we shall be considering including the option to produce smaller presentations of H for use in those cases when G/K is too large to compute the regular permutation representation. This would probably make it quicker and less demanding on storage to compute at least the order of these large quotients.

Returning to the Reidemeister-Schreier process, if we assume that H is normal in G, then the elements $g_l^{-1}h(j, k)g_l$ will lie in H. By writing the $h(j, k)$ as words in the generators g_i, we can regard these expressions as relators of G, and then express them as words in the generators of H as described above. In other words, we can compute the conjugation action of the generators of G on those of H. This, in turn, enables us to compute the conjugation action of G/H on $H/H^{(p)}$ as an induced matrix group over the field K of order p. We can do the same for the action of G/H on any of the sections $\gamma_i^p(H)/\gamma_{i+1}^p(H)$ of the lower exponent-p-central series of H as computed by the p-quotient algorithm. We can then use the submodule lattice package, which was made available to us from Aachen, to compute the $K(G/H)$-submodules of this induced module. These correspond precisely to the normal subgroups of G that lie between H and $H^{(p)}$, or between $\gamma_i^p(H)$ and $\gamma_{i+1}^p(H)$. Since the sublattice package can output generators (as vector spaces) of each of the submodules found, there is no difficulty in computing the regular permutation representation of the corresponding quotient of G (provided, of course, that it has small enough order). In fact, since there may be a large number of these submodules, the user must choose at the end of the calculation how many, if any, to plot on the screen.

We turn now to the lattice operations, intersection and join. Suppose that we have permutation representations of G on sets Ω_1 and Ω_2, with kernels M and N respectively. We replace Ω_2 by a set that is disjoint with Ω_1, and then the natural induced action of G on $\Omega_1 \cup \Omega_2$ is a permutation representation of G with kernel $M \cap N$. We have provided a program **convrep** which computes the permutation representation of a permutation group P in its action by right multiplication on the right cosets of a subgroup Q of P and, by taking $Q = 1$, we compute the regular permutation representation of the quotient $G/M \cap N$ if its order is sufficiently small. By comparing the order of this quotient with those of G/M and G/N, we can decide whether either of M and N is contained in the other, or if $M = N$. In case of strict containment, an edge is plotted between the two vertices if there is none there already. In case of equality, the two vertices will be identified in the graph (which might then result in the identification of further edges and vertices).

For the join operation, we first calculate the action on Ω_2 of the kernel M of the action of G on Ω_1. To do this, we find bases $b_1, b_2, \dots, b_r$ and $c_1, c_2, \dots, c_s$ of the two permutation representations. Then clearly $b_1, \dots, b_r, c_1, \dots, c_s$ will be a base of the action of G on $\Omega_1 \cup \Omega_2$, and a strong generating set with respect to this base will contain generators of the subgroup M in its action on Ω_2. Then we apply **convrep** to G^{Ω_2}, using M^{Ω_2} as the subgroup. The result will be the regular permutation representation of G/MN, which is what we want.

Finally, the program **ordcrs**, which counts the number of elements of each order in a regular permutation group P on a set Ω, is straightforward. For each i in Ω, it constructs the unique permutation in P that takes 1 to i, and computes the length of the orbit of 1 in this permutation, to give its order.

4. Examples and illustrations

We start with four examples which we were asked to analyze by Jeff Weeks. They arose as fundamental groups of closed manifolds, and he wished to know whether they were finite, to help determine whether or not the manifolds were hyperbolic. These were:

$$G_1 = \langle a, b | ab^3a = b^2, a^3 = b^5 \rangle, G_2 = \langle a, b | a = b^3a^2b^3, a^3 = b^7 \rangle,$$

$$G_3 = \langle a, b | a = b^3ab^3, ba^2(b^{-2}a^2ba^2)^2 = 1 \rangle, G_4 = \langle a, b | a^2ba^2 = b^3, a^3b^4 = 1 \rangle.$$

Using **quotpic**, we managed to determine the precise structures of G_1 and G_4, both of which turned out to be finite. In fact $G_1 \cong SL(2,5) \times C_{13}$ (where C_{13} is the cyclic group of order 13), and $G_4 \cong \hat{S}_4 \times C_{11}$, where $\hat{S}_4$ is the 2-fold covering group of S_4 that is *not* isomorphic to $GL(2,3)$. (The 'orders' option to count the number of elements of each order in a group was used to distinguish between the two nonisomorphic 2-fold covering groups of S_4.) Having obtained the above as quotients of G_1 and G_4, we succeeded in recognizing these quotients as the full group in both cases, by calculating presentations of the relevant normal subgroups and finding them to be trivial presentations. (Of course, the orders of these groups can also be determined very quickly by a Todd-Coxeter coset enumeration.)

On the other hand, G_2 and G_3 both turned out to be infinite with very many finite quotients. For example, G_2 has a map onto $PSL(2,7)$ of which the kernel has $C_{100} \times C_\infty^6$ as maximal abelian quotient, and G_3 has a map onto A_5 of which the kernel has $C_2^2 \times C_\infty^{29}$ as maximal abelian quotient.

As an illustration, Fig. 1 shows two views of what the user sees when carrying out this computation in G_1. Although it cannot be seen in these pictures, selecting the 'ab. invariants' menu option shows us that the maximal abelian quotient of G_1 itself has order 13. This proves that G_1 has the 2-fold cover of A_5 (i.e. $SL(2,5)$) as a direct factor, rather than $A_5 \times C_2$, and so the structure of the group is unambiguous. The vertex representing the quotient of order 1560 is triangular (or yellow) in the first picture. The menu item 'conv. to reg. rep.' was then selected to compute the regular permutation representation of this quotient, which automatically converted the vertex to circular (green), and then the full menu of operations became available. It was then possible to select the menu option 'rep. kernel', which computed a presentation of the normal subgroup corresponding to this vertex, and this presentation turned out to be trivial.

Our next illustrative example is the Hurwitz group $G = \langle x, y | x^2 = y^3 = (xy)^7 = 1\rangle$. In Fig. 2, we have found the (unique) epimorphisms of G onto the simple groups $PSL(2,7)$ and $PSL(2,8)$. The maximal abelian quotients of their kernels were C_∞^6 and C_∞^{14} respectively. We have then plotted the maximal elementary abelian p-quotients of these kernels, for $p = 2$ and 3 in the first case and $p = 2$ in the second, and used the 'intermediate quotient' ('Meataxe') option to find and plot all intermediate quotients. The original edges representing the full elementary abelian quotients have been removed where possible, in order to clarify the diagram. Notice that many of the vertices are octagonal (red). Finally, in order to illustrate the parameters section of the menu, we have taken the picture while it was in the process of calculating the maximal elementary abelian 5-quotient of one of the normal subgroups.

The following interesting example was proposed some years ago by Heineken. $G = \langle x, y, z | [[x,[x,y]] = z, [[y,[y,z]] = x, [z,[z,x]] = y\rangle$. Neubüser in Aachen discovered that it had an epimorphism onto A_5 of which the kernel had a quotient

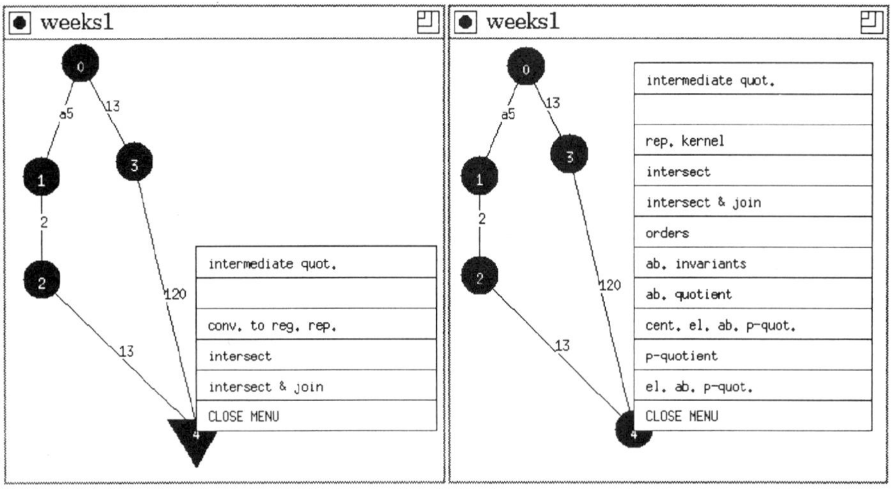

FIGURE 1.

of class 5 and order 2^{24}. No other quotients of this group are known. In Fig. 3, we have applied the p-quotient algorithm to the kernel of the map onto A_5 to obtain the terms in the descending 2-central series. These occur down the main diagonal in the picture. In addition, we have found those intermediate quotients that lie between the successive terms of the series, and plotted these to the left of the main diagonal. In particular, we notice that the two layers of order 2^5 have different submodule structures. (In fact, there are two nonisomorphic A_5-modules of order 2^4 involved in this sequence, and the 'Meataxe' package does distinguish between them, but we have not yet made it possible to display this information.) Finally, we have opened the global menu.

As a final example, we should like to mention a group that was produced at the DIMACS conference itself by Leonard Soicher and Jon Hall. This has 5 involutory generators and far too many relations to write down here. (The interested reader could obtain the presentation from us, or from them.) The conjecture is that it defines the largest 3-transposition group generated by 5 transpositions, such that no two transpositions commute. If so, then it should be an extension of a large 3-group by a group of order 2. We discovered, by using the p-quotient algorithm from within **quotpic**, that it has a quotient that is an extension of a group of order 3^{49} (and exponent-3-class 6) by a group of order 2, and that this is the largest quotient of this form. We found no other finite quotients of this group. This information is therefore consistent with the conjecture.

5. Future development

quotpic is still very much under development. We have plans to

(a) include a low index subgroup algorithm (see Section 6 of [**9**]). This would

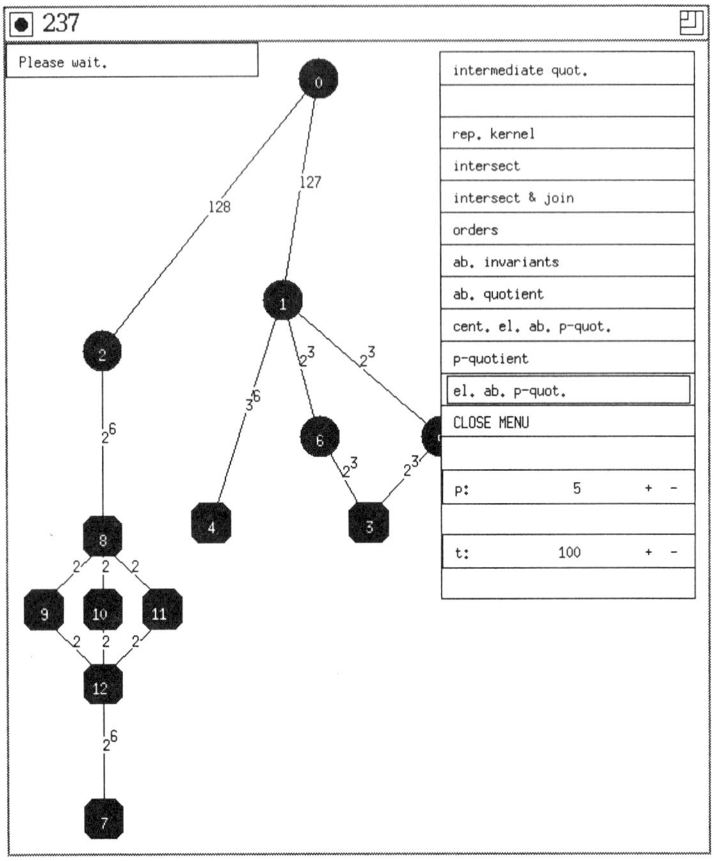

FIGURE 2.

give an alternative approach to that of **permim**, and would allow us to hold several permutation representations of the same quotient.

(b) allow the possibility of applying the Reidemeister-Schreier related programs to the point stabilizers in non-regular permutation representations (which are in general not normal subgroups). The examination of non-normal subgroups has been used in the past both to prove infiniteness (see [**6**]), and to distinguish between non-isomorphic groups (see [**2**]).

(c) replace our own implementation of the Canberra p-quotient program by the much more efficient implementation of Eamonn O'Brien and Mike Newman.

(d) include the general nilpotent quotient algorithm, as implemented by Werner Nickel. This will enable us to gather information also about infinite nilpotent sections.

(e) make it possible for **quotpic** to look for all maps onto *any* permutation group on which the user has supplied information, rather than restrict to a specific list of permutation groups held by the program.

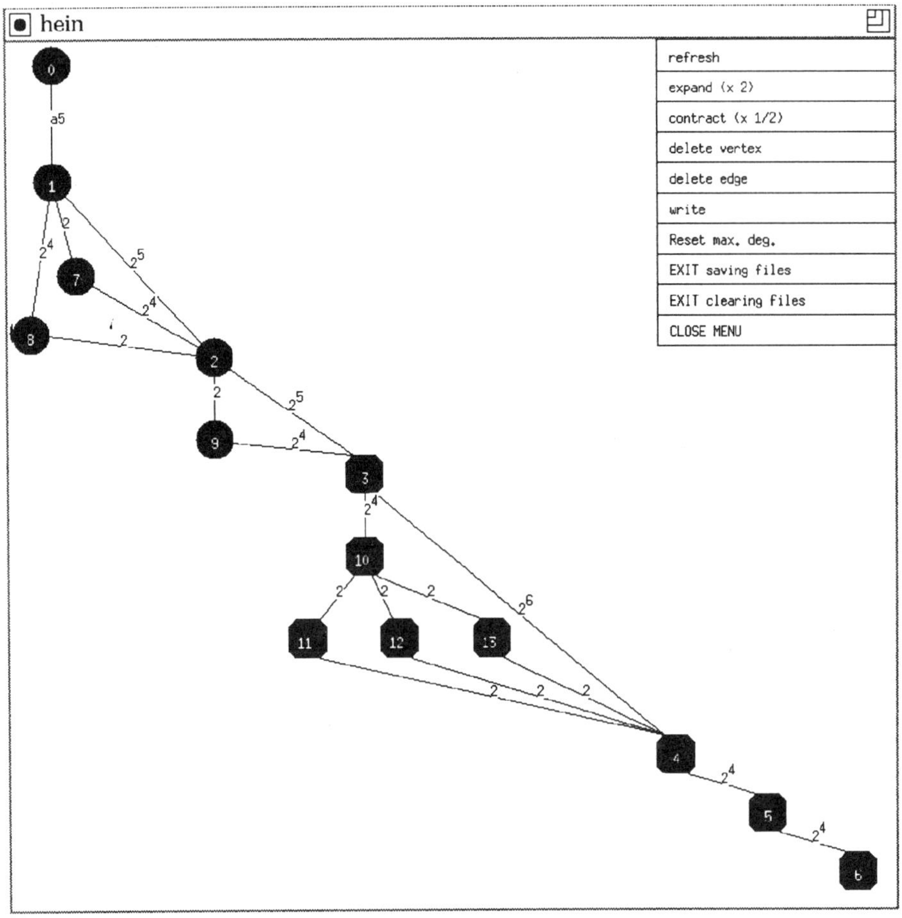

FIGURE 3.

We are keen to receive comments from users about our program, and will be very pleased to give copies (via magnetic tape or FTP) to anyone who asks for it. We invite suggestions for improvements, and would like any bugs, or odd behaviour to be reported; ideally a bug report should be accompanied by a copy of the presentation input to the program, together with the log file created by **quotpic**.

The files that can be collected via anonymous FTP are compressed UNIX tarred files named isom$i$.tar.Z for $i = 1, 2, 3, \ldots$. They can be found either in the directory pub/local/nser on the machine tuda.ncl.ac.uk or in the directory pub/sources/geom on the machine amazon.warwick.ac.uk.

REFERENCES

1. G. Havas, *A Reidemeister-Schreier program*, Proceedings of the Second International Conference on the Theory of Groups, Canberra 1973, Lecture Notes in Math., vol. 372, (M.F.

Newman, ed.), Springer-Verlag, Berlin and New York, 1974, pp. 347–356.
2. G. Havas and L.G.Kovács, *Distinguishing eleven crossing knots*, Computational Group Theory (M. Atkinson, ed.), Academic Press, London, 1984, pp. 367–73.
3. G. Havas and M.F. Newman, *Applications of computers to questions like those of Burnside*, Burnside Groups, Lecture Notes in Math., vol. 806, (J.L. Mennicke, ed.), Springer-Verlag, Berlin and New York, 1980, pp. 211–30.
4. G. Havas and L. Sterling, *Integer matrices and abelian groups*, Symbolic and Algebraic Computation, Lecture Notes in Computer Science, vol. 72, (E.W. Ng, ed.), Springer-Verlag, Berlin and New York, 1979, pp. 431–451.
5. D. F. Holt and W. Plesken, *Perfect Groups*, Oxford University Press, 1989.
6. D. F. Holt and W. Plesken, *A cohomological criterion for a finitely presented group to be infinite* (to appear).
7. D. F. Holt and Sarah Rees, *Testing for isomorphism between finitely presented groups*, Groups, Combinatorics and Geometry, LMS Lecture Note Series 165, (M.W. Liebeck and J. Saxl, eds.), Cambridge University Press, 1992, pp. 459–475.
8. K. Lux, J. Müller and M. Ringe, *Peakword condensation and submodule lattices, an application of The Meat-Axe* (to appear).
9. J. Neubüser, *An elementary introduction to coset table methods in computational group theory*, Groups - St Andrews 1981, LMS Lecture Note Series 71, (C. M. Campbell and E. F. Robertson, eds.), Cambridge University Press, 1982, pp. 1–45.
10. R. Parker, *The computer calculation of modular characters. (The Meat-Axe)*, Computational Group Theory (M. Atkinson, ed.), Academic Press, London, 1984, pp. 267–74.

Mathematics Institute, University of Warwick, Coventry CV4 7AL, U.K.
E-mail: dfh@maths.warwick.ac.uk

Mathematics Department, University of Newcastle, Newcastle NE1 7RU, U.K.
E-mail: sarah.rees@newcastle.ac.uk

DIMACS Series in Discrete Mathematics
and Theoretical Computer Science
Volume 11, 1993

Random Remarks on Permutation Group Algorithms

WILLIAM M. KANTOR

I

Quite a few papers in these Proceedings concern goal–oriented Computational Group Theory, aimed at producing software and leading to important applications. There are also a number of papers concerned with algorithms and their inherent limitations, from a somewhat more theoretical point of view; their goal is new data structures and new mathematical approaches more than the immediate production of software. The present note is partly in the latter direction, but also partly points elsewhere: the production of new, purely mathematical theorems and directions, having algorithmic applications but capable of standing by themselves without any algorithmic components. It is my contention that such theorems can not only be of value within algorithmic contexts, but also within other areas of mathematics.

Let me start with some examples of theorems I have in mind.

- Orders of primitive subgroups G of S_n: $|G| < n^{C \log n}$ *for some constant* C, *unless* $n = \binom{m}{k}^{\ell}$ *and* G *is a subgroup of* $S_m \,\mathrm{wreath}\, S_\ell$ *containing* $(A_m)^\ell$ *for some* m, k, ℓ, *where* S_m *acts on the* k*–sets of an* m*–set and the wreathed product has the product action.* This result is due to Cameron [**Cam**], based on work on permutation representations of classical groups [**Ka1**] as well as the classification of finite simple groups. (In [**Li**] it is shown that one can use $C = 9$ here; in fact, $C = 5$ will do.) The theorem, and refinements of it, have been of great importance to some of the work described in other papers in these Proceedings (by Babai–Luks–Seress, Seress–Weisz and Sarawagi–Cooperman–Finkelstein), and would have been discovered due to their need. However, the

1991 *Mathematics Subject Classification.* Primary 20B40, 20P05.

Research supported in part by the NSF and the NSA.

This is the final version of this paper.

results in [**Cam**] and [**Ka1**] arose in the context of very classical questions concerning primitive groups.

- Orders of primitive solvable subgroups G of S_n: $|G| < n^{3.25}$ (Pálfy [**Pá**], Wolf [**Wo**]). While Pálfy was motivated by possible algorithmic applications, Wolf was not. For him this was a natural question in the context of properties of solvable groups he was studying.

- Orders of primitive subgroups G of S_n: *If every noncyclic composition factor of G has order bounded by some constant b, then $|G| = O(n^{c(b)})$ for some constant $c(b)$* (Babai-Cameron-Pálfy [**BCP**]). While this theorem certainly stands by itself, it was motivated by algorithmic applications.

- *The classification of subgroups of $GL(n,q)$ containing an irreducible element subject to suitable additional assumptions.* Results of this type are discussed in the paper by Praeger in these Proceedings. Other results of the same sort were obtained by Hering, initially for applications to projective planes, and later in order to study 2–transitive permutation groups (cf. [**Ka3**] for a discussion of those and more recent theorems). Penttila and Praeger are working on extensions of research that was initiated by Neumann and Praeger for algorithmic purposes; and these extensions will very likely lead to other purely geometric applications as well.

There are other examples. An additional one will be discussed later. For now, the preceding examples should suffice to make the point concerning "stand alone" theorems that have algorithmic implications. The direction of all of the above results is, to some extent, towards Asymptotic Group Theory; compare Pyber's contribution to these Proceedings (and also [**Ka4**]).

II

If $G \leq S_n$, what can be said about a random element of G? What can be said about a random subgroup G of S_n? What can be said about a random transitive subgroup G of S_n?

Answers to questions like these could be of great value to Computational Group Theory. However, it may well be that random subgroups of S_n have an uninteresting structure. Namely, whether "random" refers to averaging over all subgroups of S_n, or only up to conjugacy in S_n, it seems likely that

- Prob($G \leq S_n$ *is nilpotent*) $\to 1$ as $n \to \infty$.

Worse than this, it may even be true that the word "nilpotent" can be replaced here by "abelian". However, it seems as if this is not the correct type of question: it appears to take one outside the realm of those permutation groups studied most often. A result with a similar flavor was conjectured at the Workshop: *a random subgroup G of S_n has exponential order.* This was proved soon after the Workshop by Pyber (see his paper in these Proceedings for a more precise statement, as well as for numerous other related results), but it also may not be sufficiently useful in computational contexts. This discussion leads to the

- Metaquestion: *What are the correct questions to ask about the average behavior of permutation groups?*

In particular, this question needs to be addressed by those interested in *using* permutation group algorithms. What are the types of groups to which algorithms are most often applied? They are certainly not just the simple groups. They usually are transitive, or even primitive. However, in the course of recursion many other groups, in particular p–groups, may occur. While these do not seem interesting in the context of permutation groups, they appear to be unavoidable and can produce algorithmic difficulties. A significant instance of this is provided by a result of Blaha: the Greedy Algorithm cannot be expected to yield a near–optimal base for a permutation group, for example if that group is the direct product of isomorphic groups, such as cyclic groups of prime order or alternating groups of degree 5 (cf. [**Bl**]). If such "boring" groups could *usually* be avoided then algorithms could be developed in that direction. If such simple–minded groups are inherently unavoidable and occur fairly often — in particular, if a random subgroup of S_n happens to be "boring" — then the average behavior of algorithms might be of less significance than worst–case behavior.

One further remark is needed concerning the notion of "random" subgroups of S_n. A lovely and well-known theorem of Dixon [**Dix**] states that a random pair of elements of S_n generates S_n or A_n with probability $\to 1$ as $n \to \infty$. There are analogous results for most (and probably all) finite simple groups [**KL**]. Therefore, it is not at all clear how one would even go about *choosing* a random subgroup of any such group.

III

A standard question in Computational Group Theory is that of finding an element of order p, where p is a prime dividing the order of G. Such elements are needed, for example, as starting points towards Sylow subgroups. The only provably "efficient" algorithm for finding such an element seems impractical [**Ka2**]. The usual approach is to pick an element of G randomly a few times, and hope that an element of order divisible by p is obtained. Choosing an element at random is quite easy, assuming that a base B is available. For example, if $B = \{1, 2, \dots, b\}$, if $G_{(i)}$ is the stabilizer of the points $1, 2, \dots, i$ (so $G_{(0)} = G$ and $G_{(b)} = 1$), and if T_i is a set of coset representatives of $G_{(i+1)}$ in $G_{(i)}$, simply choose a random element $t_i \in T_i$ for each i; then $g = t_{b-1} \cdots t_1 t_0$ is a random element of G.

J. Cannon observed that, when randomly picking an element of the subgroup $\mathrm{PSL}(2, p)$ of S_{p+1}, Cayley often takes a long time to produce an element of order p. He hoped something could be done to handle this difficulty. First, observe that $\mu_p(\mathrm{PSL}(2, p)) = 2/p = 2/(n-1)$, where for any group G let $\mu_p(G)$ denote the probability that an element of G has order divisible by p (i.e., $\mu_p(G)$ is the proportion of such elements of G). Note that the stabilizer G_1 of a point in $\mathrm{PSL}(2,p)$ is a Frobenius group of order $p \cdot \frac{1}{2}(p-1)$, and $\mu_p(G_1) = 2/p$. However, it is certainly easy to use a commutator or two of any given generators of G_1 in order to obtain an element of order p.

Clearly $\mathrm{PSL}(2,p)$ is a standard type of group occurring in group-theoretic computations. There are many examples of permutation groups exhibiting bad behavior from the present point of view. Some further examples are as follows:

$G = \mathrm{PSL}(2,p) < S_{\binom{p+1}{2}}$, where $\mu_p(G) \sim \sqrt{2/n}$;

$G = \mathrm{PSL}(2,2^e) \leq S_{2^e+1}$, where $\mu_2(G) = 1/(n-1)$;

$G = S_p$, $p = n$, where $\mu_p(G) = 1/p = 1/n$; and

G is sharply 2-transitive and n is a power of p, where $\mu_p(G) = 1/n$.

These should be compared with the following

THEOREM [**IKS**]. *If $G \leq S_n$ and p is a prime dividing $|G|$, then $\mu_p(G) \geq 1/n$. Equality holds only in the last two examples given above.*

Clearly, this is another example of a "stand alone" theorem. From a computational point of view, there are two ways to view this theorem: (i) *the optimistic view*: no more than n random choices should be needed in order to obtain an element of order divisible by p; and (ii) *the pessimistic view*: there can be some situations where random selection will work too poorly to be of value. As noted above, there are many other "bad" situations for the random method — perhaps not as bad as in the theorem, but bad enough. Of course, there are other ways of dealing with some of them — the case $G = S_p$, $p = n$, being especially easy (just add a subroutine to write a cycle). However, all of this points towards a need to develop algorithms that will try random selection, but will then do something else if not successful after a few tries.

It should be possible to go somewhat further than the Theorem. There is no doubt that all permutation groups of degree n such that $\mu_p(G) \leq 2/n$ can be classified. It may be possible to give a reasonable description of those groups G for which $\mu_p(G) \leq c/n$ for some constant c. For purposes of applications it would be ideal to be able to show that $\mu_p(G) \leq 1/\log n$ for "most" groups, whatever that means; but this seems extremely difficult to state, much less prove.

Nevertheless, it would be interesting to know from a user's point of view what types of theorems in this direction might be desirable.

The proof of the above theorem uses the classification of finite simple groups. First there is a reduction to the primitive case, then to the (almost) simple case, and finally to simple groups of Lie type in characteristic $\neq p$. These reductions are relatively straightforward, and not especially enlightening. The remainder of the proof is more interesting, leading to the following

THEOREM [**IKS**]. *If G is a simple group of Lie type of characteristic $\neq p$, then $\mu_p(G) \geq 1/p^2$ — independent of the group G — and also $\mu_p(G) \geq (1-1/p)/2h$.*

(Here, h is the Coxeter number of the corresponding group over an algebraically closed field; for example, $h = n$ when $G = \mathrm{PSL}(n, q)$.) This requires a counting argument involving groups of Lie type, fairly general considerations using partitions in the case of classical groups, and a case by case calculation with Weyl groups and cyclotomic polynomials for the exceptional groups of Lie type. (Cyclotomic polynomials arise as follows: p divides the order of G, which is essentially a product of cyclotomic polynomials evaluated at a prime power q. The proof then considers which of these polynomials a given prime p might divide.)

The following amusing result is easily deduced from the preceding theorem:

COROLLARY **[IKS]**. *If G is any group having a simple homomorphic image that is neither cyclic nor Lie type of characteristic 2, then $\mu_2(G) \geq 1/4$.*

Thus, for such a group G it is "easy" to find involutions.

All of this leaves open a much more important question: is there any way to choose random subgroups in order to obtain nice ones — such as Sylow subgroups? Once again, **[Dix]** as well as related results **[KL]** present obstacles to the most obvious notions of randomness.

REFERENCES

[BCP] L. Babai, P. J. Cameron and P. Pálfy, *On the order of primitive groups with restricted nonabelian composition factors*, J. Algebra **79** (1982), 161–168.

[Bl] K. D. Blaha, *Minimum bases for permutation groups: The greedy approximation*, J. Algorithms **13** (1992), 297–306.

[Cam] P. J. Cameron, *Finite permutation groups and finite simple groups*, Bull. London Math. Soc. **13** (1981), 1–22.

[Dix] J. D. Dixon, *The probability of generating the symmetric group*, Math. Z. **110** (1969), 199–205.

[IKS] I. M. Isaacs, W. M. Kantor, and N. Spaltenstein, *On the probability that a group element is p–singular* (to appear in J. Algebra).

[Ka1] W. M. Kantor, *Permutation representations of the finite classical groups of small degree or rank*, J. Algebra **60** (1979), 158–168.

[Ka2] ———, *Polynomial-time algorithms for finding elements of prime order and Sylow subgroups*, J. Algorithms **6** (1985), 478–514.

[Ka3] ———, *2–Transitive and flag-transitive designs* (to appear in Proc. Marshall Hall Conf. 1990).

[Ka4] ———, *Some topics in asymptotic group theory*, Groups, Combinatorics and Geometry (eds. M. W. Liebeck and J. Saxl), LMS Lecture Notes 165, 1992, pp. 403–421.

[KL] W. M. Kantor and A. Lubotzky, *The probability of generating a finite classical group*, Geom. Ded. **36** (1990), 67–87.

[Li] M. W. Liebeck, *On the minimal degrees and base sizes of primitive groups*, Arch. Math. **43** (1984), 11–15.

[Pá] P. P. Pálfy, *A polynomial bound for the orders of primitive solvable groups*, J. Algebra **77** (1982), 127–137.

[Wo] T. R. Wolf, *Solvable and nilpotent subgroups of $GL(n, q^m)$*, Can. J. Math. **34** (1982), 1097–1111.

DEPARTMENT OF MATHEMATICS, UNIVERSITY OF OREGON, EUGENE, OR 97403

E-mail address: kantor@bright.uoregon.edu

DIMACS Series in Discrete Mathematics
and Theoretical Computer Science
Volume **11**, 1993

Application of group theory to combinatorial searches

CLEMENT W. H. LAM

ABSTRACT. This paper discuss why algorithms for permutation groups are important in combinatorial searches. Its aim is to describe this particular application to the designers of algorithms. Its hope is to encourage the development of better algorithms. The paper defines what is a search problem and explains how symmetry can be used to reduce the size of a search. It also outlines how algorithms for permutation groups are used and highlights where improvements would be desirable.

1. Introduction

In recent years, several famous combinatorial problems have been "solved" after thousands of hours of computing. Note-worthy examples are the four-color theorem [**1**] and the non-existence of a finite projective plane of order 10 [**7**]. These are examples of combinatorial search problems where efficient permutation group algorithms are used to reduce the searches to a manageable size. Yet, we may be seeing only the tip of the iceberg. The objective of this paper is to explain how permutation group algorithms arise in these applications. Our hope is to encourage the development of better algorithms, which in turn would mean that even more combinatorial search problems are within the computer's reach.

Since the intended audience consists of experts in permutation groups, we shall skip over the basic definitions from group theory. We shall also assume that concepts such as base and strong generators are known. Readers interested in more background information are referred to [**5**].

Since our emphasis is on practical permutation group algorithms, we should keep in mind that a typical value of n, the degree of the permutation group, is about 100. For smaller values of n, say 10, algorithms do not have to be very efficient, but very few interesting search problems are left. For larger values

---

1991 *Mathematics Subject Classification.* Primary 05B05, 20B25; Secondary 05B25.

The author was supported in part by NSERC Grants #9393, #9413, and FCAR Grant #0518.

This paper is in final form and no version of it will be submitted for publication elsewhere.

of n, the search problems are so big that even the most efficient algorithms do not help. Fortunately, with $n \approx 100$, there are still many unsolved interesting problems.

Given a typical value of $n \approx 100$, we can translate the complexity of an algorithm roughly into actual computing time. An $O(n)$ algorithm would take less than a millisecond and can be used frequently. An $O(n^3)$ method would take seconds and should be used with some care, while an $O(n^5)$ one would require hours and should be used rarely. Some of the basic algorithms we need have complexities in the range between $O(n)$ and $O(n^5)$.

In Section 2 we shall define a combinatorial search problem and its symmetry group. In Section 3, we shall explain how several permutation group algorithms arise naturally in this context.

2. Search and Symmetry

First, let us define a combinatorial search problem.

> **Search problem:**
> Given a collection of sets of candidates $C_1, C_2, C_3, \ldots, C_m$ and a boolean *predicate* function $\mathcal{P}$ defined on the Cartesian product $C_1 \times C_2 \times \ldots \times C_m$, find an *m-tuple* $(x_1, \ldots, x_m)$ such that $\mathcal{P}((x_1, \ldots, x_m))$ is true.

For example, the adjacency matrix of a loopless undirected graph with n vertices, when written in row-major order, is an n^2-tuple. In this case, the candidate sets C_i are all equal to $\{0, 1\}$. The boolean predicate function would be one that ensures the original adjacency matrix is symmetrical and has zeros in the diagonal.

The predicate $\mathcal{P}$ is often called a *property*. An m-tuple $(x_1, \ldots, x_m)$ is called a *combinatorial object*. An m-tuple $(x_1, \ldots, x_m)$ satisfying $\mathcal{P}((x_1, \ldots, x_m))$ is called a *solution*. A brute-force approach to find all solutions is to generate all objects and test which one satisfies the predicate. Of course, this method would be too slow for most problems.

There are two methods to speed up the search — backtracking and isomorph rejection.

2.1 Backtrack Search.

In a backtrack approach, we work with *k-tuples* with $k \leq m$. We require a *partial predicate* which is defined for all k-tuples and agrees with the predicate $\mathcal{P}$ when $k = m$. Thus, there is little ambiguity in using $\mathcal{P}$ to denote both the partial and complete predicates. A k-tuple $(x_1, \ldots, x_k)$ is a *partial solution* at level k if the partial predicate $\mathcal{P}((x_1, \ldots, x_k))$ is true. The basic idea of the backtrack approach is: given a partial solution at level k, extend it to one at level $k+1$, and if this extension is impossible, then go back to the partial solution at level $k-1$ and attempt to generate a different partial solution at level k. However, we have to ensure that this process does not lead to a loss of solutions. The following *Consistency Criterion* ensures that this backtrack process is correct.

Consistency Criterion
If $\mathcal{P}((x_1, \ldots, x_k))$ is true, then $\mathcal{P}((x_1, \ldots, x_i))$ is also true for all $i \leq k$.

In other words, once a partial predicate $\mathcal{P}((x_1, \ldots, x_i))$ is false, then it is not necessary to extend $(x_1, \ldots, x_i)$, because it will never lead to a complete solution.

2.2 Isomorph Rejection. In order to discuss isomorph rejection, there must be some notion of permutations acting on the set of combinatorial objects. A permutation $p \in S_{C_1 \times C_2 \times \ldots \times C_m}$, is *property preserving* if $\mathcal{P}((x_1, \ldots, x_m)^p) = \mathcal{P}((x_1, \ldots, x_m))$ for all m-tuples $(x_1, \ldots, x_m)$. The degree of this permutation is, however, very large. Computing with these permutations is practically impossible.

Therefore, we typically work with permutations defined over a much smaller Ω. In the example with graphs, such an Ω may be the set of vertices. We say that a $p \in S_\Omega$ *induces* an action on $C_1 \times C_2 \times \ldots \times C_m$ if $(x_1, \ldots, x_m)^p$ is another well defined m-tuple. A permutation p is a *symmetry* if its induced action on $C_1 \times C_2 \times \ldots \times C_m$ is property preserving.

One can show that the set of symmetries forms a group. It is called the *symmetry group* and is denoted by $\mathcal{S}(\mathcal{P})$. In other words,

$$\mathcal{S}(\mathcal{P}) = \{p \in S_\Omega | \mathcal{P}(A^p) = \mathcal{P}(A) \ \forall \text{ objects } A\}.$$

Two combinatorial objects A and B are said to be *isomorphic* if there exists a permutation $p \in \mathcal{S}(\mathcal{P})$ such that $A^p = B$. An *automorphism* of an object A is a permutation $p \in \mathcal{S}(\mathcal{P})$ such that $A^p = A$. The set of automorphisms of an object A forms a subgroup of $\mathcal{S}(\mathcal{P})$ and is called its *automorphism group*, $Aut(A)$.

In Section 2.1, we defined the general search problem and the notion that a complete solution can be generated by recursively extending a partial solution. Isomorph rejection is a process which applies isomorphism testing to a partial solution and returns two possible results — either it is *admitted*, or it is *rejected*. If it is admitted, then it is also said to be *admissible*. Only admissible partial solutions are furthered extended. The test has to ensure that if a partial solution $(x_1, \ldots, x_k)$ is rejected, it does not lead to a loss of non-isomorphic solutions. In order to establish the correctness of an isomorph rejection process, we have to show that it satisfies the following two criteria.

Consistency Criteria of Isomorph Rejection

(a) Suppose $(x_1, \ldots, x_m)$ is a complete solution. In the orbit generated by $(x_1, \ldots, x_m)$ under the action of $\mathcal{S}(\mathcal{P})$, there exists exactly one admissible solution.

(b) If a complete solution $(x_1, \ldots, x_m)$ is admissible, then for $j \leq m$, the partial solutions $(x_1, \ldots, x_j)$ are all admissible.

The first condition ensures that at most one representative is generated from each orbit of complete solutions. The second condition guarantees that this representative is actually generated. Together, they imply that all non-isomorphic solutions are obtained.

If we are working on an existence type of problem, we may not care whether the solutions are isomorphic. In this case, we can relax the first condition and require that, in each orbit, at least one solution is admissible. This relaxation may lead to a faster isomorph rejection test.

3. Relationship with Permutation Group Algorithms

Isomorphism testing is a basic step in any search incorporating isomorph rejection. It is here that good permutation group algorithms are required.

It is often noted that the complexity of isomorphism testing is equivalent to the complexity of finding an automorphism group[**8**]. Suppose we wish to test whether objects A and B are isomorphic, we can form a new object $\{A, B\}$ and find its automorphism group. If there is an automorphism switching A and B, then the two objects are isomorphic.

How do we find the automorphism group? It would take too long to test every $p \in \mathcal{S}(\mathcal{P})$. The idea of *automorphism pruning* is based on the fact that if g and h are automorphisms of an object A, then $\langle g, h\rangle$ is a subgroup of $Aut(A)$. Thus, we can find $Aut(A)$ incrementally. Initially, we set $Aut(A)$ to be the trivial group. We have to test the coset representatives of $\mathcal{S}(\mathcal{P})/Aut(A)$. If a coset representative u is found to be an automorphism, then we replace $Aut(A)$ by $\langle Aut(A), u\rangle$ and repeat the testing of coset representatives.

The implementation of automorphism pruning gives rise to our first algorithmic question in permutation groups — how to calculate $\langle Aut(A), u\rangle$? Here, the representation of a permutation group by a base and a set of strong generators is helpful[**10**]. However, even a fast heuristic $O(n^4)$ algorithm[**4**] to update the automorphism group is too expensive for most applications. Most practical isomorphism testing programs arrange to test the coset representatives in an order which fixes fewer and fewer base points. Thus, the generators for $Aut(A)$ are discovered in a bottom-up manner and it is not necessary to sieve the Schreier generators.

However, there are situations where it is unnatural to test the coset representatives in this bottom-up manner. It would be nice to have a faster algorithm to find the strong generators so that the user would be free to choose any order of testing the coset representatives which fits the problem.

Our next algorithmic question is on the efficient generation of coset representatives. This step is embedded in the innermost loop of our isomorphism testing routine. In a typical run, we may have to generate and test hundreds of thousands of coset representatives. Because we want to test the coset representatives in a bottom-up order, they are usually organized in a tree-like fashion such that all the coset representatives which fix 1 will be in the first subtree, all those which map 1 to 2 will be in the second subtree and so on. Suppose we are investigating the second subtree. We have to find the subgroup of $Aut(A)$ stabilizing the point 2. This brings us to the third algorithmic question — base change. The use of base change was first reported in [**3**]. The recent development of fast base change algorithms [**2**] has meant significant improvement in our isomorphism testing routines.

Since both base change and the generation of coset representatives are in the innermost loop of an isomorphism testing routine, efficiency is extremely important. Here, even an improvement by a constant factor is worthwhile.

Most practical isomorphism testing programs use invariant analysis to reduce the number of coset representatives that have to be tested[**9**]. Given an object A, an *invariant* is a function, v, mapping each point $\beta \in \Omega$ into a set of values satisfying the condition that if $\exists p \in Aut(A)$, then $v(\beta) = v(\beta^p)$. An example of such an invariant is the degree of a vertex of a graph. If two points have different invariants, then there cannot exist an automorphism mapping one point to the other. Hence, the point set Ω is partitioned into equivalence classes,

$$\Omega = \bigcup_{i=1}^{r} B_i,$$

according to their invariants. The automorphism group of A must be a subgroup of $S_{B_1} \times S_{B_2} \ldots \times S_{B_r}$. This implies that we can replace $\mathcal{S}(\mathcal{P})$ by $\mathcal{S}(\mathcal{P}) \cap S_{B_1} \times S_{B_2} \ldots \times S_{B_r}$ when we are trying to find $Aut(A)$. Since we are using a smaller symmetry group, the number of coset representatives to be tested is correspondingly smaller.

This method of invariant analysis raises the fourth algorithmic question — group intersection. If the initial symmetry group is the complete symmetric group on Ω, then the group intersection problem is easy, because it reduces to merely a problem of refining a partition. In fact, most isomorphism testing programs implicitly assume that the initial symmetry group is the symmetric group. However, when the symmetry group is a general permutation group, this group intersection problem is practically as hard as the isomorphism testing problem. The current practice is to combine this group intersection calculation with the generation of coset representatives. Are there better approaches?

The problem of group intersection also arises in another context. Isomorph rejection requires that isomorphism testing be done on partial objects $(x_1, \ldots, x_k)$ with $k < m$. Thus, the m-tuple has its first k entries defined but the last $m-k$ entries undefined. However, a symmetry operation of the complete object may scramble the defined and undefined parts of the partial object. One way to avoid this mad scrambling is to find the subgroup of $\mathcal{S}(\mathcal{P})$ that preserves the boundary between the defined and undefined parts. The computation of this boundary preserving subgroup often reduces to another group intersection routine[**6**].

4. Conclusion

In summary, we have highlighted a few of the most often used permutation group algorithms in combinatorial searches. They are:

(1) finding a strong generating set,
(2) finding all the coset representatives,
(3) base change, and
(4) group intersection.

Any improvement in these algorithms, such as the ones in the article by Cooperman and Finkelstein elsewhere in this proceeding, would have a significant impact on which search problems are practically computable.

REFERENCES

1. K. Appel and W. Haken, *Every planar map is four-colorable*, Bull. Amer. Math. Soc. **82** (1976), 711–712.
2. C. A. Brown, L. Finkelstein and P. W. Purdom, Jr., *A New Base Change Algorithm For Permutation Groups*, SIAM J. of Comp. **18** (1989), 1034–1047.
3. G. Butler and C. W. H. Lam, *A general Backtrack Algorithm for the Isomorphism Problem of Combinatorial Objects*, J. Symbolic Comp. **1** (1985), 363–381.
4. G. Cooperman, L. Finkelstein and P. W. Purdom. Jr., *Fast Group Membership using a Strong Generating Test for Permutation Groups*, Computers and Mathematics (E. Kaltofen and S. M. Watt, eds.), Springer-Verlag, New York, 1989, pp. 27–36.
5. C. M. Hoffman, *Group-Theoretic Algorithms and Graph Isomorphism*, Lecture Notes in Computer Science, vol. 136, Springer-Verlag, 1982.
6. C. W. H. Lam and L. H. Thiel,, *Backtrack Search with Isomorph Rejection and Consistency Check*, J. Symbolic Comp. **7** (1989), 473–485.
7. C. W. H. Lam, L. H. Thiel and S. Swiercz, *The Non-existence of Finite Projective Planes of Order 10*, Canad. J. of Math. **XLI** (1989), 1117–1123.
8. R. A. Mathon, *A Note on the Graph Isomorphism Counting Problem*, Inf. Proc. Letters **8** (1979), 131–132.
9. B. D. McKay, *nauty User's Guide (Version 1.5)*, Aust. Nat. Univ. Dept. Comp. Sci. Tech. Report TR-CS-90-02, 1990.
10. C. C. Sims, *Computational methods in the study of permutation groups*, Computational Problems in Abstract Algebra (J. Leech, eds.), Pergamon, Elmsford, N.Y., 1970, pp. 169–183.

DEPARTMENT OF COMPUTER SCIENCE, CONCORDIA UNIVERSITY, MONTREAL, QUEBEC, CANADA H3G 1M8

E-mail: lam@vax2.concordia.ca

DIMACS Series in Discrete Mathematics
and Theoretical Computer Science
Volume **11**, 1993

Permutation Groups and Polynomial-Time Computation

EUGENE M. LUKS

ABSTRACT. We discuss aspects of computation in permutation groups assuming polynomial time as a measure of efficiency. Of particular interest are problems, such as finding the intersection of two groups, that resemble or generalize the problem of testing graph isomorphism. We also summarize the instances where the problems are known to be solvable in polynomial time and indicate methods that accomplish this. As with graph isomorphism, the computational complexities of the general problems are open, though we can demonstrate polynomial-time reductions and equivalences among them. A typical approach to such issues is shown to involve an NP-complete problem. Several open questions are listed.

1. Introduction

Our focus is on polynomial-time computability. Naturally, in employing so broad a brush, we do not pretend to delineate the present practical frontiers of computational group theory. Polynomial time is, on the other hand, a widely-recognized standard of tractability as well as a robust model in which to measure and compare efficiency. But, we leave even that point to be defended, or disputed, elsewhere. From our perspective, polynomial time provides, independently, a productive and elegant domain in which to study the structure of group-theoretic computation, while the group-theoretic setting provides insight into the structure of unresolved issues in computational complexity. Furthermore, this interface with theoretical computer science motivates attractive problems for the group theorist, a haunting example stemming from the failure of all efforts to develop a *provably* efficient method for testing graph isomorphism. Thus, where the state-of-knowledge about polynomial-time efficiency does not conform to current

1991 *Mathematics Subject Classification.* Primary 20B40,68Q25; Secondary 68Q15.
Research supported in part by NSF grant CCR-9013410.
This is the final version of this paper.

perceptions of "practical" efficiency, there lie the most tantalizing of the open questions.

We discuss permutation groups $G \leq \mathrm{Sym}(\Omega)$ that are input via generators. It is reasonable to insist that the generating set is "small"; e.g., of cardinality $< |\Omega|^2$. With this understanding, polynomial time "in the input" translates to an $O(|\Omega|^c)$ bound, for some constant c, on the number of steps required in the worst-case. In this paper, we are not concerned with optimizing the exponent c. Such "low-level" complexity matters are, of course, of great interest and are closer to, even when not identical with, implementational concerns. Nevertheless, these are, in the present context, extraneous issues and we gain more insight into polynomial-time matters by ignoring them. Thus, we *avoid* specification and justification of precise constants in the exponents, favoring clarity of the polynomial-time status over optimization of time or space requirements.

We sample not only what *is* in polynomial time but also a range of problems that, to date, have not met this standard. Our concentration is on issues that are motivated by graph-isomorphism testing. Such issues include important and standard group-theoretic problems, including the computation of intersections of permutation groups, centralizers of elements and stabilizers of subsets. As with graph isomorphism, these problems are not considered hard in practice. Nevertheless, no algorithm has been shown to require less than exponential time in the worst case. On the other hand, there seems some evidence that "decision" versions of the problems are not NP-complete. If that is the case, then another level of difficulty (assuming P≠NP) is represented by the related problem of finding lexicographically least elements in double cosets, for we show (the decision version of) this one is NP-complete.

On the positive side, we offer various proofs of polynomial time. While there is a large polynomial-time library (see [**18**]), our emphasis again is on instances of the problems that are related to graph isomorphism and its group-theoretic analogues. For most of the problems, there are efficient procedures for solvable groups.

Our discussion also brings out several open questions.

In Section 3, we review basic polynomial-time tools for dealing with permutation groups. Consistent with computational experience, these efficient techniques are rooted in methods of C.C. Sims. However, in Section 4 we move on to some problems for which efficiency has not been theoretically substantiated. Although their complexity is unknown, they can be shown to be polynomial-time equivalent. The NP-hardness of the aforementioned lex-least problem is proved in Section 5. To better understand the difficulty, we offer two proofs. One of these involves abelian groups with small orbits. In that case, we can explain away the difficulty in terms of the choice of linear orderings on the permutation domain. To be precise, we go on, in Section 6, to show that, with an ordering based upon the orbit/imprimitivity-block structure of the group, the lex-least problem is in polynomial time for groups with restricted noncyclic composition factors, thus automatically including all solvable groups. This result, in turn, recovers instances where the graph-isomorphism-inspired problems are in polynomial time. Other polynomial-time instances are discussed in Section 7, where it is seen that the search for subgroups is apparently made easier if the targeted subgroup is normal. For instance, in Sections 7 and 8, we show that the cores of intersections, centralizers and set stabilizers can be found in polynomial time. In Section 8, we give samples of results of Kantor and Luks that indicate how the procedures, as well as the open problems, are extended to quotients of permutation groups, the theme being that, as far as polynomial time is concerned, the problems are no harder when dealing with quotients. There is, however, remarkable additional overhead in generalizing to quotient groups, for some problems that previously had elementary solutions now seem to require much deeper theory. In particular, the solutions make use of Sylow subgroups which are accessible in polynomial time only through results of Kantor that use consequences of the classification of finite simple groups. By contrast, we present, in Section 9, an approach to finding p-cores (maximal normal p-subgroups). While implemented solutions to this problem typically use Sylow subgroups, we describe an elementary, self-contained method that bypasses these. Some other related problems are listed in Section 10, none of which are known to be in polynomial time. In fact, to date, they seem to represent various levels of difficulty, thereby opening up questions even about the existence of polynomial-time reductions between the problems.

2. Notation and Preliminaries

Let G be a group. We write $H \leq G$, respectively $H \trianglelefteq G$, to indicate H is a subgroup of G, respectively a normal subgroup; $H < G$ and $H \triangleleft G$, respectively, indicate strict inclusion. We say H is *subnormal* in G, denoted $H \triangleleft\triangleleft G$, if there exist groups $H_1, ..., H_m$ such that $H \trianglelefteq H_1 \trianglelefteq \cdots \trianglelefteq H_m \trianglelefteq G$. If $H \leq G$, a *right (left) transversal* for H in G is a complete set of right (left) coset representatives for H in G. A *right (left) subcoset* of G is a coset Hx (xH) of a subgroup $H \leq G$. For $A \subseteq G$, $\langle A \rangle$ denotes the subgroup generated by A. For $g, h \in G$, let $h^g = g^{-1}hg$, the *conjugate* of h by g, and let $[g, h] = g^{-1}h^{-1}gh$, the *commutator* of g and h. For $A \subseteq G$, $A^g = \{a^g \mid a \in A\}$; if $H \leq G$, the *centralizer* of A in H is $\mathrm{C}_H(A) = \{h \in H \mid a^h = a, \forall a \in A\}$; for $a \in G$, $\mathrm{C}_H(a) = \mathrm{C}_H(\{a\})$. For subgroups $H, K \leq G$, the *normalizer* of H in K is $\mathrm{N}_K(H) = \{k \in K \mid H^k = H\}$; we let $H^K = \langle \bigcup_{k \in K} H^k \rangle$, this is the *normal closure* of H in $\langle H, K \rangle$, namely the smallest normal subgroup of $\langle H, K \rangle$ that contains H. For $H \leq G$, the *core* of H in G is $\mathrm{Core}_G(H) = \bigcap_{g \in G} H^g$, it is the largest subgroup of H that is normalized by G.

We denote by $\mathrm{Sym}(\Omega)$ the symmetric group on the finite set Ω. Suppose G acts on Ω, that is, there is a homomorphism $\phi : G \to \mathrm{Sym}(\Omega)$; if $G \leq \mathrm{Sym}(\Omega)$, ϕ is understood to be the natural injection. For $\omega \in \Omega$, $g \in G$, ω^g denotes the image of ω under $\phi(g)$; for $\Delta \subseteq \Omega$, $\Delta^g = \{\omega^g \mid \omega \in \Delta\}$; for $\omega \in \Omega$, the *orbit* of ω is $\{\omega^g \mid g \in G\}$ and is denoted ω^G. For $\omega \in \Omega$, G_ω denotes the subgroup fixing ω, namely $\{g \in G \mid \omega^g = \omega\}$; for $\Delta \subseteq \Omega$, G_Δ denotes the *pointwise stabilizer* of Δ, namely $\bigcap_{\omega \in \Delta} G_\omega$; if an ordering $\omega_1, \ldots, \omega_n$ of Ω is understood, and $\Delta_i = \{\omega_1, \ldots, \omega_{i-1}\}$ then $G^{(i)} = G_{\Delta_i}$ (in particular, $G^{(1)} = G$ and $G^{(n)} = 1$). For $\Delta \subseteq \Omega$, the *(set) stabilizer* of Δ in G, denoted $\mathrm{Stab}_G(\Delta)$, is $\{g \in G \mid \Delta^g = \Delta\}$; G *stabilizes* Δ if $\mathrm{Stab}_G(\Delta) = G$. For $a \in \mathrm{Sym}(\Omega), \Delta \subseteq \Omega$, let a^Δ denote the induced function $\Delta \to \Delta^a$, and for $A \subseteq \mathrm{Sym}(\Omega)$, $A^\Delta = \{a^\Delta \mid a \in A\}$. In particular, if $G \leq \mathrm{Sym}(\Omega)$ and G stabilizes Δ then $G^\Delta \leq \mathrm{Sym}(\Delta)$. A subset $\Delta \subseteq \Omega$ is called a *block* for G if, for all $g \in G$, either $\Delta^g = \Delta$ or $\Delta^g \cap \Delta = \emptyset$. We say that G acts *transitively* on Ω if Ω consists of a single orbit; G acts *primitively* if it acts transitively and there is no block Δ for G with $1 < |\Delta| < |\Omega|$; G acts *regularly* if it acts transitively and $G_\omega = 1$ for any (all) $\omega \in \Omega$.

We refer to Chapter 1 of [**30**] for elementary results in permutation groups that are not specifically recalled herein.

Unless otherwise indicated, we suppose $n = |\Omega|$. It is useful to recall that an increasing chain of subgroups of $\mathrm{Sym}(\Omega)$ has polynomially-bounded length; for example, Lagrange's Theorem implies that the length cannot exceed $\log_2 n! = O(n \log n)$ (in fact, a linear bound can be proved [**1**]). Though not explicitly stated, this is often essential to the verification of polynomial running times. For algorithmic purposes, unless indicated otherwise, it is assumed that subgroups of $\mathrm{Sym}(\Omega)$ are specified (input or output) by generating sets.

A problem is said to be in *polynomial time* if it is solvable in $O(m^c)$ steps where m is the size of a reasonable encoding of the input. In saying that a problem $\mathcal{A}$ is *polynomial-time reducible* to a problem $\mathcal{B}$, we mean that if $\mathcal{B}$ is in polynomial time then $\mathcal{A}$ is in polynomial time. However, in the case of reductions between two *decision* problems, i.e., those with a "yes"/"no" answer, we always intend *Karp reductions*, that is, there is a polynomial-time-computable mapping of instances of $\mathcal{A}$ to instances of $\mathcal{B}$, so that "yes", "no" instances map, respectively, to "yes", "no" instances. Two problems are *polynomial-time equivalent* (in either sense) if there are reductions in both directions. See, for example, [**12**], for elaboration of these issues, including the need for Karp reductions, as well as formal definitions of the classes P, NP, and NP-Complete. Note that these particular terms apply only to decision problems.

3. Basic Polynomial-Time Tools

In this section, we recall some elementary problems that are solvable in polynomial time and that we need to reference later. The techniques will be quite familiar to most readers. Still, it is worth reviewing a few of these not only to emphasize polynomial-time thinking, but also to distinguish these problems from those in Sections 4 and 5, for which presently implemented methods do not meet our measure of efficiency.

We assume that $G = \langle X \rangle \leq \mathrm{Sym}(\Omega)$.

Some efficient procedures follow a divide-and-conquer approach that exploits the orbit and imprimitivity structure of the group. We observe that standard computations of orbits and imprimitivity blocks run in polynomial time.

(3.1) *Given $\omega \in \Omega$, find ω^G, the orbit of ω under G, and for each $\psi \in \omega^G$, find $g \in G$ such that $\omega^g = \psi$.*

A naive transitive closure algorithm involves, at worst, applying each generator to each element of Ω, for a worst-case time of $O(|X|n)$. □

(3.2) *If G acts transitively on Ω, test whether G acts primitively and, if not, find a non-trivial block system.*

For example, the (unique) smallest block containing any given $\alpha, \beta \in \Omega$ is the component of α in the undirected graph $(\Omega, \{\alpha, \beta\}^G)$. □

We often concatenate polynomial-time procedures, up to a polynomial number in fact, to obtain, thereby, a polynomial-time procedure. To do so freely, however, we must be sure that the size of the output of each procedure is bounded by some fixed polynomial in the size of the permutation domain. (Technically, the output of a quadratic procedure could use space that is quadratic in its input size; the concatenation of an unbounded number of such procedures is prohibitive.) In particular, since many procedures involve finding generators for some targeted subgroup, we need to exercise some control on the sizes of generating sets. For example, there is an polynomial-time procedure for

(3.3) *Find a set of $< n^2$ generators for G.*

The underlying logic for this is at the heart of Sims's methods [**29**]. We review the idea, from which polynomial time is then a straightforward observation.

for $i = 1$ **to** $n - 1$ **do**
  while $\exists$ distinct $x, y \in X \cap G^{(i)} \setminus G^{(i+1)}$ such that $\omega_i^x = \omega_i^y$ **do**
    replace such a pair x, y by x, yx^{-1}.

Discarding duplicates, the modified X does not contain distinct elements of any $G^{(i)}$ that lie in the same right coset of $G^{(i+1)}$. Hence, the final X has size at most $\sum_{i=1}^{n-1} |G^{(i)} : G^{(i+1)}| \leq \sum_{i=1}^{n-1}(n - i + 1) < n^2$. □

Remark. Henceforth, we assume that all polynomial-time algorithms that output generators for a subgroup return fewer than n^2 generators. We also assume that $|X| < n^2$.

It is often the case that a subgroup H of our given G is specified only by some testable condition, i.e., there is a polynomial-time procedure which, for any $g \in G$, determines whether $g \in H$. (The subgroups $G^{(i)}$ serve as examples.) In such case, we say that H is *(polynomial-time) recognizable.*

(3.4) *Find generators for a polynomial-time-recognizable subgroup H for which $|G\colon H| = O(n^c)$ and determine $|G\colon H|$.*

A right transversal T for H in G and a set Y of Schreier generators [**14**] of H are both constructed in:

$T := \{1\};\ Y := \emptyset;\ Q := \{1\}$
while $Q \neq \emptyset$ **do**
  remove q from Q;
  for all $x \in X$ **do**
    if $\exists t \in T : qxt^{-1} \in H$ **then** add such qxt^{-1} to Y
    else add qx to T and to Q.

Thus, T has been constructed by a transitive closure method, having right multiplied every element of T by all generators of G to see if this produces any new cosets. In the end, $TX \subseteq \langle Y\rangle T$, so that $G = \langle Y\rangle T$, whence $\langle Y\rangle = H$.

The running time is $O(|G\colon H|^2 n^{c'+c''})$, where $O(n^{c'})$ is the time for a membership-test in H and c'' is an absolute constant, so that an upper bound is $O(n^{2c+c'+c''})$. □

Remarks. It is often the case that H-recognizability also involves some natural interpretation of the cosets, obviating the search through all $t \in T$ to find whether some $qxt^{-1} \in H$; this could eliminate as much as $c+c'$ from the exponent in the timing. The prototypical example is $H = G^{(2)}$, wherein T is keyed by the orbit of ω_1.

It is assumed that the bound $|G\colon H| = O(n^c)$, for some constant c, is known, though the method can also be interpreted as testing this bound in polynomial time; if the procedure takes longer than the predicted number of steps, the bound does not hold.

Applying (3.4) iteratively yields a polynomial-time solution to

(3.5) *Find generators for a subgroup H given that $H = H_m \le H_{m-1} \le \cdots \le H_0 = G$ where the H_i are each polynomial-time-recognizable and $|H_i\colon H_{i+1}| = O(n^c)$ for $0 \le i < m$.* □

In particular, as each $G^{(i)}$ is polynomial-time recognizable, we can solve in polynomial time

(3.6) *Given any $\Delta \subset \Omega$, find generators for G_Δ (the pointwise stabilizer of Δ).* □

Since $|G| = \prod_{i=1}^{n-1} |G^{(i)}\colon G^{(i+1)}|$, we can, in polynomial time,

(3.7) *Find $|G|$.* □

Noting that $x \in G$ iff $|G| = |\langle G, x\rangle|$, we have a polynomial-time algorithm for membership-testing:

(3.8) *Given $x \in \mathrm{Sym}(\Omega)$, test whether $x \in G$.* □

Remarks. We should emphasize that we are not recommending this indirect approach to membership-testing in practice. We are only reminding the reader of a particular logical interconnection of these problems, through which polynomial time is made clear.

The observation that Sims's methods run in polynomial time was made by Furst, Hopcroft and Luks [**11**].

More generally we refer later to a polynomial-time procedure for

(3.9) *Given $\psi_1, \ldots, \psi_m \in \Omega$, with $m \leq n = |\Omega|$, test whether $\exists x \in G$ such that $\omega_i^x = \psi_i$, for $1 \leq i \leq m$, and, if so, find (the subcoset of) all such x.*

By (3.1), we can find $y \in G$ such that $\omega_1^y = \psi_1$, if any such y exists. Recursively find the subcoset Hz of G_{ω_1} mapping $\omega_i \mapsto \psi_i^{y^{-1}}$, for $2 \leq i \leq m$; return Hzy. (Note that the single recursive call involves a permutation group on at most $n-1$ letters.) □

As an immediate consequence of (3.8), we have a polynomial-time algorithm for

(3.10) *Given $H = \langle Y \rangle \leq \mathrm{Sym}(\Omega)$, test whether $H \leq G$.*

I.e., test $Y \subseteq G$. □

Several applications require normal closures.

(3.11) *Given $H = \langle Y \rangle \leq \mathrm{Sym}(\Omega)$, find H^G.*

To get generators, $\bar{Y}$ of H^G: initialize $\bar{Y} = Y$; while there exist $x \in X, y \in \bar{Y}$ such that $y^x \notin \langle \bar{Y} \rangle$, add such y^x to $\bar{Y}$. □

Remark. We remind the reader that we use (from now on implicitly) the polynomial constraint on the length of any increasing chain of subgroups of $\mathrm{Sym}(\Omega)$.

As the derived group G' is the normal closure in G of $\langle [x,y] \mid x, y \in X \rangle$, we have a polynomial-time procedure for

(3.12) *Find the derived series $G \geq G' \geq (G')' \geq \cdots$. Hence, test whether G is solvable.* □

To compute the lower central series of G, let $L_1(G) = G = \langle X \rangle$; then if $L_i = \langle X_i \rangle$, $L_{i+1}(G) = \langle [x,y] \mid x, \in X, y \in X_i \rangle^G$. Thus we have a polynomial-time procedure for

(3.13) *Find the lower central series of G. Hence, test whether G is nilpotent.* □

An alternative polynomial-time nilpotence test is to check that G is a direct product of p-groups: For each $x \in X$ and each prime p dividing $|G|$, let $\langle x_p \rangle$ be the Sylow p-subgroup of $\langle x \rangle$; for example, $x_p = x^m$ where $|G| = mp^k$ with $(m,p) = 1$ (such large powers of x are computable by "repeated squaring" tricks, though an even faster approach could be to compute the power in each cycle of x, first reducing m modulo the cycle length). Letting $G_p = \langle x_p \mid x \in X \rangle$, verify that $|G_p|$ is a power of p for each p and that the generators of G_p commute with the generators of G_q for $p \neq q$. □

4. Not Known to be in Polynomial Time

There is general agreement that, by all measures, the problems in the preceding section have efficient solutions. We turn now to some which seem to have satisfactory implementations but for which all known algorithms have exponential worst-case complexity. Were there no other reason for looking at them, these would be of theoretical interest because of their relation to the graph-isomorphism problem:

PROBLEM. GRAPH ISOMORPHISM (GRAPH-ISO)
INPUT: *Graphs* $\mathcal{G}_1 = (V_1, E_1)$, $\mathcal{G}_2 = (V_2, E_2)$.
QUESTION: *Are* $\mathcal{G}_1, \mathcal{G}_2$ *isomorphic?*

It is generally felt that GRAPH-ISO is not a hard problem in practice (see, e.g., [**23**] for an implemented procedure that many have found satisfactory). Nevertheless, although the problem has been extensively studied, nothing close to polynomial time has been proved. Indeed, there is no known approach that has proved to be *subexponential* (say, for example, in $O(n^{\log^c n})$ time) in the worst case. (See remarks at the end of Section 6). On the other hand, there is evidence that GRAPH-ISO is not NP-complete, otherwise there would be a collapse of the "polynomial-time hierarchy" [**13**]. Indeed, from the earliest expositions of NP-completeness (e.g., see discussion in [**12**]), there has been speculation that GRAPH-ISO may be one of the few classical decision problems that is neither in polynomial time nor NP-complete.

We recall polynomial-time reductions of GRAPH-ISO to permutation-group problems. To introduce the groups, we consider

PROBLEM. GRAPH AUTOMORPHISM-GROUP (GRAPH-AUTO)
INPUT: *Graph* $\mathcal{G} = (V, E)$.
FIND: *Generators for* $\mathrm{Aut}(\mathcal{G})$*, the automorphism group of* $\mathcal{G}$.

The following is well known.

PROPOSITION 4.1. *GRAPH-ISO and GRAPH-AUTO are polynomial-time equivalent problems.*

PROOF. To reduce GRAPH-ISO to GRAPH-AUTO, we first note that it suffices to consider the GRAPH-ISO case where the graphs $\mathcal{G}_1, \mathcal{G}_2$ are connected, for, in general, one may test all pairs of connected components. Given connected $\mathcal{G}_1 = (V_1, E_1), \mathcal{G}_2 = (V_2, E_2)$, form the disjoint union $\mathcal{G} = (V_1 \dot{\cup} V_2, E_1 \dot{\cup} E_2)$ and suppose $\mathrm{Aut}(\mathcal{G}) = \langle X \rangle$. Then $\mathcal{G}_1 \cong \mathcal{G}_2$ iff $\exists x \in X : V_1^x = V_2$.

We turn to the reverse reduction. For this, we first observe that GRAPH-ISO would enable us to solve

PROBLEM. RESTRICTED GRAPH AUTOMORPHISM (RES-GRAPH-AUTO)
INPUT: *Graph* $\mathcal{G} = (V, E)$ *and, for some* $m \leq |V|$;
sequences $v_1, v_2, \ldots, v_m$ *and* $w_1, w_2, \ldots, w_m$ *of distinct vertices in* V.
QUESTION: *Is there some* $g \in \mathrm{Aut}(\mathcal{G})$ *such that* $v_i^g = w_i$, *for* $1 \leq i \leq m$?

Reducing RES-GRAPH-AUTO to GRAPH-ISO: Attaching distinguishable "gadgets" to the v_i forming a graph $\mathcal{G}_1$ and similar gadgets to the respective w_i forming a graph $\mathcal{G}_2$, RES-GRAPH-ISO reduces to testing isomorphism of the modified $\mathcal{G}_1$, $\mathcal{G}_2$. A suitable gadget at v_i, respectively w_i, could be a new cycle of length $|V| + i$ through the vertex.

Reducing GRAPH-AUTO to RES-GRAPH-AUTO: Repeated application of a procedure for the decision problem RES-GRAPH-AUTO facilitates the actual construction of a suitable g; for, having received a "yes", we go on to find a possible v_{m+1}^g (using a RES-GRAPH-AUTO procedure to test all candidates) then v_{m+2}^g, etc. Note this will have used $O(|V|^2)$ calls to RES-GRAPH-AUTO.

In this fashion, $O(|V|^3)$ applications of RES-GRAPH-AUTO determine the orbit of v_1 under $\mathrm{Aut}(\mathcal{G})$ and a right transversal for $\mathrm{Aut}(\mathcal{G})^{(2)}$ (the stabilizer of v_1) in $\mathrm{Aut}(\mathcal{G})$. Similarly, we get transversals for each $\mathrm{Aut}(\mathcal{G})^{(i+1)}$ in $\mathrm{Aut}(\mathcal{G})^{(i)}$. The union of such transversals generate $\mathrm{Aut}(\mathcal{G})$. Thus, GRAPH-AUTO has been recovered from $O(|V|^4)$ applications of GRAPH-ISO (to graphs of polynomial size $O(|V|^2)$). □

Observing then that GRAPH-AUTO is the problem that we would have to solve, we consider the natural action of $G = \mathrm{Sym}(V)$ on the set of unordered pairs in V, and see that $\mathrm{Aut}(\mathcal{G})$ is precisely the subgroup that stabilizes E. With this in mind, we define the problem

PROBLEM. SET-STABILIZER (STAB)
INPUT: $G \leq \mathrm{Sym}(\Omega); \Delta \subseteq \Omega$.
FIND: $\mathrm{Stab}_G(\Delta) = \{g \in G \mid \Delta^g = \Delta\}$.

Thus, the above argument showed

PROPOSITION 4.2. *GRAPH-ISO is polynomial-time reducible to STAB.* □

There are two other important formulations of STAB. Consider

PROBLEM. INTERSECTION (INTER)
INPUT: $G, H \leq \mathrm{Sym}(\Omega)$.
FIND: $G \cap H$.

PROBLEM. CENTRALIZER (CENT)
INPUT: $G \leq \mathrm{Sym}(\Omega)$; $x \in \mathrm{Sym}(\Omega)$.
FIND: *The centralizer,* $\mathrm{C}_G(x)$, *of* x *in* G.

PROPOSITION 4.3. *The problems STAB, INTER and CENT are polynomial-time equivalent.*

PROOF. Suppose we are given an instance (G, Ω, Δ) of STAB. To reduce this to INTER or CENT, let G act in the diagonal on the disjoint union $\widehat{\Omega} = \Omega_1 \dot{\cup} \Omega_2$ of two copies of Ω (i.e., $(\omega_i)^g = \omega_i^g$, $\forall \omega \in \Omega$, $i = 1, 2$, $\forall g \in G$, where ω_i denotes the Ω_i copy of $\omega \in \Omega$). Let x be the involution in $\mathrm{Sym}(\widehat{\Omega})$ specified by: $(\omega_i)^x = \omega_i$ if $\omega \notin \Delta$ and $(\omega_i)^x = \omega_{3-i}$ if $\omega \in \Delta$. Then $\mathrm{Stab}_G(\Delta) = G \cap G^x = \mathrm{C}_G(x)$.

We indicate reductions in the other direction. For INTERS: let $G \times H$ act on $\Omega \times \Omega$ in the natural way, and set $\Delta = \{(\omega, \omega) \mid \omega \in \Omega\}$; then $\mathrm{Stab}_{G \times H}(\Delta) = \{(g, g) | g \in G \cap H\}$. FOR CENT: $g \in G$ commutes with x iff g, acting diagonally on $\Omega \times \Omega$, stabilizes $\{(\omega, \omega^x) \mid \omega \in \Omega\}$. □

Remarks. In a panel discussion at the DIMACS workshop that gave rise to these Proceedings, the sentiment was generally expressed that STAB, INTER and CENT are "not hard in practice." Thus one should ask:

QUESTION 1. *Are STAB (or INTER or CENT) in polynomial time? Are there even subexponential methods?*

Of course, affirmative answers would carry over to GRAPH-ISO. Until such time as this is resolved, implemented methods that rely on *general* procedures for STAB, INTER or CENT cannot be proved efficient. In particular, they must be excluded from the polynomial-time toolkit.

We emphasize "general" in the last paragraph, for it is entirely plausible that the problems can often be solved efficiently. The challenge that we put forth, therefore, is to back this up with theory.

QUESTION 2. *For what classes of inputs do implemented procedures, or modifications thereof, for STAB, INTER or CENT, have polynomial (or subexponential) worst case performance?*

If the question seems vague, we welcome reformulation, even to the exclusion of polynomial time as a targeted criterion. Assuming there is an acknowledged class of "interesting" groups, what can you *guarantee* about the running time over that class? A proffered system should be able to promise efficiency beyond the observation that a procedure took S seconds for group G on machine M.

At first glance, the sentiment that STAB, etc., are "not hard in practice," seems entirely consistent with feelings about GRAPH-ISO. However, in the latter case, one can provide some theoretical justification, since there are well-defined and accepted notions of *random graphs*, with respect to which naive isomorphism-testing procedures are provably fast on average (e.g., [**5**]). Can one do the same for groups?

QUESTION 3. *What is the* average *running time of implemented procedures for STAB, etc.?*

We leave open the choice of probability distributions from which to approach this problem. A uniform distribution over all permutation groups is just one possibility. One could also look at conjugacy classes (in $\mathrm{Sym}(\Omega)$) or even isomorphism classes. One should, as well, look at this problem for restricted classes of groups.

In Sections 6 and 7, we do give examples of procedures for STAB, INTER, CENT that are provably in polynomial time for specified input classes, including, for example, solvable groups.

The similarity of these problems to GRAPH-ISO carries over to analogues of the GRAPH-AUTO/GRAPH-ISO relationship. Namely, there is, in each case, an equivalent decision problem in NP. Essentially, these may be obtained by substituting cosets for (one or both of) the groups in the problem and asking whether the targeted set is nonempty. E.g., corresponding to INTER, we ask whether $Gx \cap H \neq \emptyset$. Then, with minor reformulations, we obtain the following problems corresponding, respectively, to STAB, INTER, CENT.

PROBLEM. SET-TRANPORTER (TRANS)
INPUT: $G \leq \mathrm{Sym}(\Omega)$; $\Delta_1, \Delta_2 \subseteq \Omega$.
QUESTION: *Is there some* $g \in G$ *such that* $\Delta_1^g = \Delta_2$?

PROBLEM. DOUBLE-COSET EQUALITY (DC-EQ)
INPUT: $G, H \leq \mathrm{Sym}(\Omega)$; $x_1, x_2 \in \mathrm{Sym}(\Omega)$.
QUESTION: *Does* $Gx_1H = Gx_2H$?

PROBLEM. CONJUGACY OF ELEMENTS (CONJ-ELT)
INPUT: $G \leq \mathrm{Sym}(\Omega)$; $x_1, x_2 \in \mathrm{Sym}(\Omega)$.
QUESTION: *Is there some* $g \in G$ *such that* $x_1^g = x_2$?

Note, we include "ELEMENTS" in the title specifically to distinguish from the corresponding question of conjugacy of groups (see Section 10.2).

We have the following analogue of Proposition 4.1.

PROPOSITION 4.4. *STAB is equivalent to each of the problems TRANS, DC-EQ and CONJ-ELT.*

PROOF. We outline this for the equivalence STAB $\equiv$ TRANS. (One can also get CENT $\equiv$ CONJ-ELT and INTER $\equiv$ DC-EQ by establishing TRANS $\equiv$ CONJ-ELT $\equiv$ DC-EQ the way one established STAB $\equiv$ CENT $\equiv$ INTER. Note that DC-EQ is trivially restated as testing non-emptiness of an intersection of a group $H^{x_1^{-1}}$ and a coset $Gx_2x_1^{-1}$.)

Reducing TRANS to STAB: Given an instance (G, Δ_1, Δ_2) of TRANS, consider the wreath product $\hat{G} = G \wr Z_2$ acting on the disjoint union $\Omega \dot{\cup} \Omega$ of two copies of Ω and let $\hat{\Delta} = \Delta_1 \dot{\cup} \Delta_2$, in which Δ_i is considered as lying in the ith copy of Ω. Then the answer to TRANS is affirmative iff some generator of $\mathrm{Stab}_{\hat{G}}(\hat{\Delta})$ switches Δ_1 and Δ_2.

For a reverse reduction, we consider the following analogue of RES-AUTO:

PROBLEM. RESTRICTED SET STABILIZER (RES-STAB)
INPUT: $G \leq \mathrm{Sym}(\Omega); \Delta \subseteq \Omega$; *sequences* $\omega_1, \ldots, \omega_m$ *and* $\psi_1, \ldots, \psi_m$ *in* Ω.
QUESTION: *Is there some* $g \in \mathrm{Stab}_G(\Delta)$ *such that* $\omega_i^g = \psi_i$ *for* $1 \leq i \leq m$.

Reducing RES-STAB to TRANS: Find, using (3.9), the subcoset Hy of G consisting of elements mapping $\omega_i \mapsto \psi_i$ for $1 \leq i \leq m$; apply TRANS to $(H, \Delta, \Delta^{y^{-1}})$.

Reducing STAB to RES-STAB: This proceeds exactly as the reduction of GRAPH-AUTO to RES-GRAPH-AUTO. □

Remark. Extending the analogy to GRAPH-ISO, Babai and Moran [**8**] have shown that TRANS (therefore DC-EQ and CONJ-ELT) could be NP-complete only if the polynomial-time hierarchy collapses to $\Sigma_2^p = \Pi_2^p$. Thus, even if GRAPH-ISO were to be resolved via other methods (there is a legion of sufferers from the "Graph-Isomorphism Disease", see [**28**] for traditional attacks) such group-theoretic problems would very possibly remain as outstanding candidates for membership in a complexity class strictly between P and NP-Complete. There is also the possibility of an affirmative answer to the following open question:

QUESTION 4. *Is DC-EQ (equivalently TRANS, CONJ-ELT) polynomial-time reducible to GRAPH-ISO?*

See Section 10 for additional open questions on where group-theoretic decision problems fit in this hierarchy.

5. Not Likely to be in Polynomial Time

A suggested approach to DC-EQ (e.g., [**9, 19**]) has been to determine, in any given double-coset GxH, its lexicographically least element, as $Gx_1H = Gx_2H$ iff the lex-least element in Gx_1H is the lex-least element in Gx_2H. This is analogous to, and a generalization of, attacking GRAPH-ISO by establishing lexicographically least representations (e.g. lex-least adjacency matrices) as canonical forms.

Of course, proponents realize that the approach has limitations. It is nevertheless worthwhile to provide theoretical substantiation of its difficulty, namely, that it involves an NP-hard problem.

We suppose that Ω is linearly ordered with respect to a relation $<$. Then $\mathrm{Sym}(\Omega)$ acquires a lexicographic ordering $\prec$, specifically, if $x, y \in \mathrm{Sym}(\Omega)$ and $x \neq y$, then $x \prec y$ iff $\omega^x < \omega^y$ for the least $\omega \in \Omega$ such that $\omega^x \neq \omega^y$.

We state a polynomial-time equivalent decision problem in order to bring the question into NP.

PROBLEM. LEXICOGRAPHIC LEADER in DOUBLE COSET (LLDC)
INPUT: *A linearly-ordered set* Ω*;* $G, H < \mathrm{Sym}(\Omega)$*;* $x, y \in \mathrm{Sym}(\Omega)$.
QUESTION: *Is there some* $z \in GxH$ *such that* $z \prec y$*?*

LLDC is in NP, for one can guess permutations g, h and verify that $g \in G$, $h \in H$ and $gxh \prec y$. Clearly, if one could find lex-least elements in polynomial time then LLDC would be solvable in polynomial time. Conversely, a polynomial number of calls to an LLDC procedure would suffice in a binary-search procedure for lex-least elements.

We show

THEOREM 5.1. *LLDC is NP-complete.*

In fact, we give two distinct reductions of known NP-complete problems to LLDC, as they involve different, yet seemingly reasonable, classes of groups and seem to display different reasons for the difficulty. The first shows that LLDC is "hard" even if G and H are symmetric groups (not, of course, in their natural actions). The second reduction shows that LLDC is "hard" even if G and H are abelian and even if the orbits of $\langle G, x, H\rangle$ are small (size 3). (However, see the remark following the proofs.)

FIRST PROOF OF THEOREM 5.1. It is known that the following problem is NP-complete (see, e.g., [**12**]).

PROBLEM. MAXIMAL CLIQUE (MAX-CLIQ)
INPUT: *Graph* $\mathcal{G} = (V, E)$*, integer* K.
QUESTION: *Does* $\mathcal{G}$ *contain a clique of cardinality* K*?*

(Recall that $W \subseteq V$ is called a *clique* in $\mathcal{G}$ if $\{w_1, w_2\} \in E$, for all $w_1, w_2 \in W$.)

We reduce MAX-CLIQ to LLDC, thereby establishing LLDC is also NP-complete:

Let $(\mathcal{G} = (V, E), K)$ be an instance of MAX-CLIQ. We may assume that V is linearly ordered so that $V = \{v_1, \ldots, v_m\}$, the subscripts reflecting the ordering. Let Ω be the set of pairs $\{\{v_i, v_j\} \mid 1 \leq i < j \leq m\}$. Then Ω is linearly ordered with $\{v_i, v_j\} < \{v_k, v_l\}$, for $1 \leq i < j \leq m$, $1 \leq k < l \leq n$, if either $j < l$ or $j = l$ and $i < k$. For any q, $1 \leq q \leq \binom{m}{2}$, let Ω_q denote the set consisting of the first q elements of Ω in this ordering.

Note that $E \subseteq \Omega$. We may assume that $|E| \geq \binom{K}{2}$. Let x be any permutation in $\mathrm{Sym}(\Omega)$ that maps $\Omega_{|E|}$ to E. Let H be the natural image of $\mathrm{Sym}(V)$ in $\mathrm{Sym}(\Omega)$ and let $G = \mathrm{Sym}(\Omega)_{\Omega \setminus \Omega_{|E|}}$ (so $G \simeq \mathrm{Sym}(\Omega_{|E|})$).

Let $y \in \mathrm{Sym}(\Omega)$ be the transposition switching the elements in positions $\binom{K}{2}$ and $\binom{K}{2}+1$. Then, for $z \in \mathrm{Sym}(\Omega)$, $z \prec y$ iff z pointwise fixes $\Omega_{\binom{K}{2}}$ We claim there is a clique of size K in $\mathcal{G}$ iff $\exists z \in GxH$ such that $z \prec y$. This follows from the fact that there is a clique in $\mathcal{G}$ of cardinality K iff there exists $h \in H(=\mathrm{Sym}(V))$ such that $(\Omega_{\binom{K}{2}})^h \subseteq E$ (the clique then being $\{v_1^h, \ldots, v_K^h\}$). But, as the permutations in Gx map $\Omega_{\binom{K}{2}}$ precisely to the subsets of cardinality $\binom{K}{2}$ in the set E, there is a clique of size K iff some permutation in H agrees with some permutation in Gx on $\Omega_{\binom{K}{2}}$, which is true iff some $z \in GxH$ pointwise fixes $\Omega_{\binom{K}{2}}$, i.e., iff $z \prec y$. □

SECOND PROOF OF THEOREM 5.1. The following is also NP-complete (see, e.g., [**12**]).

PROBLEM. EXACT 3-COVER (X3C)

INPUT: *A set Σ together with a collection Θ of size-3 subsets of Θ.*

QUESTION: *Is there a subcollection $\Theta' \subseteq \Theta$ with $|\Theta'| = |\Sigma|/3$ such that $\Sigma = \bigcup_{\theta\in\Theta'} \theta$.*

Reduction of X3C to LLDC:

Given an instance (Σ, Θ) of X3C, we construct an instance (Ω, G, H, x, y) of LLDC as follows.

We may assume the triples in Θ are distinct. Let $\Psi = \{\{\theta, \theta'\} \mid \theta, \theta' \in \Theta, \theta \cap \theta' \neq \emptyset\}$, the collection of unordered-pairs of intersecting triples. Let $\Phi = \Sigma \cup \Theta \cup \Psi$. The desired permutation domain is

$$\Omega = \Phi \times \{1, 2, 3\}$$

and we fix any linear ordering of Ω subject only to the condition that $\Phi \times \{1\}$ precedes $\Phi \times \{2\}$ and $\Phi \times \{2\}$ precedes $\Phi \times \{3\}$.

For any $\phi \in \Phi$, we let $a_\phi \in \mathrm{Sym}(\Omega)$ be the 3-cycle $((\phi, 1), (\phi, 2), (\phi, 3))$ (i.e., $(\phi, 1) \to (\phi, 2) \to (\phi, 3) \to (\phi, 1)$), and let $b_\phi \in \mathrm{Sym}(\Omega)$ be the transposition $((\phi, 1), (\phi, 2))$. (Note that $\langle\{a_\phi \mid \phi \in \Phi\}\rangle$ is an elementary abelian 3-group and $\langle\{b_\phi \mid \phi \in \Phi\}\rangle$ is an elementary abelian 2-group.) For $\theta = \{\sigma_1, \sigma_2, \sigma_3\} \in \Theta$, define $c_\theta \in \mathrm{Sym}(\Omega)$ by

$$c_\theta = a_{\sigma_1} a_{\sigma_2} a_{\sigma_3} a_\theta \cdot \prod_{\theta \in \psi \in \Psi} a_\psi \ .$$

Now let

$$\begin{aligned} G &= \langle\{c_\theta \mid \theta \in \Theta\}\rangle; \\ H &= \langle\{b_\phi \mid \phi \in \Theta \cup \Psi\}\rangle; \\ x &= \prod_{\sigma \in \Sigma} a_\sigma^{-1}. \end{aligned}$$

Further, let $y \in \mathrm{Sym}(\Omega)$ be the transposition switching the last point in $\Phi \times \{1\}$ with the first point in $\Phi \times \{2\}$. Then, for $z \in \mathrm{Sym}(\Omega)$, $z \prec y$ iff z fixes $\Phi \times \{1\}$ pointwise.

To establish the reduction we show that LLDC with input (Ω, G, H, x, y) has an affirmative answer iff X3C with input (Σ, Θ) has an affirmative answer.

First suppose that $\Theta' \subseteq \Theta$ is an exact cover of Σ. Let

$$\begin{aligned} g &= \prod_{\theta \in \Theta'} c_\theta \ \in G \\ h &= \prod_{\theta \in \Theta'} (b_\theta \cdot \prod_{\theta \in \psi \in \Psi} b_\psi) \ \in H. \end{aligned}$$

We claim that gxh pointwise fixes $\Phi \times \{1\}$ (whence $gxh \prec y$). To see this: for $\sigma \in \Sigma$, σ is in θ for exactly one $\theta \in \Theta'$ so that $(\sigma,1)^{gxh} = (\sigma,2)^{xh} = (\sigma,1)^h = (\sigma,1)$; for $\theta \in \Theta \setminus \Theta'$, $(\theta,1)$ is fixed by each of g,x,h; for $\theta \in \Theta'$, $(\theta,1)^{gxh} = (\theta,2)^{xh} = (\theta,2)^h = (\theta,1)$; finally, for $\psi \in \Psi$, if $\psi \cap \Theta' = \emptyset$ then $(\psi,1)$ is fixed by each of g,x,h, otherwise $\theta \in \psi$ for precisely one $\theta \in \Theta'$ so that $(\psi,1)^{gxh} = (\psi,2)^{xh} = (\psi,2)^h = (\psi,1)$.

Conversely, suppose that $gxh \prec y$, for $g \in G, h \in H$, so that gxh pointwise fixes Φ_1. We can express

$$\begin{aligned} g &= \prod_{\theta \in \Theta} c_\theta^{e_\theta}, \text{ where } e_\theta = 0, 1, \text{ or } 2; \\ h &= \prod_{\theta \in \Theta'} b_\theta \cdot \prod_{\psi \in \Psi'} b_\psi, \text{ where } \Theta' \subseteq \Theta, \Psi' \subseteq \Psi. \end{aligned}$$

We show that Θ' is an exact cover of Σ: For $\theta \in \Theta$, $(\theta,1)^{gxh} = (\theta,1)$ implies $e_\theta = 0$ if $\theta \notin \Theta'$ and $e_\theta = 1$ if $\theta \in \Theta'$. Hence, $g = \prod_{\theta \in \Theta'} c_\theta$. For $\sigma \in \Sigma$, $(\sigma,1) = (\sigma,1)^{gxh} = (\sigma,1)^{ga_\sigma^{-1}}$, and so $\sigma \in \bigcup_{\theta \in \Theta'} \theta$. Finally, we must show that Θ' does not contain θ, θ' with $\theta \cap \theta' \neq \emptyset$. Suppose, to the contrary, that such $\theta, \theta' \in \Theta'$ and let $\psi = \{\theta, \theta'\} \in \Psi$; then $(\psi,1)^{gxh} = (\psi,1)^{c_\theta c_{\theta'} h} = (\psi,3)^h = (\psi,3)$, contradicting the fact that gxh fixes $(\psi,1)$. □

Remarks. The construction in the second proof should be compared with the result of Theorem 6.2, where it is shown that, with a judicious choice of ordering of Ω (determined by G alone), the problem is actually in polynomial time for interesting classes of groups G. This includes all solvable groups as well as all groups with bounded orbits. Either of these conditions are satisfied by the groups of the above reduction, in fact for $\langle G, x, H\rangle$.

On the other hand, we conjecture that there is no analogous fix for the situation encountered in the first proof. Therein Ω is the set of size-2 subsets of a linearly ordered set V, H the natural image of $\mathrm{Sym}(V)$ in $\mathrm{Sym}(\Omega)$ and G is the subgroup of $\mathrm{Sym}(\Omega)$ fixing all but the first q points for some $q \leq \binom{|V|}{2}$. In this special setting, we ask

QUESTION 5. *Given such Ω, G, H, is there a reordering of Ω with respect to which the lexicographically least elements in GxH, for any $x \in \mathrm{Sym}(\Omega)$, can be found in polynomial time?*

With such an ordering in hand, one could define a polynomial-time computable *canonical form* for graphs $\mathcal{G} = (V, E)$ with $|E| = q$: take any $x \in \mathrm{Sym}(\Omega)$ such that $\Omega_q^x = E$ (Ω_q remains the first q elements in the original ordering); find z, the lex-least element (with respect to the new ordering) in GxH; set $\mathrm{CF}(\mathcal{G}) = (V, \Omega_q^z)$. The graph $\mathrm{CF}(\mathcal{G})$ is independent of the choice of x; even more significantly, it is a complete isomorphism invariant (hence a canonical form), that is, $\mathcal{G} = (V, E)$ is isomorphic to $\mathcal{G}' = (V, E')$ iff $\mathrm{CF}(\mathcal{G}) = \mathrm{CF}(\mathcal{G}')$.

As we do not expect an affirmative answer to Question 5, we suggest looking for evidence to the contrary. Could one show, for example, that with such G and H, LLDC remains NP-complete for *any* prescribed ordering of Ω?

An important restricted case of LLDC occurs when $G = H$ (e.g., [**19**]). Thus, it is worth observing that this case is equally "hard".

COROLLARY 5.2. *LLDC remains NP-complete when $G = H$.*

PROOF. We describe a polynomial-time reduction of the general LLDC problem to this special case.

Suppose (Ω, G, H, x, y) is an instance of LLDC.

Set $\hat{\Omega} = \Omega \times \{1, 2\}$ and linearly order $\hat{\Omega}$ so that $(\omega, i) \prec (\psi, j)$ if $i < j$ or if $i = j$ and $\omega \prec \psi$. Let $\hat{G} = G \times H$ act on $\hat{\Omega}$ via $(\omega, 1)^{(g,h)} = (\omega^g, 1)$, $(\omega, 2)^{(g,h)} = (\omega^h, 2)$ for $\omega \in \Omega, g \in G, h \in H$ (thus G acts naturally on $\Omega \times \{1\}$ and H acts naturally on $\Omega \times \{2\}$). Let $\hat{x}, \hat{y} \in \mathrm{Sym}(\hat{\Omega})$ satisfy $(\omega, 1)^{\hat{x}} = (\omega^x, 2)$, $(\omega, 1)^{\hat{y}} = (\omega^y, 2)$ and $(\omega, 2)^{\hat{x}} = (\omega, 2)^{\hat{y}} = (\omega, 2)$ for $\omega \in \Omega$.

To establish the reduction, we show that the instance (Ω, G, H, x, y) of LLDC has an affirmative answer iff the instance $(\hat{\Omega}, \hat{G}, \hat{G}, \hat{x}, \hat{y})$ has an affirmative answer.

Suppose $\exists g \in G, h \in H$ satisfying $gxh \prec y$. Then $(g, 1)\hat{x}(1, h) \prec \hat{y}$. This follows from that the fact that, for $\omega \in \Omega$, $(\omega, 1)^{(g,1)\hat{x}(1,h)} = (\omega^{gxh}, 2)$.

Conversely, suppose $\exists (g_1, h_1), (g_2, h_2) \in \hat{G}$ satisfying $(g_1, h_1)\hat{x}(g_2, h_2) \prec \hat{y}$. Then $g_1 x h_2 \prec y$. To see this: the first point in $\hat{\Omega}$ at which $(g_1, h_1)\hat{x}(g_2, h_2)$ and $\hat{y}$ differ must lie in $\Omega \times \{1\}$ for no permutation agrees with $\hat{y}$ on $\Omega \times \{1\}$ and strictly precedes it; the result now follows from the observation $(\omega, 1)^{(g_1, h_1)\hat{x}(g_2, h_2)} = (\omega^{g_1 x h_2}, 2)$. □

Remark. A context in which LLDC arises is the actual enumeration of all double cosets GxH of subgroups G, H of some $L \in \mathrm{Sym}(\Omega)$ (see, e.g., [**9**]). (For, assuming LLDC is not significantly harder than DC-EQ for the particular groups, it would be most efficient to compute and store the lex-least elements as canonical representatives of their double cosets.) Since the answer is not necessarily of polynomial size, the natural question to ask in this case is how much work has to be done beyond that which is dictated by the output.

QUESTION 6. *Given* $G, H, L \in \mathrm{Sym}(\Omega)$ *with* $G, H \le L$, *can the double cosets,* GxH, *for* $x \in L$, *be enumerated in time* $O((\nu + n)^c)$, *where* ν *is the number of such double cosets?*

The proof of Theorem 5.1 contains the ingredients of another NP-completeness result. Consider the class of problems

PROBLEM. MEMBERSHIP IN PRODUCT OF m GROUPS (m-MEMB)
INPUT: $A_1, A_2, \ldots, A_m \le \mathrm{Sym}(\Omega)$; $x \in \mathrm{Sym}(\Omega)$.
QUESTION: *Is* $x \in A_1 A_2 \ldots A_m$?

Of course, 1-MEMB is in polynomial time by (3.8). The problem 2-MEMB is simply a restatement of DC-EQ, which is of unknown complexity.

What happens for $m \ge 3$? Clearly, m-MEMB is in NP, for one can guess and verify a factorization $x = a_1 \cdots a_m$ with $a_i \in A_i$. Now, in the reduction in each proof of Theorem 5.1, the element y was taken to be a transposition of two adjacent elements ω_{i-1}, ω_i in the linearly ordered Ω. Then, for $z \in \mathrm{Sym}(\Omega)$, $z \prec y$ iff $z \in \mathrm{Sym}(\Omega)^{(i)}$. Hence, there exists $z \in GxH$ preceding y iff $GxH \cap \mathrm{Sym}(\Omega)^{(i+1)} \ne \emptyset$, which occurs iff $x \in G\mathrm{Sym}(\Omega)^{(i+1)}H$. It follows that

PROPOSITION 5.3. *m-MEMB is NP-complete for* $m \ge 3$. □

Remark. Variations on these reductions can be used to show that even the special case of testing membership in GHG, for $G, H \le \mathrm{Sym}(\Omega)$, is NP-complete. Furthermore, this remains NP-complete even when G and H are both abelian.

6. A Polynomial-Time Instance of LLDC

Following techniques introduced in [**20**] and [**6**], we show that the difficulty of the LLDC instance used in the second proof of Theorem 5.1 is attributable entirely to the particular linear ordering of Ω. That is, for a more general choice of G, one can define (in polynomial time) a linear ordering of Ω so that the lexicographically-least element in any GxH is obtainable in polynomial time. This will, in turn, yield polynomial-time solutions to special cases of INTER, STAB and CENT.

The restriction to be placed, on G alone, is a limit on the sizes of the noncyclic composition factors. Specifically, for any fixed integer d, let Γ_d denote the class of groups all of whose non-cyclic composition factors are isomorphic to subgroups of S_d. So, in particular, Γ_d includes all solvable groups. The following is proved in [**2**].

LEMMA 6.1 (BABAI, CAMERON, PÁLFY). *There is a constant c such that if G is a primitive subgroup of* $\mathrm{Sym}(\Psi)$ *and* $G \in \Gamma_d$ *then* $|G| = O(|\Psi|^{cd})$.

Remark. Since many problems on permutation groups have natural reductions to the primitive case, results that bound the size of primitive groups under various conditions are often essential to the analysis (see, e.g., [**7**] for other examples). Indeed, the investigations leading to Lemma 6.1 were inspired by computational complexity applications. In particular, the lemma enables a simplification as well as a wider applicability of the set-stabilizer algorithm in [**20**].

In this section, the lemma comes into play in the base case of a "divide-and-conquer" algorithm that is guided by the orbit/imprimitivity structure of G. For convenience, we keep track of this in a *structure forest* $\mathcal{F}$ for G. Such a forest includes one *structure tree* for each orbit. The (rooted) tree $\mathcal{T}_\Delta$ on the orbit Δ has leaf set Δ and is such that the action of G on Δ can be lifted to automorphisms of $\mathcal{T}_\Delta$, with the further property that the subgroup of G that fixes any node acts primitively on the children of that node. The polynomial-time construction of a suitable $\mathcal{T}_\Delta$ is an easy consequence of (3.2): if G is primitive then simply attach all points to a root, else find any non-trivial block Δ_1 and recursively construct a structure tree for the action of G on $\{\Delta_1\}^G$ and a tree for the action of $\mathrm{Stab}_G(\Delta_1)$ on Δ_1, using G to copy the latter to the other blocks (in this case, $\mathrm{Stab}_G(\Delta_1)$ is the stabilizer of a single "point" in the action on $\{\Delta_1\}^G$).

We establish some additional notation that is convenient for a recursive exploitation of orbits and blocks. Let Ω be a fixed linearly-ordered set. For any $\Delta \subseteq \Omega$, $\mathrm{Sym}(\Omega)^\Delta$ acquires a lexicographic linear ordering (via $f_1 \prec f_2$ iff $f_1(\delta) < f_2(\delta)$ for the least $\delta \in \Delta$ such that $f_1(\delta) \neq f_2(\delta)$). Define $\mu : \mathrm{Sym}(\Omega) \times \mathrm{Sym}(\Omega) \to \mathrm{Sym}(\Omega)$ by $\mu(g,h) = g^{-1}h$ and let $\mathrm{pr}_i \colon \mathrm{Sym}(\Omega) \times \mathrm{Sym}(\Omega) \to \mathrm{Sym}(\Omega)$, for $i = 1, 2$ be the projections onto the first and second factors, respectively. For $A \subseteq \mathrm{Sym}(\Omega) \times \mathrm{Sym}(\Omega)$, $\Delta \subseteq \Omega$, let $\ell\ell_\Delta(A)$ denote the lexicographically-least element in $\mu(A)^\Delta$; observe that the lex-least element of GxH is $\ell\ell_\Omega((1,x)G \times H)$. We also consider the elements that induce $\ell\ell_\Delta(A)$, namely $\mathrm{LL}_\Delta(A) = \{z \in A \mid \mu(z)^\Delta = \ell\ell_\Delta(A)\}$. We need the following facts.

FACT 1. *If $\Delta = \Delta_1 \dot\cup \Delta_2$ with the elements of Δ_1 strictly preceding all those of Δ_2 then*

$$\mathrm{LL}_\Delta(A) = \mathrm{LL}_{\Delta_2}(\mathrm{LL}_{\Delta_1}(A)).$$

FACT 2. *If $A = A_1 \cup A_2 \cup \cdots \cup A_m$ then*

$$\begin{aligned} \ell\ell_\Delta(A) &= \text{lex-least}\{\ell\ell_\Delta(A_i) \mid 1 \le i \le m\} \\ \mathrm{LL}_\Delta(A) &= \bigcup \{\mathrm{LL}_\Delta(A_i) \mid \ell\ell_\Delta(A_i) = \ell\ell_\Delta(A),\ 1 \le i \le m\} \end{aligned}$$

FACT 3. *If A is a left coset of M and Δ is invariant under* $\mathrm{pr}_1(M)$, *then* $\mathrm{LL}_\Delta(A)$ *is a left coset of* $\mathrm{Stab}_M(\{(\delta, \delta^{\ell\ell_\Delta(A)}) \mid \delta \in \Delta\})$.

The proofs of Facts 1 and 2 are straightforward. Fact 3 follows from the observation that, for $u, v \in \mathrm{Sym}(\Omega) \times \mathrm{Sym}(\Omega)$, if $\mathrm{pr}_1(u^{-1}v)$ stabilizes Δ then $\mu(u)^\Delta = \mu(v)^\Delta$ iff $u^{-1}v$ stabilizes $\{(\delta, \delta^{\mu(u)}) \mid \delta \in \Delta\}$. $\square$

These facts are used in the main theorem of this section:

THEOREM 6.2. *Let d be fixed. Given $G < \mathrm{Sym}(\Omega)$, with $G \in \Gamma_d$, in polynomial time one can establish a linear ordering of Ω with respect to which one can then find, in polynomial time, the lexicographical least element in GxH for any given $x \in \mathrm{Sym}(\Omega)$ and any given $H \leq \mathrm{Sym}(\Omega)$.*

PROOF. Let $\mathcal{F}$ be a fixed structure forest for G. Taking any planar layout of $\mathcal{F}$, with the leaves (i.e., the set Ω) situated at the same level, order the leaves left-to-right.

To accommodate recursion, we describe a general procedure for finding $\mathrm{LL}_\Omega(zM)$ where zM is any left coset of $M \leq \mathrm{Sym}(\Omega)\times\mathrm{Sym}(\Omega)$ and $\mathrm{pr}_1(M) < G$. (Our overall goal is the special case $\ell\ell_\Omega((1,x)G\times H)$.) Note that we may consider M as acting on Ω via either pr_1 or pr_2, and $\mathrm{pr}_1(M)$ also acts on $\mathcal{F}$.

Assuming the orbits of G occur in the order $\Omega_1, \ldots, \Omega_m$, we have, by Fact 1,

$$\mathrm{LL}_\Omega(zM) = \mathrm{LL}_{\Omega_m}(\cdots(\mathrm{LL}_{\Omega_1}(zM))\cdots).$$

By Fact 3, the intermediate answers are always left cosets.

Thus it suffices to describe the construction of $\mathrm{LL}_\Phi(zM)$ where Φ is the set of roots descendent from a node ν in $\mathcal{F}$ and ν (therefore Φ) is fixed by $\mathrm{pr}_1(M)$.

If ν is a leaf, then $\Phi = \{\phi\}$. In this case, $\ell\ell_\Phi(zM)$ is the least element in the orbit of $\phi^{\mu(z)}$ under $\mathrm{pr}_2(M)$. If wL is the subcoset of M mapping $\phi^{\mu(z)}$ to $\ell\ell_\Phi(zM)$ (via the pr_2 action) then $\mathrm{LL}_\Phi(zM) = zwL$.

If ν is not a leaf, then let L be the subgroup of M that fixes the immediate children, $\nu_1, \ldots, \nu_r$ (listed left-to-right), of ν (L is found by an application of (3.6)) and find a left transversal $\{w_1, \ldots, w_{|M:L|}\}$ for L in M, so

$$M = \bigcup_{i=1}^{|M:L|} w_i L.$$

By Fact 2, computation of $\mathrm{LL}_\Phi(zM)$ follows from computation of $\mathrm{LL}_\Phi(zw_iL)$ for $1 \leq i \leq |M : L|$. By Fact 3, each contributing subanswer, i.e., each $\mathrm{LL}_\Phi(zw_iL)$, for which $\ell\ell_\Phi(zw_iL) = \ell\ell_\Phi(zM)$, is a coset v_iK of the same subgroup $K = \mathrm{Stab}_L(\{(\delta, \delta^{\ell\ell_\Delta(zM)}) \mid \delta \in \Delta\})$, so that the subanswers $v_{i_1}K, \ldots, v_{i_s}K$ glue together to a coset as in:

$$v_{i_1}K \cup \cdots \cup v_{i_s}K = v_{i_1}\langle K, \{v_{i_1}^{-1}v_{i_t}\}_{2\leq t\leq s}\rangle.$$

Finally, to compute $\mathrm{LL}_\Phi(zw_iL)$, we exploit the fact that $\mathrm{pr}_1(L)$ stabilizes each ν_i, in the iterative approach

$$\mathrm{LL}_\Phi(zw_iL) = \mathrm{LL}_{\Phi_r}(\cdots(\mathrm{LL}_{\Phi_1}(zw_iL))\cdots),$$

where Φ_i denotes the set of leaves descendent from ν_i.

For the purpose of timing, we observe that, that $|M : L|$ is bounded by the size of the primitive group in the action of G_ν on $\{\nu_1, \ldots, \nu_r\}$. Suppose now that $G \in \Gamma_d$. Then $|M : L| \leq O(r^{cd})$. Thus, the one problem on Φ has involved at most $O(r^{cd+1})$ recursive calls to problems on sets of size $|\Phi|/r$. It follows that the timing for the entire procedure is $O(n^{cd+c'})$. $\square$

Since double cosets can be compared when lex-least elements are available, Theorem 6.2 has immediate applications to the problems of Section 4.

COROLLARY 6.3. *Let d be fixed. Given $G < \mathrm{Sym}(\Omega)$, with $G \in \Gamma_d$, in polynomial time one can*

(i) *for any $\Delta_1, \Delta_2 \subseteq \Omega$, test whether there exists $g \in G$ such that $\Delta_1^g = \Delta_2$;*
(ii) *for any given $H \leq \mathrm{Sym}(\Omega)$ and any $x_1, x_2 \in \mathrm{Sym}(\Omega)$ test whether $Gx_1H = Gx_2H$;*
(iii) *for any $x_1, x_2 \in \mathrm{Sym}(\Omega)$, test whether there exists $g \in G$ such that $x_1^g = x_2$.* $\square$

The methods of Section 4 yield polynomial-time equivalent "AUTO" versions for the statements in Corollary 6.3.

COROLLARY 6.4. *Let d be fixed. Given $G < \mathrm{Sym}(\Omega)$, with $G \in \Gamma_d$, in polynomial time one can*

(i) *for any $\Delta \subseteq \Omega$, find $\mathrm{Stab}_G(\Delta)$;*
(ii) *for any given $H \leq \mathrm{Sym}(\Omega)$, find $G \cap H$;*
(iii) *for any $x \in \mathrm{Sym}(\Omega)$, find $\mathrm{C}_G(x)$.* $\square$

Remarks. The timing in all of these results, as implied by the proof of Theorem 6.2, can be expressed in the form $O(|\Omega|^{cd})$, for constant c. An improvement described in [**4**] results in the timing $O(|\Omega|^{cd/\log d})$.

If $\mathcal{G} = (V, E)$ is a connected graph of valence d, and $e \in E$, then $\mathrm{Aut}(\mathcal{G})_e \in \Gamma_{d-1}$. This observation, together with the result in Corollary 6.4(i), was used in [**20**] to establish a polynomial-time isomorphism test for graphs of bounded valence. Using the improved timing as above, one gets isomorphism-testing for valence-d graphs in time $O(|V|^{cd/\log d})$ (so the exponent is $o(d)$). Together with the "valence-reduction" trick of Zemlyachenko [**31**], this, in turn, yields the best-known timing for general graph isomorphism, $O(n^{\sqrt{cn/\log n}})$ [**4**].

The result of Corollary 6.4(i) also underlies polynomial-time isomorphism tests for a broader class of graphs generalizing both bounded valence and bounded genus [**25, 26**].

We remark, finally, that Corollary 6.4 can be approached directly, and possibly a bit more compactly, than via Theorem 6.2. However, there is some dividend in the lex-least approach. For example, one can apply it to find *canonical forms* in the above graph classes [**6**].

7. Exploiting Normality

Problems that involve finding *normal* subgroups often have efficient solutions according to the criterion of this paper. We illustrate the point, in this section, with the problems of Section 4. Other examples are given in Sections 8 and 9.

We first consider INTER.

PROPOSITION 7.1. *Given $G, H \leq \mathrm{Sym}(\Omega)$, where G normalizes H, then $G \cap H$ can be found in polynomial time.*

PROOF. This is an application of (3.5), for we have the tower

$$G \cap H = G \cap G^{(n)}H \leq G \cap G^{(n-1)}H \leq \cdots \leq G \cap G^{(2)}H \leq G \cap G^{(1)}H = G.$$

Generators for $G^{(i)}H$ are available (union of generators for $G^{(i)}$ and generators for H) and so membership-testing in both G and $G^{(i)}H$, therefore in $G \cap G^{(i)}H$, is in polynomial time. Moreover,

$$|G \cap G^{(i-1)}H \colon G \cap G^{(i)}H| \leq |G^{(i-1)}H \colon G^{(i)}H| \leq |G^{(i-1)} \colon G^{(i)}| \leq n - i.$$

Hence, (3.5) applies. □

The result generalizes to

COROLLARY 7.2. *Given G and H such that $H \lhd\lhd \langle G, H \rangle$. Then $G \cap H$ can be found in polynomial time.*

PROOF. If G normalizes H then apply Proposition 7.1. Otherwise, since $H \lhd\lhd \langle G, H \rangle$, $H^G < \langle G, H \rangle$ and so $G \cap H^G < G$. It suffices then to observe that $G \cap H = H \cap (G \cap H^G)$, which we compute recursively ($G \cap H^G$ being obtained by the proposition). □

Remark. In particular, Corollary 7.2 offers an alternative approach for intersecting subgroups of a nilpotent group (wherein all subgroups are subnormal). The method appears substantially different from the orbit and imprimitivity-blocks divide-and-conquer that led to Corollary 6.4(ii).

If a targeted normal subgroup $N \trianglelefteq G$ can be interpreted as the kernel of some induced action $\pi : G \to \mathrm{Sym}(\Psi)$, then $N = G_\Psi$ (obtainable in polynomial time by (3.6)). We use this in several places, including the following.

PROPOSITION 7.3. *Given $G, H \leq \mathrm{Sym}(\Omega)$, where G normalizes H, then $\mathrm{C}_G(H)$ can be found in polynomial time.*

PROOF. We describe an action $\pi : G \to \mathrm{Sym}(\Psi)$, with $|\Psi| \leq |\Omega|$. Then if $K = \ker(\pi)$, we describe a new action $\phi : K \to \mathrm{Sym}(\Omega)$ such that $\mathrm{C}_G(H) = \ker(\phi)$.

Let Ψ be the set of equivalence classes in Ω relative to the relation defined by $\alpha \sim \beta \Leftrightarrow H_\alpha = H_\beta$. Let $\pi : G \to \mathrm{Sym}(\Psi)$ be the action of G induced by conjugation. Note that $\mathrm{C}_G(H) \leq K = \ker(\pi)$.

To define ϕ, fix a point α_Δ in each orbit Δ of H. Then for $k \in K$ let

$$(\alpha_\Delta^h)^{\phi(k)} = \alpha_\Delta^{h^k}.$$

Since $H_{\alpha_\Delta}^k = H_{\alpha_\Delta}$, ϕ is well-defined, whence it is immediate that ϕ is a homomorphism. We need to verify only that $\mathrm{C}_G(H) = \ker(\phi)$

Clearly, if k centralizes H then $k \in \ker(\phi)$. Conversely, suppose $k \in \ker(\phi)$. Let $h \in H$; we must show $hk = kh$. For any $\omega \in \Omega$, $\omega = \alpha_\Delta^{h_1}$ for some Δ and some $h_1 \in H$. Since $k \in \ker(\phi)$,

$$\omega^h = \alpha_\Delta^{h_1 h} = \alpha_\Delta^{(h_1 h)^k} = (\alpha_\Delta^{h_1^k})^{h^k} = (\alpha_\Delta^{h_1})^{h^k} = \omega^{h^k}.$$

Hence $h^k = h$. □

Remarks. In [**21**], it is observed that, when G normalizes H, $\mathrm{C}_G(H)$ can be directly interpreted as a kernel, though the action is on a set of size $O(|\Omega^2|)$. The above approach avoids this blowup in space demands.

Proposition 7.1 offers still another approach, as centralizers in $\mathrm{Sym}(\Omega)$ can be found in polynomial time (see, e.g., [**10**] or [**15**]). With that in mind, we can employ $\mathrm{C}_G(H) = G \cap \mathrm{C}_{\mathrm{Sym}(\Omega)}(H)$.

Proposition 7.3 has the immediate corollary

COROLLARY 7.4. *Given $G \leq \mathrm{Sym}(\Omega)$, then the center of G can be found in polynomial time.* □

In practice, centers are typically computed by cutting down to the centralizers of successive generators. Since the elements to centralize are chosen in a special way, for example, the first one from within the group itself, one might ask whether there may be a polynomial-time approach of this sort, notwithstanding the uncertain complexity of general CENT. However, we observe that in the first round, one is already solving a problem as hard as CENT. Consider

PROBLEM. INTERNAL-CENTRALIZER (INT-CENT)
INPUT: $G \leq \mathrm{Sym}(\Omega)$; $x \in G$.
FIND: $\mathrm{C}_G(x)$.

Unfortunately,

PROPOSITION 7.5. *INT-CENT is polynomial-time equivalent to STAB.*

PROOF. Suppose INT-CENT is in polynomial time. Then, with notation as in the reduction of STAB to CENT (in proof of Proposition 4.3), in the faithful action of $\langle G, x\rangle$ on $\widehat{\Omega}$ we could find $\mathrm{C}_{\langle G,x\rangle}(x)$ Observe, however, that $\langle G, x\rangle$ also acts on the system $\overline{\Omega} = \{\{\omega_1, \omega_2\} \mid \omega \in \Omega\}$, which may be identified with Ω (via $(\omega_1, \omega_2) \leftrightarrow \omega$); the resulting action of $\mathrm{C}_{\langle G,x\rangle}(x)$ on Ω is $\mathrm{C}_G(x)$. □

Remarks. We note that the proof shows finding $C_G(x)$ is "no easier" when $x \in G$ is an involution.

CONJ-ELT (Section 4) has an analogous "internal" case in which x_1, x_2 are assumed to be in G. Again, this is polynomial-time equivalent to the general problem.

Corollary 7.2 inspires the question of whether the following is also in polynomial time.

PROBLEM. SUBNORMAL-CENTRALIZER (SUBNORM-CENT)
INPUT: $G, H \leq \mathrm{Sym}(\Omega)$ *with* $H \lhd\lhd G$.
FIND: $C_G(H)$.

However, this problem, too, is no easier then CENT.

PROPOSITION 7.6. *SUBNORM-CENT is polynomial-time equivalent to STAB.*

PROOF. In the above discussion of INT-CENT, $\langle x\rangle^G$ is an elementary abelian 2-group, so that $\langle x\rangle \trianglelefteq \langle x\rangle^G \trianglelefteq \langle G, x\rangle$, whence $\langle x\rangle \lhd\lhd \langle G, x\rangle$. □

Proposition 7.3 does give a bit of information about general centralizers.

COROLLARY 7.7. *Given* $G, H \leq \mathrm{Sym}(\Omega)$, *then* $\mathrm{Core}_G(C_G(H))$ *can be found in polynomial time.*

PROOF. $\mathrm{Core}_G(C_G(H)) = C_G(H^G)$. □

Following the reduction of STAB to CENT, this immediately yields

COROLLARY 7.8. *Given* $G \leq \mathrm{Sym}(\Omega)$ *and* $\Delta \subseteq \Omega$, *then* $\mathrm{Core}_G(\mathrm{Stab}_G(\Delta))$ *can be found in polynomial time.* □

Corollaries 7.7 and 7.8 prompt the question of whether $\mathrm{Core}_G(G \cap H)$ can be computed in polynomial time. It can. However, we do not know an "elementary" proof (see Proposition 8.6).

Remarks. The polynomial-time methods for Propositions 7.1, 7.3 ultimately utilize the fact that the targeted subgroup $H \leq G$ lies in a chain

$$H = H_m \leq H_{m-1} \leq \cdots \leq H_0 = G \tag{1}$$

with $|H_i : H_{i+1}|$ "small". In fact, this is true for any $H \lhd\lhd G$, where "small" can be interpreted as $\leq n$. To show this, it suffices to consider to assume $H \trianglelefteq G$, in which case

$$H = HG^{(n)} \leq \cdots HG^{(2)} \leq HG^{(1)} = G.$$

This suggests that (3.5) should provide the tool for finding targeted normal subgroups much more generally. The difficulty that arises, however, is that we do not have, a priori, ways of "recognizing" the intermediate groups. (See Question 10, for example.)

One expects, also, to find normal subgroups as kernels of actions. However, for arbitrary $N \lhd G \leq \mathrm{Sym}(\Omega)$, G/N may not be representable on a polynomial-size set [**27**]. One knows, however, for $H \lhd\lhd G$, there is a chain

$$H = L_m \trianglelefteq L_{m-1} \trianglelefteq \cdots \trianglelefteq L_0 = G$$

with $L_i/L_{i+1} \hookrightarrow \mathrm{Sym}(\Omega)$ for each i. (To show this, we may assume $H \trianglelefteq G$; using the chain in (1), inductively let L_{i+1} be the kernel of the right-multiplication action of L_i on the right cosets of H_{i+1} in H_i.) Call the minimal such m the *depth* of L in G. It is not hard to show that if $L \trianglelefteq G$ then $m = O(\log^2 |\Omega|)$. (This reduces easily to the primitive case, wherein one uses the Cameron classification of primitive groups, see, e.g., [**7**].)

Though we are not sure of polynomial-time implications, the following question seems of interest.

QUESTION 7. *What is the least upper bound on the depths of normal and subnormal subgroups in permutation groups?*

The proof of Proposition 7.3 shows, for example, that the depth of the centralizer of a normal subgroup is at most 2.

8. Quotient Groups

In [**18**], Kantor and Luks suggest the thesis that problems that are in polynomial time for permutation groups remain in polynomial time for quotients of permutation groups. The justification is not, however, via routine consideration of the quotients as permutation groups, as is often the case in available systems, inasmuch as quotients may not have any small (polynomial-size) faithful permutation representations [**27**]. The generalizations of problems INTER and CENT provide good illustrations of the techniques that are brought to bear in [**18**].

PROBLEM. QUOTIENT-INTERSECTION (Q-INTER)
INPUT: $G, H, K \leq \mathrm{Sym}(\Omega)$ *with* $K \trianglelefteq G$, $K \trianglelefteq H$.
FIND: $G/K \cap H/K$.

PROBLEM. QUOTIENT-CENTRALIZER (Q-CENT)
INPUT: $G, K \leq \mathrm{Sym}(\Omega)$; $x \in \mathrm{Sym}(\Omega)$, *with* $K \trianglelefteq G$ *and* x *normalizing* K.
FIND: $\mathrm{C}_{G/K}(xK/K)$.

($G/K, H/K$ may be considered as contained in the group $\langle G, H\rangle/K$ and G/K, xK/K in the group $\langle G, x\rangle/K$.)

Since permutation representations of the quotients may be infeasible, the question arises of whether these problems present a still higher level of challenge. However,

PROPOSITION 8.1. *Q-INTER and Q-CENT are polynomial-time equivalent to STAB.*

PROOF. It is obvious that Q-INTER is no harder than INTER, since $G/K \cap H/K = (G \cap H)/K$.

Reduction of Q-CENT to STAB: Let (G, K, x) be an instance of Q-CENT. Let $\hat{G} = \{(g, gk) \mid g \in G, k \in K\}$ acting on $\Omega \times \Omega$. For $x \in G$ let $\Delta(x) = \{(\omega, \omega^x) \mid \omega \in \Omega\} \subseteq \Omega \times \Omega$. Then for $(g, gk) \in \hat{G}$, $\Delta(x)^{(g,gk)} = \Delta(x^g k)$, so that (g, gk) stabilizes $\Delta(x)$ iff $[x, g] = k^{-1}$. But $gK \in \mathrm{C}_{G/K}(xK/K)$ iff there exists $k \in K$ such that $[g, x] = k$. Hence, if we compute $\mathrm{Stab}_{\hat{G}}(\Delta(x))$ and let H be its first coordinate projection, we have $\mathrm{C}_{G/K}(xK/K) = H/K$. □

Remark. Similarly, the quotient versions of DC-EQ and ELT-CONJ are polynomial-time equivalent to the permutation-group cases.

Thus, it seems, from a polynomial-time perspective, that these problems do not get any harder for quotients. In a positive direction, we next show that the instances where INTER and CENT are in polynomial time generalize to quotient groups.

The following is just a repeat of Proposition 7.1.

PROPOSITION 8.2. *Given $G, H, K \leq \mathrm{Sym}(\Omega)$, with $K \trianglelefteq G$ and $K \trianglelefteq H$, and where G/K normalizes H/K, then $G/K \cap H/K$ can be found in polynomial time.*

PROOF. $G/K \cap H/K = (G \cap H)/K$ and the hypotheses imply G normalizes H. □

Corollary 7.2 generalizes immediately, as well.

Generalizations of Corollary 6.4(ii,iii) and Proposition 7.3 require a surprising amount of additional machinery.

The following is proved in [**16, 17**].

LEMMA 8.3 (KANTOR). *Given $G \leq \mathrm{Sym}(\Omega)$ then*

(i) *For any prime p dividing $|G|$, a Sylow p-subgroup of G can be found in polynomial time.*

(ii) *Given Sylow p-subgroups P_1, P_2 of G, some $g \in G$ such that $P_1^g = P_2$ can be found in polynomial time.*

(iii) *Given $K, P \leq \mathrm{Sym}(\Omega)$ with P a Sylow p-subgroup of K and $K \trianglelefteq G$, then $\mathrm{N}_G(P)$ can be found in polynomial time.*

Quite unlike the methods being described in this paper, which have relied on elementary group theory, the algorithms and proofs underlying Lemma 8.3 use substantial consequences of the classification of finite simple groups, including detailed knowledge of simple-group types. Nevertheless, that having been done, it is demonstrated in [**18**] that one can effectively use the result as a "black-box" in further, once again elementary, arguments. We illustrate first with a generalization of Corollary 6.4(iii). This involves a constructive version of the well-known

FRATTINI ARGUMENT. *Let $P \leq K \trianglelefteq G$ with P a Sylow p-subgroup of K. Then $G = K\mathrm{N}_G(P)$.*

PROPOSITION 8.4. *Given $G, K \leq \mathrm{Sym}(\Omega)$ and $x \in \mathrm{Sym}(\Omega)$, where $K \trianglelefteq G$, x normalizes K, and with $G/K \in \Gamma_d$, then $\mathrm{C}_{G/K}(xK/K)$ can be found in polynomial time.*

PROOF. Note that the Γ_d hypothesis applies only to G/K. If, however, $G \in \Gamma_d$ then the reduction in the proof of Proposition 8.1 would lead to an instance of STAB with a group, $\hat{G}$, in Γ_d, whence we could apply 6.4(i).

So suppose $G \notin \Gamma_d$. Then $K \notin \Gamma_d$. In particular, K is not nilpotent, so that, for some prime p dividing $|K|$, any Sylow p-subgroup, P, of K is not normal. Find generators for such a P as well as $G_1 = \mathrm{N}_G(P)$ and $K_1 = \mathrm{N}_K(P)$, and find $k \in K$ such that $P^k = P^x$ (Lemma 8.3). Let $y = xk^{-1}$, so y normalizes P and therefore normalizes K_1.

Recursively compute $L/K_1 = \mathrm{C}_{G_1/K_1}(yK_1/K_1)$. Then $\mathrm{C}_{G/K}(xK/K) = LK/K$.

The recursive procedure runs in polynomial time since $G_1 < G$. The correctness is a consequence of a Frattini argument: Since $G = G_1K$, it suffices to show, for $g_1 \in G_1$ that g_1K centralizes xK/K (in $\langle G, x\rangle/K$) iff g_1K_1 centralizes yK_1 (in $\langle L, y\rangle/K_1$). But g_1K centralizes xK/K iff $[g_1, x] \in K$ iff $[g_1, y] \in K$ iff $[g_1, y] \in K_1$ (since $K_1 = \mathrm{N}_K(P)$ and both g_1 and y normalize P) iff g_1K_1 centralizes yK_1. $\square$

A similar Frattini argument (see [**18**]) is used for the following extension of Corollary 6.4(ii).

PROPOSITION 8.5. *Given $G, H, K \leq \mathrm{Sym}(\Omega)$, where $K \trianglelefteq G$, $K \trianglelefteq H$, and with $G/K \in \Gamma_d$, then $G/K \cap H/K$ can be found in polynomial time.* $\square$

We reiterate that, while Corollary 6.4(ii,iii) has been extended to quotient groups, the fact that the extensions are dependent upon Lemma 8.3, means that we have now had to invoke the classification of simple groups. On the other hand, in the special case when $\langle G/K, H/K\rangle$ is solvable, there are "elementary" proofs of Propositions 8.4 and 8.5 based upon results in [**22**].

QUESTION 8. *Is there an "elementary" construction of $\mathrm{C}_{G/K}(H/K)$ and/or $G/K \cap H/K$ if only G/K is assumed to be solvable?*

Our extension of Proposition 7.3 to quotient groups requires the ability to compute cores of given subgroups of permutation groups. In practice, this is commonly done by intersecting conjugates until the resulting group is normal. Since intersections are not presently available, this approach is not yet feasible in polynomial time. Nevertheless, cores are attainable. Following the theme of Section 7, we observe that the normality of the targeted group facilitates this.

PROPOSITION 8.6. *Given* $G, H \leq \mathrm{Sym}(\Omega)$, *then* $\mathrm{Core}_G(G \cap H)$ *can be found in polynomial time.*

PROOF. For each prime p, find a Sylow p-subgroup, P_p, of G. Since $\mathrm{Core}_G(G \cap H) \trianglelefteq G$,

$$\mathrm{Core}_G(G \cap H) = \langle \{P_p \cap \mathrm{Core}_G(G \cap H) \mid p \text{ divides } |G|\} \rangle.$$

It suffices to determine $P_p \cap \mathrm{Core}_G(G \cap H)$ for each p. This is made feasible by the fact that we can test membership in $\mathrm{Core}_G(G \cap H)$, that is, if $g \in G$ then $g \in \mathrm{Core}_G(G \cap H)$ iff $\langle g \rangle^G \leq H$. Thus, initially setting $T = P_p$, test whether $T^G \leq H$ and, if so, output T; else we can find $g \in G$ such that $T^g \not\leq H$ (the computation of generators for the normal closure, (3.11), can maintain generators as conjugates of the generators of T) and repeat with $T := T \cap H^{g^{-1}}$ (intersect by Corollary 6.4(ii)).

The procedure succeeds because the relation $T \cap \mathrm{Core}_G(G \cap H) = P_p \cap \mathrm{Core}_G(G \cap H)$ is maintained. □

It is immediate that

COROLLARY 8.7. *Given* $G, H \leq \mathrm{Sym}(\Omega)$ *with* $H \leq G$, *then* $\mathrm{Core}_G(H)$ *can be found in polynomial time.* □

Remark. The proof of Proposition 8.6 provides a striking counterpoint to that of Corollaries 7.7 and 7.8. While the latter two were elementary, the former uses Lemma 8.3 which, in turn, uses the classification of finite simple groups. On the other hand, we observe in Section 9 that another problem (finding p-cores) which, in practice, has seemed to require construction of Sylow subgroups, has a direct and elementary approach. Once again we are led to questions about the existence of non-classification-dependent arguments.

QUESTION 9. *Is there an "elementary" approach to finding* $\mathrm{Core}_G(G \cap H)$ *or even for finding* $\mathrm{Core}_G(H)$ *when* $H \leq G$*?*

In particular, considering the remarks at the end of Section 7,

QUESTION 10. *Is there an "elementary" construction of a chain*

$$\mathrm{Core}_G(H) = N_0 \trianglelefteq N_1 \trianglelefteq \cdots \trianglelefteq N_m = G$$

in which N_{i+1} *is the kernel of a "small" degree representation of* N_i*?*

Of course, these issues may lie with Lemma 8.3 itself.

QUESTION 11. *Is there an "elementary" approach to finding Sylow subgroups?*

Can one even get started?

QUESTION 12. *Is there an "elementary" method for locating an element of order* p *where* p *is a prime dividing* $|G|$*?*

Returning to the main issue, we complete this section with the following extension of Proposition 7.3.

COROLLARY 8.8. *Given* $G, H, K \leq \mathrm{Sym}(\Omega)$, *with* $K \trianglelefteq G$ *and* $K \trianglelefteq H$, *and where* G *normalizes* H, *then* $\mathrm{C}_{G/K}(H/K)$ *can be found in polynomial time.*

PROOF. Consider $G \times G$ acting naturally on the disjoint union, $\Omega \dot{\cup} \Omega$, of two copies of Ω. Let $L = \{(g, gk) \mid g \in G, k \in K\}$ and $M = \{(g, gh) \mid g \in G, h \in H\}$. Find $\mathrm{Core}_M(L)$ (Corollary 8.7) and let C be the group obtained by restricting $\mathrm{Core}_M(L)$ to the first copy of Ω. Output C/K.

We show that $C/K = \mathrm{C}_{G/K}(H/K)$, i.e., that for $g \in G$, $(g, gk) \in \mathrm{Core}_M(L)$ for some $k \in K$ iff gK centralizes H/K (in $\langle G, H\rangle/K$). Since $K \times K \trianglelefteq M$, we have $K \times K \leq C$ so that $K \leq C$. Then, for $(g, gk) \in L$,

$$
\begin{aligned}
(g, gk) = (g, g)(1, k) \in \mathrm{Core}_M(L) \quad &\text{iff} \quad (g, g)^M \subseteq L \\
&\text{iff} \quad (g, g)^{(1,h)} \in L, \ \forall h \in H \\
&\text{iff} \quad g^{-1} g^h \in K, \ \forall h \in H \\
&\text{iff} \quad gK \text{ centralizes } H/K. \qquad \square
\end{aligned}
$$

9. p-Cores

For any prime p and group G, the *p-core* of G is the (unique) maximal normal p-subgroup of G and is denoted $\mathrm{O}_p(G)$.

THEOREM 9.1. *Given* $G \leq \mathrm{Sym}(\Omega)$, *then* $\mathrm{O}_p(G)$ *can be found in polynomial time.*

A suggested method for computing the p-core of a permutation group has been to find a Sylow p-subgroup $P \leq G$ and then use

$$\mathrm{O}_p(G) = \mathrm{Core}_G(P).$$

This does give a polynomial-time solution. However, the conceptual overhead in this approach to $\mathrm{O}_p(G)$ is that the known method for finding P (Lemma 8.3) uses the classification of finite simple groups. Nevertheless, unlike the situation for general cores, we offer a self-contained elementary proof of Theorem 9.1, giving another measure of support for the theme that *normal* targets are easier to locate. (See **[27]** for another direct approach to p-cores.)

A few lemmata are required.

LEMMA 9.2. *Given a transitive* $G \leq \mathrm{Sym}(\Omega)$ *with* $|G| > n$, *in polynomial time one can find a proper normal subgroup or else establish that* G *does not have a* regular abelian *normal subgroup.*

PROOF. If $G^{(3)} = 1$ then $|G| < n^2$. (Recall that $G^{(3)}$ is the subgroup fixing ω_1 and ω_2.) In that case, the elements of G can be listed and the normal closure of the group generated by each can be computed in polynomial time. If none of these yield a proper normal subgroup then G does not have a regular abelian normal (or *any* proper normal) subgroup.

Assume $G^{(3)} \neq 1$ and let $\Psi = \{(G^{(3)})^g \mid g \in G\}$. Then $1 < |\Psi| \leq \binom{n}{2}$ and G acts transitively (via conjugation) on Ψ. Let $\mathcal{B}$ be a minimal G-block system in Ψ (i.e., start with $\mathcal{B} = \Psi$ and while G does not act primitively on $\mathcal{B}$, replace $\mathcal{B}$ by a nontrivial partition of $\mathcal{B}$ into blocks of imprimitivity). Output $G_{\mathcal{B}}$ (the kernel of the action of G on $\mathcal{B}$) if it is proper, else declare that G does not have a regular abelian normal subgroup.

We must show, under the assumption that G has a regular abelian normal subgroup A that G does not act faithfully on $\mathcal{B}$. Since A is regular, there is a unique $a \in A$ such that $\omega_1^a = \omega_2$. Such a normalizes, in fact centralizes, $G^{(3)}$, for if $x \in G^{(3)}$, both a and $x^{-1}ax$ are elements of A mapping ω_1 to ω_2 so that $a = x^{-1}ax$. Hence a fixes the block in $\mathcal{B}$ containing $G^{(3)}$. We conclude that A does not act regularly on $\mathcal{B}$. But then A cannot act faithfully on $\mathcal{B}$, for a normal subgroup of a primitive group is transitive and so, if it is abelian, it is regular. □

Remarks. The above algorithm simplifies one with an analogous purpose in [**21**]. The modification is due to Á. Seress. (See also [**7**].)

Note that the output of a proper normal subgroup leaves open the question of whether there is a regular abelian normal subgroup, thus leading us to ask

QUESTION 13. *Given $G \leq \mathrm{Sym}(\Omega)$, can one determine, in polynomial time, whether G has a regular abelian normal subgroup and, if so, find one?*

More generally,

QUESTION 14. *Given $G \leq \mathrm{Sym}(\Omega)$, can one determine, in polynomial time, whether G has a regular normal subgroup and, if so, find one?*

Both of these issues are in polynomial time for primitive groups: if a primitive group G has an abelian normal subgroup N, then $N = \mathrm{O}_p(G)$ for (the unique) prime p dividing n; in general, if H is the smallest nontrivial term in a composition series for G ([**21**]), then G has a regular normal subgroup iff H^G is regular.

LEMMA 9.3. *Given $G \leq \mathrm{Sym}(\Omega)$, in polynomial time one can find a proper normal subgroup of G or else establish that $\mathrm{O}_p(G) = 1$.*

Remark. Output of a proper normal subgroup does not yet mean $\mathrm{O}_p(G) \neq 1$.

PROOF OF LEMMA. Let Δ be any nontrivial orbit of G and construct a minimal G-block system $\mathcal{B}$ in Δ (so that G acts primitively on $\mathcal{B}$). Let $\phi : G \to \mathrm{Sym}(\mathcal{B})$ be the induced action. If $K = \ker(\phi) \neq 1$, output generators for K. Otherwise, the primitive group $\phi(G)$ is isomorphic to G. We may assume $|G| > n$ else G is listable and the lemma resolvable by brute force. Apply Lemma 9.2 to $\phi(G)$. If $\phi(G)$ has no abelian regular normal subgroup then $\mathrm{O}_p(G) = 1$ (else, if A is the the last nontrivial term in the derived series for $O_p(G)$, $\phi(A)$ would be a regular normal subgroup of the primitive group). Otherwise, the call to Lemma 9.2 produces Y, generating a proper normal subgroup, in which case return $N = \phi^{-1}(Y)$. (The lifting $\phi^{-1}(y)$ for $y \in Y$ is computed, for example, (3.9); alternatively, while computing Y, keep track of liftings of elements yielding $Y' \subseteq G$ with $\phi(Y') = Y$; then $N = \langle Y', K\rangle$.) $\square$

LEMMA 9.4. *Given $G \leq \mathrm{Sym}(\Omega)$, in polynomial time one can find a nontrivial normal p-subgroup of G or else establish that $O_p(G) = 1$.*

PROOF. We describe a procedure p-NORM(G) with output as indicated.

Apply Lemma 9.3. If we discover $O_p(G) = 1$ then return that information. Otherwise we have $1 \neq N \lhd G$ and we proceed as follows.

Recursively call p-NORM(N). If the call returns $P \trianglelefteq N$, then output P^G. Else ($O_p(N) = 1$) recursively call p-NORM($C_G(N)$) (using Proposition 7.3 to find $C_G(N)$). If the call returns for $P \trianglelefteq C_G(N)$ then output P^G. Else report "$O_p(G) = 1$."

The procedure succeeds since $O_p(N) = 1$ implies $O_p(G) \cap N = 1$, whence $O_p(G) \leq C_G(N)$.

Timing concern: What if both recursive calls are made? That only happens when $O_p(N) = 1$ so that p does not divide $|N \cap C_G(N)|$, whence

$$|N|_p|C_G(N)|_p = |NC_G(N)|_p \leq |G|_p$$

where sub-p denotes p-part. Thus, except for multiplicative contributions from known polynomial timings, the time is linear in $\log |G|_p$. $\square$

LEMMA 9.5. *Given $P, G \leq \mathrm{Sym}(\Omega)$ with $P \lhd G$ where P is a p-group and G is not a p-group, one can construct in polynomial time another action $\phi : G \to \mathrm{Sym}(\Omega)$, where $\ker(\phi)$ is a nontrivial normal p-subgroup of G.*

PROOF. Replacing P, if necessary, by the last nontrivial term in its derived series, we may assume that P abelian. Let $\{\Delta_i\}_{i \in I}$ be the set of orbits of P and let $\pi : G \to \mathrm{Sym}(I)$ be the naturally induced action, i.e., $\Delta_i^g = \Delta_{i^{\pi(g)}}$ for $i \in I, g \in G$. Choose $\delta_i \in \Delta_i$ for each $i \in I$. Then ϕ is defined via

$$(\delta_i^x)^{\phi(g)} = \delta_{i^{\pi(g)}}^{x^g},$$

for $i \in I, x \in P$. (The superscripts x, x^g denote the given action.) That ϕ is well defined follows from the fact that P^{Δ_i} is regular (since it is abelian), for, if $\delta_i^x = \delta_i^y$ for $x, y \in P$, then x and y act identically on Δ_i so that x^g and y^g act identically on $\Delta_{i^{\pi(g)}}$. From this it is straightforward to see ϕ is a homomorphism. Since $\ker(\phi)$ stabilizes each Δ_i and commutes with the action of P thereon, $\ker(\phi)^{\Delta_i} = P^{\Delta_i}$. Hence $\ker(\phi)$ is an abelian p-group containing P. It is proper in G as G is not a p-group. $\square$

PROOF OF THEOREM 9.1. We may assume G is not a p-group. By Lemma 9.4, we establish immediately that $O_p(G) = 1$ or else obtain a proper normal p-subgroup K. In the latter case, we apply Lemma 9.5 to obtain an action $\phi : G \to \mathrm{Sym}(\Omega)$ with $1 < K = \ker(\phi) \lhd G$. Recursively, compute $\langle Y \rangle = O_p(\phi(G))$. Then $O_p(G) = \phi^{-1}(Y)$ (computed, say, via (3.9), wherein it is convenient to consider $\phi(G)$ as acting on a disjoint copy of Ω).

For correctness, we observe that, since $\ker(\phi)$ is a p-group, $\phi(O_p(G)) = O_p(\phi(G))$.

For the timing, note that the recursive call involves a smaller group $\phi(G)$ on a permutation domain of the same size. □

Remark. In **[18]** it is pointed out that, in polynomial time, one can construct the maximal normal subgroup with composition factors in any specified collection of simple groups, but the general result ultimately makes use of the classification of finite simple groups.

10. Other Problems and their Relationships

We comment on several other problems resembling GRAPH-ISO and STAB, etc. There are open questions, not only about when they are in polynomial time, but in the relationships among them.

10.1. Finding Subgroups. Possibly presenting a challenge beyond STAB is

PROBLEM. NORMALIZER (NORM)
INPUT: $G, H \leq \mathrm{Sym}(\Omega)$.
FIND: $N_G(H)$.

Techniques announced in **[22]** show that NORM is in polynomial time when $\langle G, H \rangle$ is solvable. Questions that immediately arise include

QUESTION 15. *Is NORM in polynomial time when only G is assumed to be solvable?*

The next step up the group ladder would appear to be

QUESTION 16. *Is NORM in polynomial time when* $\langle G, H \rangle$ *is in* Γ_d *(See Section 6).*

How is NORM related to the problems of Section 4? STAB reduces to NORM, either by Proposition 7.1, or, following the reduction of STAB to CENT in Proposition 4.3, the fact that x is an involution implies $N_g(\langle x \rangle) = C_G(x)$. But is NORM, in general, "harder" than STAB, etc.?

QUESTION 17. *Is there a polynomial-time reduction of NORM to STAB?*

For this question, notice that it would suffice to find a polynomial-time solution to the special case

PROBLEM. NORMALIZER IN SYMMETRIC GROUP (NORM-SYM)
INPUT: $G \leq \mathrm{Sym}(\Omega)$.
FIND: $\mathrm{N}_{\mathrm{Sym}(\Omega)}(G)$.

Recall that *centralizers* in the symmetric group are computable in polynomial time (see, e.g., [**10**]). However, the complexity of NORM-SYM is open.

QUESTION 18. *Is NORM-SYM in polynomial time? Is there even a subexponential solution?*

If NORM-SYM were in polynomial time, then NORM would reduce to INTER (since $\mathrm{N}_G(H) = \mathrm{N}_{\mathrm{Sym}(\Omega)}(H) \cap G$). In fact, even if polynomial-time algorithms are not available, reductions between the problems are of interest.

QUESTION 19. *Is NORM-SYM polynomial-time reducible to STAB? Is STAB polynomial-time reducible to NORM-SYM?*

Affirmative answers would, respectively, put NORM equivalent to STAB or NORM-SYM.

One of the reasons that Questions 18 and 19 are particularly intriguing is that GRAPH-AUTO *is* polynomial-time reducible to NORM-SYM (as well as to STAB). *Reduction:* Given a graph $\mathcal{G} = (V, E)$, we construct Ω, $G \leq \mathrm{Sym}(\Omega)$ and describe an epimorphism $\phi : \mathrm{N}_{\mathrm{Sym}(\Omega)}(G) \to \mathrm{Aut}(\mathcal{G})$. Let $I = \{1, 2, \ldots, 2|V|\}$. Set $\Omega = V \times I \,\dot\cup\, E \times \{1, 2\}$ (so $|\Omega| = 2(|V|^2 + |E|)$). For each $v \in V$, let g_v be the involution in $\mathrm{Sym}(\Omega)$ that transposes $(v, 2i-1)$ with $(v, 2i)$, for $1 \leq i \leq |V|$, and transposes $(e, 1)$ with $(e, 2)$, for every $e \in E$ having endpoint v, while leaving other points fixed; thus g_v moves precisely $2(|V| + \mathrm{degree}(v))$ points. Set $G = \langle \{g_v\}_{v \in V} \rangle$ (an elementary abelian 2-group). Within G the only non-identity elements that move $< 4|V|$ points are the g_v. Hence, permutations in $\mathrm{N}_{\mathrm{Sym}(\Omega)}(G)$ permute the g_v, so that there is an induced homomorphism $\phi : \mathrm{N}_{\mathrm{Sym}(\Omega)}(G) \to \mathrm{Sym}(V)$. Since $\{v, w\} \in E$ iff g_v and g_w move the same point (i.e., the point $(\{v, w\}, 1)$), it is clear that $\phi(G) \leq \mathrm{Aut}(\mathcal{G})$. It is straightforward to show that ϕ is surjective. $\square$

Notice that the above reduction involved an elementary abelian 2-group. Thus, Question 18 is interesting *and open* even in this case.

For any finite field $\mathrm{GF}(q)$, there is a natural action of $\mathrm{Sym}(\Omega)$ on $\mathrm{GF}(q)^{\Omega}$ via permutation of coordinates. Then $g \in \mathrm{Sym}(\Omega)$ stabilizes $\Delta \subseteq \Omega$ iff g stabilizes the vector $(a_\omega)_{\omega \in \Omega}$ with $a_\omega = 1$ for $\omega \in \Delta$ and $a_\omega = 0$ otherwise. Thus STAB is polynomial-time reducible to

PROBLEM. VECTOR STABILIZER (VEC-STAB)
INPUT: $G \leq \mathrm{Sym}(\Omega)$*; a representation* $\phi : G \to \mathrm{GL}(V)$*, where* V *is a finite dimensional vector space over* $\mathrm{GF}(q)$*;* $v \in V$
FIND: $G_v = \{g \in G \mid v^g = v\}$.

Here, we assume that V is specified via a basis and ϕ is specified on the given generators of G. By results of [**22**], VEC-STAB is solvable in polynomial time if G is solvable. ("Polynomial in the input" is taken to be $O((|\Omega|+\dim(V)+\log q)^c)$.) In fact, it is also indicated there that the following is solvable in polynomial time if G is solvable.

PROBLEM. SUBSPACE STABILIZER (SUBSP-STAB)
INPUT: $G \leq \mathrm{Sym}(\Omega)$; *a representation* $\phi : G \to \mathrm{GL}(V)$, *where* V *is a finite dimensional vector space over* $\mathrm{GF}(q)$; *a subspace* $W \leq V$
FIND: $\mathrm{Stab}_G(W) = \{g \in G \mid W^g = W\}$.

QUESTION 20. *Is VEC-STAB in polynomial time for* $G \in \Gamma_d$? *Is SUBSP-STAB in polynomial time for* $G \in \Gamma_d$?

VEC-STAB is polynomial-time reducible to SUBSP-STAB. The "obvious" reduction seems to be to stabilize first the 1-dimensional W =Span(v), after which we only need the kernel of a homomorphism $\phi : \mathrm{Stab}_G(W) \to \mathrm{GF}(q)^*$ (multiplicative group). While this is not difficult to complete, we refer instead to the reduction between the corresponding decision problems (VEC-TRANS $\leq$ SUBSP-TRANS) in Section 10.2.

Remark. One can also show that the problem of finding $\mathrm{N}_G(H)$ when $H \lhd\lhd \langle G, H\rangle$ is polynomial-time reducible to SUBSP-STAB.

Another question that arises is whether normality helps for some of these problems. Techniques of [**18**] (in particular the method of Theorem 8.6 of this present paper), can be used to find $\mathrm{Core}_G(G_v)$ and $\mathrm{Core}_G(\mathrm{Stab}_G(W))$ (where v, W are a vector and subspace, respectively). We wonder, however, about

QUESTION 21. *Given* $G, H \leq \mathrm{Sym}(\Omega)$, *can one find* $\mathrm{Core}_G(\mathrm{N}_G(H))$ *in polynomial time?*

Note that we have found $\mathrm{Core}_G(G \cap H)$, which is the kernel of the right-multiplication-action of G on right cosets of H by G, while $\mathrm{Core}_G(\mathrm{N}_G(H))$ is the kernel of the conjugacy action of G on the conjugates of H by G.

10.2. Decision Questions. The problems of Section 10.1 suggest decision analogues.

Corresponding to NORM:

PROBLEM. CONJUGACY OF GROUPS (CONJ-GROUP)
INPUT: $G, H_1, H_2 \leq \mathrm{Sym}(\Omega)$.
QUESTION: *Is there some* $g \in G$ *such that* $H_1^g = H_2$?

As in the STAB $\equiv$ TRANS equivalence, NORM is polynomial-time equivalent to CONJ-GROUP.

The right analogue of NORM-SYM would seem to be

PROBLEM. CONJUGACY IN THE SYMMETRIC GROUP (CONJ-SYM)
INPUT: $H_1, H_2 \le \mathrm{Sym}(\Omega)$.
QUESTION: *Is there some* $x \in \mathrm{Sym}(\Omega)$ *such that* $H_1^x = H_2$*?*

Here, we do not see the equivalence. While a reduction of CONJ-SYM to NORM-SYM is not difficult, we do not have a reverse reduction. Thus, we ask

QUESTION 22. *Is NORM-SYM polynomial-time equivalent to CONJ-SYM?*

VEC-STAB and SUBSP-STAB are, respectively, polynomial-time equivalent to

PROBLEM. VECTOR TRANSPORTER (VEC-TRANS)
INPUT: $G \le \mathrm{Sym}(\Omega)$*; a representation* $\phi : G \to \mathrm{GL}(V)$*, where* V *is a finite dimensional vector space over* $\mathrm{GF}(q)$*;* $v_1, v_2 \in V$
QUESTION: *Is there some* $g \in G$ *such that* $v_1^g = v_2$*?*

and

PROBLEM. SUBSPACE TRANSPORTER (SUBSP-TRANS)
INPUT: $G \le \mathrm{Sym}(\Omega)$*; a representation* $\phi : G \to \mathrm{GL}(V)$*, where* V *is a finite dimensional vector space over* $\mathrm{GF}(q)$*; subspaces* $W_1, W_2 \le V$
QUESTION: *Is there some* $g \in G$ *such that* $W_1^g = W_2$*?*

The reduction of VEC-TRANS to SUBSP-TRANS is worth noting. While a polynomial-time reduction can be completed along the lines begun in the VEC-STAB to SUBSP-STAB discussion, that would not then be a Karp reduction (yes/no instance mapping to yes/no instance). Here then is another approach: Let (G, ϕ, V, v_1, v_2) be an instance of VEC-TRANS; we may assume that $v_2 \neq 0$. Then G acts naturally on the tensor product $V \otimes V$ (so that $(v \otimes w)^g = v^g \otimes w^g$) and therefore there is an induced action $G \to \mathrm{GL}(V \oplus (V \otimes V))$. Let $W_i = \mathrm{Span}((v_i, v_i \otimes v_i))$, for $i = 1, 2$. We claim that for $g \in G$, $v_1^g = v_2$ iff $W_1^g = W_2$. The only-if direction is clear. Assume $W_1^g = W_2$. Then, for some $c \in \mathrm{GF}(q)$, $(v_1, v_1 \otimes v_1)^g = c(v_2, v_2 \otimes v_2)$, so that $v_1^g = cv_2$ and $(v_1 \otimes v_1)^g = c(v_2 \otimes v_2)$. Thus, $c^2(v_2 \otimes v_2) = (cv_2 \otimes cv_2) = (v_1^g \otimes v_1^g) = (v_1 \otimes v_1)^g = c(v_2 \otimes v_2)$. It follows that $c = 1$, proving the claim. □

We add two more problems that seem of particular interest.

PROBLEM. GROUP ISOMORPHISM (GROUP-ISO)
INPUT: *Cayley tables for groups* G, H.
QUESTION: *Are* G *and* H *isomorphic?*

Here "polynomial in the input" translates to polynomial in $|G|$ (presumably $|G| = |H|$). It is not hard to reduce GROUP-ISO to GRAPH-ISO. (See **[24]** for a discussion of this and related issues.) But is this problem easier? In particular,

QUESTION 23. *Is GROUP-ISO in polynomial time?*

It can be solved in subexponential $O(|G|^{c+\log_2 |G|})$ time since there is a set of $\leq \log_2 |G|$ generators, and a homomorphism $G \to H$ is determined by the images of the generators. This, however, appears to be the best result known for general groups.

Possibly on the "harder" side is

PROBLEM. PERMUTATION-GROUP ISOMORPHISM (PERM-GROUP-ISO)
INPUT: $G, H \leq \mathrm{Sym}(\Omega)$.
QUESTION: *Are G and H isomorphic?*

PERM-GROUP-ISO is in NP: Supposing $G = \langle X \rangle$, one can guess an isomorphism $f : G \to H$ by guessing $f(x)$ for all $x \in X$ and then verifying that f is indeed an isomorphism by checking that $|G| = |H| = |\langle\{(x, f(x))\}_{x\in X}\rangle|$ (the latter being considered as a subgroup of $G \times H$ acting, say, on $\Omega\dot{\cup}\Omega$). □

It is shown in **[3]** that CONJ-GROUP is polynomial-time reducible to PERM-GROUP-ISO.

To summarize the known relationships, letting "$\leq$" denote "is polynomial-time Karp-reducible to" we have:

$$
\begin{aligned}
\text{GROUP-ISO} &\leq \text{GRAPH-ISO}\\
\text{GRAPH-ISO} &\leq \text{TRANS}\\
\text{GRAPH-ISO} &\leq \text{CONJ-SYM}\\
\text{TRANS} &\leq \text{CONJ-GROUP}\\
\text{TRANS} &\leq \text{VEC-TRANS}\\
\text{CONJ-SYM} &\leq \text{CONJ-GROUP}\\
\text{CONJ-GROUP} &\leq \text{PERM-GROUP-ISO}\\
\text{VEC-TRANS} &\leq \text{SUBSP-TRANS}
\end{aligned}
$$

(And recall, TRANS $\equiv$ DC-EQ $\equiv$ CONJ-ELT.)

QUESTION 24. *Are there any other reductions between these problems except as implied by the above?*

We do not anticipate seeing a negative answer very soon (as that would necessarily include a proof of P $\neq$ NP), but we believe a search for other relationships could shed additional light on these problems.

References

1. L. Babai, *On the length of subgroup chains in the symmetric group,* Comm. in Alg. 14 (1986), 1729–1736.
2. L. Babai, P. Cameron and P.J. Pálfy, *On the order of primitive groups with restricted nonabelian composition factors,* J. Algebra 79 (1982), 161–168.
3. L. Babai, S. Kannan and E.M. Luks, *A bounded round interactive proof of permutation group non-isomorphism,* in preparation.
4. L. Babai, W.M. Kantor and E.M. Luks, *Computational complexity and the classification of finite simple groups,* Proc. 24th IEEE FOCS (1983), 162–171.
5. L. Babai and L. Kučera, *Canonical labelling of graphs in linear average time,* Proc. 20th IEEE FOCS (1979), 49–46.
6. L. Babai and E.M. Luks, *Canonical labeling of graphs,* Proc. 15th ACM STOC (1983), 171–183.
7. L. Babai, E.M. Luks and Á. Seress, *Computing composition series in primitive groups,* in these Proceedings.
8. L. Babai and S. Moran, *Arthur-Merlin games: A randomized proof system and a hierarchy of complexity classes.* J. Comp. Sys. Sci. 36 (1988), 254–276.
9. G. Butler, *On computing double coset representatives in permutation groups,* in Computational Group Theory, M.D. Atkinson, ed., Academic Press, 1984, 283–290.
10. G. Cooperman, L. Finkelstein and E. Luks, *Reduction of group constructions to point stabilizers,* Proc. 1989 ACM-SIGSAM Intern. Symp. on Symbolic and Algebraic Comp., (1989), 351-356.
11. M.L. Furst, J. Hopcroft and E.M. Luks, *Polynomial time algorithms for permutation groups,* Proc. 21st IEEE FOCS (1980), 36–41.
12. M.R. Garey and D.S. Johnson, Computers and Intractability, A Guide to the Theory of NP-completeness, W.H. Freeman, 1979.
13. O. Goldreich, S. Micali and A. Wigderson, *Proofs that yield nothing but their validity and a methodology of cryptographic protocol design,* Proc. 27th IEEE FOCS, 1986, 174–187.
14. M. Hall, Jr., The Theory of Groups, Macmillan, New York 1959.
15. C.M. Hoffmann, Group Theoretic Algorithms and Graph Isomorphism, Lect. Notes in Comp. Sci. 136, Springer 1982.
16. W.M. Kantor, *Sylow's theorem in polynomial time,* J. Comp. Syst. Sci. 30 (1985) 359-394.
17. W.M. Kantor, *Finding Sylow normalizers in polynomial time* J. Algorithms 11 (1990), 523–563.
18. W.M. Kantor and and E.M. Luks, *Computing in quotient groups,* Proc. 22nd ACM STOC, 1990, 524–563.
19. J. Leon, *Computing automorphisms of combinatorial objects,* in Computational Group Theory, M.D. Atkinson, ed., Academic Press, 1984, 321–335.
20. E.M. Luks, *Isomorphism of graphs of bounded valence can be tested in polynomial time,* J.Comp. Sys. Sci. 25 (1982), 42–65.
21. E.M. Luks, *Computing the composition factors of a permutation group in polynomial time,* Combinatorica 7 (1987), 87–99.
22. E.M. Luks, *Computing in solvable matrix groups,* Proc. 33rd IEEE FOCS, 1992, 111–120.
23. B. McKay, *Nauty Users Guide (Version 1.5),* Tech. Rep. TR-CS-90-02, Dept. Comp. Sci., Austral. Nat. Univ. 1990.
24. G.L. Miller, *Graph isomorphism, general remarks,* J. Comp. Sys. Sci. 18 (1979), 128–142.
25. G.L. Miller, *Isomorphism of k-contractible graphs, a generalization of bounded valence and bounded genus,* Inform. and Control 56 (1983), 1–20.
26. G.L. Miller, *Isomorphism of graphs which are pairwise k-separable,* Inform. and Control 56 (1983), 21–33.
27. P.M. Neumann, *Some algorithms for computing with finite permutation groups,* Proc. of Groups-St Andrews 1985 (Eds. E.F. Robertson and C.M. Campbell), London Math. Soc. Lect. Note 121, Cambridge U. Press 1987, 59–92.
28. R.C. Read and D.G. Corneil, *The graph isomorphism disease,* J. Graph Theory, 3 (1979), 339–363.

29. C.C. Sims, *Some group-theoretic algorithms,* Springer Lect. Notes in Math. 697 (1978), 108–124.
30. H. Wielandt, Finite Permutation Groups, Acad. Press, New York 1964.
31. V. Zemlyachenko, N. Kornienko, R. Tyshkevich, *Graph isomorphism problem* (Russian), The Theory of Computation I, Notes Sci. Sem. LOMI 118 (1982).

DEPARTMENT OF COMPUTER AND INFORMATION SCIENCE, UNIVERSITY OF OREGON, EUGENE, OR 97403
E-mail address: luks@cs.uoregon.edu

DIMACS Series in Discrete Mathematics
and Theoretical Computer Science
Volume **11**, 1993

Parallel Computation of Sylow Subgroups in Solvable Groups

PETER D. MARK

ABSTRACT. Sylow's theorem is a fundamental tool in group-theoretic investigations. In computational group theory there is an important role for efficient constructive analogs of Sylow's theorem. For computational purposes, we assume that a group is given by a set of permutations that generate it. This leads to the following problems:

SYLFIND(p, G)

GIVEN: a prime p and generators for a group G,
FIND: generators for a Sylow p-subgroup $P \leq G$.

SYLCONJ(P_1, P_2, G)

GIVEN: generators for G and two of its Sylow p-subgroups P_1 and P_2,
FIND: an element $g \in G$ for which $P_1^g = P_2$.

SYLEMBED(P, G)

GIVEN: generators for G and for a p-subgroup $P \leq G$,
FIND: generators for a Sylow p-subgroup of G containing P.

This paper shows SYLFIND, SYLCONJ, and SYLEMBED for solvable groups are in the complexity class NC; namely, they are solvable in polylogarithmic time ($O(\log^c n)$ steps) using a polynomial number of processors working in parallel. A future paper by Kantor, Luks, and Mark extends these results to general groups.

1. Introduction

Over the last two decades, sequential, polynomial-time algorithms have been found for a variety of basic problems concerning permutation groups, including determining the order, testing membership, and finding some of the important subgroups such as centers, commutator subgroups, and Sylow subgroups. Sylow

---

1991 *Mathematics Subject Classification.* Primary 20B40,68Q25; Secondary 20D20,68Q22.
The author was supported in part by NSF Grant CCR-9013410 while he was a research assistant at the University of Oregon.
This is the final version of this paper.

subgroups are a crucial tool in [**6**] for finding centers in quotient groups and for various other problems (see also [**11**]). Also, a restricted version of SYLFIND played a role in Luks's polynomial time algorithm for bounded valence graph isomorphism testing [**8**]. In a recent series of papers, Kantor gave polynomial-time solutions for the Sylow problems SYLFIND and SYLCONJ [**3, 4, 5**]. His methods exploit a well-developed library of polynomial-time procedures for permutation groups and consequences of the classification of finite simple groups.

Inspired by new generations of machines, new theoretical models were developed to describe and analyze parallel computation. In particular, the class NC gives a useful framework in which to study the inherent parallelizability and logical structure of computations independent of any interprocessor connection network. In the case of permutation group algorithms, many of the known techniques appeared to depend on inherently sequential methods. In a series of papers [**1, 10, 12, 14**], entirely new machinery was developed for permutation group computation. It ultimately brought a sizable portion of the collection of polynomial-time problems, including finding composition factors, into NC. However, a number of critical questions remained open. As pointed out in [**1**], a leading one of these was the parallelization of Sylow subgroup computations. This paper presents efficient parallelizations of these computations for solvable groups. These are special cases of methods of Kantor, Luks, and Mark that solve the same problems for general permutation groups [**7, 13**].

Most of the sequential permutation group algorithms exploit a tower of subgroups, $G = G_0 \geq G_1 \geq \cdots \geq 1$, that is either a tower of pointwise set stabilizers or a composition series for G [**2, 3, 4, 19**]. Membership testing, for example, uses a tower of point stabilizers and reduces membership testing for G_i to membership testing for G_{i+1} [**19**]. Kantor's polynomial-time sequential Sylow algorithms [**4**] use a composition series [**9**] for G.

Both of these towers can have length linear in the degree n of G, and hence may be too long for computation in NC. The parallelization of order and membership testing as well as the new NC Sylow algorithms utilize a different (normal) tower $G = K_0 \rhd K_1 \rhd \cdots \rhd K_r = 1$ for G whose length r is polylogarithmic in n. The quotients K_i/K_{i+1} of successive groups in the tower are semisimple (direct products of simple groups). The existence and NC-constructibility of such a normal tower are demonstrated in [**1**] and [**10**].

Groups in this tower arise from two different divide and conquer strategies. The first involves only a naive construction of a *structure forest* [**12**], a data structure based on the orbits and imprimitivity blocks of the group (see also [**11**]). This reduces the tower construction to the primitive group case, which requires a second divide and conquer approach based on the internal structure of primitive groups, Luks's composition factors algorithm, and other consequences of the classification of finite simple groups. The tower may be refined so that successive quotients K_i/K_{i+1} are direct products of simple groups that are either all abelian or all nonabelian. In the abelian levels (and hence for solvable groups, in which all levels are abelian) the Sylow problems reduce to the problem of solving systems of linear equations over finite fields, which has been shown to be in NC by Mulmuley [**15**]. In nonabelian levels, parallelism arises for many problems in a natural way by working within each simple group independently.

In this paper, we focus on the Sylow algorithms for solvable groups. However, we follow an approach that works for Sylow subgroup computations in solvable quotient groups as well. This plays an important role in the extension of the present results to general permutation groups [**7, 13**]. Moreover, although the nonsolvable case demands deeper group theory, including case analyses based on the classification of the finite simple groups, it exhibits much of the logical structure of the present paper.

One can describe the algorithm for SYLFIND as the computation of a tower of subgroups $G = P_0 \geq P_1 \geq \cdots \geq P_r$ where P_i/K_i is a Sylow p-subgroup of G/K_i and $G = K_0 \rhd K_1 \rhd \cdots \rhd K_r = 1$ is a semisimple tower of polylogarithmic length as described above. Such a semisimple tower is available, indeed it is intrinsic to the basic machinery for NC computation, including membership-testing [**1, 12**]. However, it simplifies our exposition to avoid repeated and explicit reference to a precomputed tower. Thus, we compute suitable K_i as they are needed.

Let $K \lhd H$ where H/K is abelian semisimple. An important ingredient in parallel permutation group computations is the ability to represent H/K in a computationally efficient manner as a product of vector spaces. Once effective representations for such quotients are available, two particular problems emerge: finding a Sylow p-subgroup P of a group H where $K \lhd H \lhd G$, K is an elementary abelian p'-group, and H/K is an elementary abelian p-group, and finding a subgroup $G^* \leq G$ that normalizes P and contains a Sylow p-subgroup of G. The latter problem is an algorithmic form of the Frattini argument. These permutation group problems are transformed into solving systems of linear equations over finite fields.

2. Preliminaries

2.1. NC Concepts and Results. The complexity class NC is the set of problems which can be solved using a polynomial ($O(n^k)$) number of processors in polylogarithmic time ($O(\log^c n)$ steps) where k and c are constants and n is the input size. Processors communicate via shared memory. This model of computation, known as a PRAM and introduced in [**16**], has become a standard tool for analyzing a problem's inherent parallelizability.

Permutation-group problems require three basic permutation operations: computing the product of two permutations, computing inverses of permutations, and computing large powers of permutations. The first two are straightforward. To form a power α^b of a permutation α, where b is represented in binary (in practice, b may be $O(n!)$ where n is the degree of α), we form α^b

independently on each cycle of α by reducing b modulo the cycle length. Note that, for sequential computation, one does not need to emphasize such powering as a "primitive" operation, for it is accomplished by repeated-squaring, therefore by a polynomial number of multiplications. However, the number of successive squarings would be prohibitive for an NC result.

We make use of NC algorithms for the following permutation group problems that are described in [**12**]. In all of the following problems, we assume G is given by a set of generating permutations: $G = \langle \mathcal{S} \rangle \leq \mathrm{Sym}(\Omega)$.

PROBLEM 2.1.1. MEMBERSHIP(G, x)

GIVEN: a permutation group $G \leq \mathrm{Sym}(\Omega)$ and an element $x \in \mathrm{Sym}(\Omega)$,
DETERMINE: whether or not $x \in G$.

PROBLEM 2.1.2. COMMUTATOR(G)

GIVEN: a permutation group $G = \langle \mathcal{S} \rangle \leq \mathrm{Sym}(\Omega)$,
FIND: the commutator subgroup G' of G.

G' is the normal closure of the set $\{[s,t] \mid s,t \in \mathcal{S}\}^G$; an NC algorithm for computing normal closures in solvable groups is given in [**12**].

PROBLEM 2.1.3. FACTOR(x, H, K)

GIVEN: $H, K \leq \mathrm{Sym}(\Omega)$ where $K = \langle Y \rangle$ is normalized by $H = \langle X \rangle$, and $x \in HK$,
FIND: elements $h \in H$ and $k \in K$ such that $x = hk$.

SKETCH OF THE ALGORITHM: The NC membership testing algorithm given in [**1**] constructs x from the set $X \cup Y$ of generators of HK. Following the same construction but substituting the identity for the elements of Y yields h.

2.2. Definitions and Basic Concepts. A group is *semisimple* if it is the direct product of simple groups; it is *abelian semisimple* if each of its simple factors is abelian. Semisimple groups play an important role in the NC algorithms for SYLFIND and SYLCONJ in general groups in [**7**]; in the present context of solvable groups, we concern ourselves only with abelian semisimple groups. Since an abelian semisimple group is a direct product of elementary abelian groups for various primes, we may view it as a direct product of vector spaces over different prime fields. For convenience, we refer to such a group as a *generalized vector space*.

A *generalized basis* for a generalized vector space $H = P_1 \times \cdots \times P_l$ (assume P_i is an elementary abelian p_i-group with $p_i \neq p_j$ for $i \neq j$) is the union $\cup_i \mathcal{B}_i$ where $\mathcal{B}_i$ is a basis for P_i. Similarly, if the quotient H/K is abelian semisimple, a subset $\mathcal{B} \subset H$ is called a *generalized basis for H modulo K* if $\{Kb \mid b \in \mathcal{B}\}$ is a generalized basis for H/K.

LEMMA 2.2.1. *Given groups $K \lhd H = \langle \mathcal{T} \rangle$ with H/K abelian semisimple, a generalized basis $\mathcal{B}$ for H modulo K can be found in NC.*

PROOF. Let $H/K = P_1/K \times \cdots \times P_l/K$ where each P_i/K is an elementary abelian p_i-group and $p_i \neq p_j$ if $i \neq j$. Let $\pi = \prod_{i=1}^{l} p_i$, let $\pi_i = \pi/p_i$, and let $\mathcal{C}_i = \{t^{\pi_i} \mid t \in \mathcal{T}\}$ for each $i = 1, \ldots, l$. Then $P_i = \langle \mathcal{C}_i, K \rangle$ To obtain a basis $\mathcal{B}_i$ of P_i modulo K for all $i = 1, \ldots, l$ in parallel, suppose $\mathcal{C}_i = \{t_1, \ldots, t_m\}$, and let $\mathcal{B}_i = \{t_j \in \mathcal{C}_i \mid t_j \notin \langle t_1, \ldots, t_{j-1}, K \rangle\}$. These membership tests are each in NC (Problem 2.1.1) and may be performed in parallel. $\mathcal{B} = \cup_{i=1}^{l} \mathcal{B}_i$ is a generalized basis for H modulo K. □

We use F_q^d to denote the d-dimensional vector space over F_q, the field of q elements, and $\{e_1, \ldots, e_d\}$ to denote the standard basis. If H/K is an abelian semisimple group, a *generalized vector space representation* for H/K is a pair (V, ϕ) consisting of a generalized vector space V together with an epimorphism $\phi : H \to V$ with kernel K. A generalized vector space representation (V, ϕ) for H/K is *NC-effective* if ϕ is an *NC-computable function*, i.e. there is an NC procedure that can compute $\phi(h)$ for any given $h \in H$.

LEMMA 2.2.2. *Given groups $K \lhd H$ where H/K is abelian semisimple, an NC-effective generalized vector space representation (V, ϕ) for H/K can be found in NC. Moreover, for any $v \in V$, a preimage of v in H, i.e. an element $h \in H$ for which $\phi(h) = v$, can be found in NC.*

PROOF. Suppose $H/K = P_1/K \times \cdots \times P_l/K$ where $P_i/K \cong V_i = F_{p_i}^{d_i}$, for each $i = 1, \ldots, l$ and the primes p_i are distinct. Let $V = V_1 \times \cdots \times V_l$. By Lemma 2.2.1, a generalized basis $\mathcal{B} = \cup_{i=1}^{l} \mathcal{B}_i$ for H modulo K may be found in NC, where $\mathcal{B}_i = \{b_{i1}, \ldots, b_{id_i}\}$ is a basis for P_i modulo K. Let $\{e_{i1}, \ldots, e_{id_i}\}$ be the standard basis of V_i. We specify an NC-computable function $\phi : H \to V$ as follows. Let h be an element of H. For all b_{ij}, $1 \leq i \leq l$, $1 \leq j \leq d_i$, in parallel: test, for all a, $0 \leq a \leq p_i - 1$, in parallel, whether $h^{-1} \cdot b_{ij}^{a} \in \langle \mathcal{B} \setminus \{b_{ij}\}, K \rangle$ (tested using Problem 2.1.1) and let a_{ij} be the unique a satisfying this condition. Then $h \equiv \prod_{i,j} b_{ij}^{a_{ij}} \pmod{K}$. Define $\phi(h) = \sum_{ij} a_{ij} e_{ij}$. Hence the map $\phi : H \to V$ is an NC-computable function with kernel K. The pair (V, ϕ) is an NC-effective representation for H/K.

Furthermore, for any element $v = \sum_{ij} a_{ij} e_{ij} \in V$, the element $h = \prod_{i,j} b_{ij}^{a_{ij}} \in H$ satisfies $\phi(h) = v$. Hence we may compute preimages in H of elements of V in NC.□

We also require the ability to perform basic operations of linear algebra within generalized vector spaces in NC. A *generalized linear transformation* of a generalized vector space $V = V_1 \times \cdots \times V_d$ is a direct product of linear transformations $L = L_1 \times \cdots \times L_d$, where each L_i is a linear transformation of V_i. Hence $Lv = (L_1 v_1, \cdots, L_d v_d)$, where $v = (v_1, \cdots, v_d) \in V$.

LEMMA 2.2.3. *Given a generalized linear transformation $L = L_1 \times \cdots \times L_d$ of a generalized vector space $V = V_1 \times \cdots \times V_d$ and an element $b = (b_1, \ldots, b_d) \in V$, the set of solutions of $Lx = b$ can be found in NC.*

PROOF. Each set $X_i = \{x_i \mid L_i x_i = b_i\}$ may be found in NC using [**15**]. Then the set $X_1 \times \cdots \times X_d \subseteq V$ is the set we seek. □

A set of equations of the form described in Lemma 2.2.3 is called a system of *generalized linear equations.*

For a prime p and a group G, let $R_p(G)$ denote the smallest normal subgroup $H \trianglelefteq G$ for which G/H is an elementary abelian p-group, and let $R_{\mathcal{A}}(G)$ denote the smallest normal subgroup $H \trianglelefteq G$ for which G/H is abelian semisimple. Note that $R_{\mathcal{A}}(G) < G$ for any solvable $G \neq 1$. Let $O^p(G)$ denote the smallest normal subgroup $H \trianglelefteq G$ such that G/H is a p-group. Let $R_p^0(G) = G$ and $R_p^{i+1}(G) = R_p(R_p^i(G))$. Let $d_p(G)$ denote the smallest integer r for which $R_p^r(G) = R_p^{r+1}(G)$. Note that $O^p(G) = R_p^r(G)$, where $r = d_p(G)$. Similarly, let $R_{\mathcal{A}}^0(G) = G$ and $R_{\mathcal{A}}^{i+1}(G) = R_{\mathcal{A}}(R_{\mathcal{A}}^i(G))$. Let $d_{\mathcal{A}}(G)$ denote the smallest integer r for which $R_{\mathcal{A}}^r(G) = R_{\mathcal{A}}^{r+1}(G)$. Note that if G is solvable, $R_{\mathcal{A}}^r(G) = 1$, where $r = d_{\mathcal{A}}(G)$.

PROPOSITION 2.2.4. *For a solvable group $G \leq \mathrm{Sym}(\Omega)$, the tower $G > R_{\mathcal{A}}(G) > R_{\mathcal{A}}^2(G) > \cdots > R_{\mathcal{A}}^s(G) = 1$ has length s* logarithmic *in $|\Omega|$, i.e. $s \leq c \log |\Omega|$ for some constant c.*

PROOF. Since $d_{\mathcal{A}}(G) \leq \max_{\Delta \in \mathcal{O}}\{d_{\mathcal{A}}(G^{\Delta})\}$ where $\mathcal{O}$ is the set of G-orbits, we may assume without loss of generality that G is transitive. If G is primitive, then $|G| < |\Omega|^{3.25}$, by [**17, 20**], hence $d_{\mathcal{A}}(G)$ is logarithmic in $|\Omega|$. If G is imprimitive, let G^* be the primitive action on $m > 1$ blocks of size n/m and let K be the kernel of this action. Inductively assuming the result is true for groups of degree smaller than $|\Omega|$, we have $d_{\mathcal{A}}(G) \leq d_{\mathcal{A}}(G^*) + d_{\mathcal{A}}(K) \leq c \log m + c \log(n/m) = c \log(n)$. □

COROLLARY 2.2.5. *For a solvable group $G \leq \mathrm{Sym}(\Omega)$, the tower*

$$G \rhd R_{\mathcal{A}}O^p(G) \rhd (R_{\mathcal{A}}O^p)^2(G) \rhd \cdots \rhd (R_{\mathcal{A}}O^p)^s(G) = 1$$

has length s logarithmic in $|\Omega|$.

PROOF. One shows that $(R_{\mathcal{A}}O^p)^i(G) \leq R_{\mathcal{A}}^i(G)$.□

COROLLARY 2.2.6. *For a solvable group $G \leq \mathrm{Sym}(\Omega)$, $d_p(G)$ is logarithmic in $|\Omega|$.*

PROOF. The quotient $G/O^p(G)$ is isomorphic to a homomorphic image $\overline{P}$ of a Sylow p-subgroup P of G. Hence $d_p(G) = d_p(G/O^p(G)) = d_p(\overline{P}) \leq d_p(P)$. But $d_p(P) = d_{\mathcal{A}}(P)$, which is logarithmic in $|\Omega|$ by Proposition 2.2.4. □

LEMMA 2.2.7. *Given a permutation group $G = \langle S \rangle \leq \mathrm{Sym}(\Omega)$, the subgroups $R_p(G), R_{\mathcal{A}}(G)$, and $O^p(G)$ can each be computed in NC.*

PROOF. $R_p(G) = \langle G', \{s^p \mid s \in \mathcal{S}\}\rangle$ and $R_{\mathcal{A}}(G) = \langle G', \{s^q \mid s \in \mathcal{S}\}\rangle$, where q is the product of the primes that divide $|G|$, so each of these groups may be found in NC (see Problem 2.1.2). Since $O^p(G) = R_p^r(G)$ where $r = d_p(G)$, $O^p(G)$ may be found by computing $R_p^i(G)$ for $i = 1, \ldots, r$ sequentially. (This is an NC computation since r is logarithmic in $|\Omega|$ by Corollary 2.2.6.) □

3. The Sylow Algorithms for Solvable Groups

3.1. A Base Case and the Frattini Argument. Finding and conjugating Sylow subgroups of solvable groups give rise to two particular subproblems: finding a Sylow p-subgroup P of a group H where $K \lhd H \lhd G$, K is an elementary abelian p'-group, and H/K is an elementary abelian p-group, and finding a subgroup $G^* \leq G$ that normalizes P and contains a Sylow p-subgroup of G. The latter problem is an algorithmic form of the "Frattini argument" ([**18**, p. 61]). This section describes procedures BASECASE1 and FRATTINI for these problems.

The essential technique in both these procedures is to transform a group theoretic condition into the problem of solving systems of generalized linear equations over finite fields, for which there exist NC algorithms (Lemma 2.2.3). Linear transformations arise where K and H are both normal subgroups of another group G, H/K is abelian semisimple, and (V, ϕ) is an NC-effective generalized vector space representation for H/K. Then each element $g \in G$ induces a generalized linear transformation on V denoted T_g, given by $\phi(x) \mapsto \phi(g^{-1}xg)$.

PROBLEM 3.1.1. *BASECASE1(H, K, L)*

GIVEN: $L \trianglelefteq K \trianglelefteq H \leq \mathrm{Sym}(\Omega)$, $L \trianglelefteq H$, H/K is an elementary abelian p-group, and K/L is an abelian semisimple p'-group,

FIND: a group P for which $L \leq P \leq H$ and P/L is a Sylow p-subgroup of H/L.

PROPOSITION 3.1.2. *BASECASE1 is in NC.*

PROOF. If P is such a group then $H = PK$ and, since $P \cap K = L$, $H/K \cong P/L$. In particular, P/L is elementary abelian. Suppose $H = \langle \mathcal{S} \rangle$. Then for each $s \in \mathcal{S}$, there exists $x_s \in K$ such that $sx_s \in P$. Hence,

(i) for all $s \in \mathcal{S}$, $(sx_s)^p \in L$

(ii) for all $s, t \in \mathcal{S}$, $[sx_s, tx_t] \in L$.

Conversely, if $\{x_s \mid s \in \mathcal{S}\} \subseteq K$ satisfies (i) and (ii), then we can take $P = \langle \{sx_s \mid s \in \mathcal{S}\} \rangle$.

Let $\phi : K \to V$ induce a generalized vector space representation of K/L (see Lemma 2.2.2) and for $g \in G$, let T_g be defined as above. Since $[sx_s, tx_t] = x_s^{-1}(x_t^{-1})^s[s,t]x_s^t x_t$,

$$\phi([sx_s, tx_t]) = -\phi(x_s) - T_s\phi(x_t) + \phi([s,t]) + T_t\phi(x_s) + \phi(x_t).$$

Thus $[sx_s, tx_t] \in L$ if and only if each pair of elements in $\{\phi(x_s) \mid s \in \mathcal{S}\}$ satisfies the following system of $|\mathcal{S}|^2$ generalized linear equations:

$$\forall\, s,t \in \mathcal{S},\ (T_t - I)X_s - (T_s - I)X_t = -\phi([s,t]).$$

Hence, a set $\{x_s \mid s \in \mathcal{S}\}$ satisfying (ii) is obtained by solving this system for $\{X_s\} \subseteq V$ (see Lemma 2.2.3) and letting x_s be a preimage of X_s in K (see Lemma 2.2.2).

We can modify this set to satisfy (i), while maintaining (ii), by replacing each x_s by $s^{-1}(sx_s)^m$ where m is chosen so that $|K/L|$ divides m and $m \equiv 1 \pmod p$. □

Recall the *Frattini argument* (see [**18**, p. 61]): if $P \leq K \trianglelefteq G$ with P a Sylow p-subgroup of K then $G = N_G(P)K$, where $N_G(P)$ is the normalizer of P in G.

PROBLEM 3.1.3. *FRATTINI(G, H, K, L, P)*

GIVEN: $L \lhd K \lhd H \trianglelefteq G \leq \mathrm{Sym}(\Omega)$, each of L, K is also normal in G, K/L is an abelian semisimple p'-group, H/K is a p-group, and P/L is a Sylow p-subgroup of H/L,

FIND: a subgroup $G^* \leq G$ that normalizes P and contains P for which G^*/L contains a Sylow p-subgroup of G/L.

PROPOSITION 3.1.4. *FRATTINI is in NC.*

PROOF. We are given $G = \langle \mathcal{S} \rangle$, $P = \langle \mathcal{T} \rangle$. By the Frattini argument, for any $s \in \mathcal{S}$ there exists $x_s \in K$ such that $sx_s \in N_G(P)$. For any such collection, $\{x_s \mid s \in \mathcal{S}\}$, we can take $G^* = \langle P, \{sx_s \mid s \in \mathcal{S}\}\rangle$; to see that G^*/L contains a Sylow p-subgroup of G/L, we observe that $|G^*K/L| = |G/L|$, but the p-part of $|G^*K/L|$ equals the p-part of $|G^*/L|$ since $(|K/L|, p) = 1$.

To find such x_s (in parallel for each $s \in \mathcal{S}$) it suffices to ensure that $t^{sx_s} \in P$ for each $t \in \mathcal{T}$. For each $t \in \mathcal{T}$, write $t^s = a_t k_t$, with $a_t \in P, k_t \in K$ (see Problem 2.1.3). The required x_s must therefore satisfy $a_t^{x_s} k_t \in P$. But, $a_t^{x_s} k_t = a_t[a_t, x_s]k_t$ and $[a_t, x_s]k_t \in K$. Since $P \cap K \leq L$, the condition on x_s is equivalent to $[a_t, x_s]k_t \in L$, or, if $\phi : K \to V$ induces a vector-space representation of K/L, to

$$\phi(x_s) - \phi(x_s^{a_t}) + \phi(k_t) = 0.$$

Therefore, $\phi(x_s)$ is a solution to the system of $|\mathcal{T}|$ generalized linear equations

$$\forall t \in \mathcal{T},\ (I - T_{a_t})X + \phi(k_t) = 0,$$

where T_{a_t} is defined just before Problem 3.1.1. Hence, x_s is obtained by solving this system for $X \in V$ (see Lemma 2.2.3) and letting x_s be a preimage of X in K (see Lemma 2.2.2). □

3.2. SYLFIND. Before giving a procedure for SYLFIND, we describe another special case.

PROBLEM 3.2.1. *BASECASE2*(G, K, L, p)

GIVEN: a solvable group G and $L \trianglelefteq K \trianglelefteq G \leq \mathrm{Sym}(\Omega)$, where $L \triangleleft G$, G/K is a p-group, and K/L is an abelian semisimple p'-group,
FIND: a subgroup $P \leq G$ containing L for which P/L is a Sylow p-subgroup of G/L.

PROPOSITION 3.2.2. *BASECASE2 is in NC.*

PROOF. Let $G_1 = R_p(G)L$ (see Lemma 2.2.7). Note that $K \leq G_1$. If $G_1 = G$, then $G = K$, and $|G/L| = |K/L|$ is relatively prime to p so G/L has no nontrivial Sylow p-subgroup. In this case, return $P = L$.

Otherwise, recursively find P_1 = BASECASE2(G_1, K, L). The group G^* = FRATTINI(G, G_1, K, L, P_1) normalizes P_1 and contains a Sylow p-subgroup of G. We seek a group P for which P/P_1 is a Sylow p-subgroup of G^*/P_1. Find the group $U = R_p(G^*)L$ (see Lemma 2.2.7) . By definition, G^*/U is an elementary abelian p-group. Also note that U/P_1 is an elementary abelian p'-group since, letting $W = U \cap K$, we have $U/P_1 = P_1W/P_1 \cong W/(P_1 \cap W) = W/L \leq K/L$, which is an elementary abelian p'-group. Hence BASECASE1 (see Problem 3.1.1) applies, and P = BASECASE1(G^*, U, P_1) is the group we seek.

In each recursive call, $L \leq G_1 < G$, so the number of recursive calls is bounded by $d_p(G/L)$. Since $d_p(G/L) \leq d_p(G)$, which is logarithmic in $|\Omega|$ by Lemma 2.2.6, the algorithm for BASECASE2 is in NC. □

PROBLEM 3.2.3. *SYLFIND*(G, p)

GIVEN: a solvable group $G \leq \mathrm{Sym}(\Omega)$ and a prime p that divides $|G|$,
FIND: a Sylow p-subgroup P of G.

THEOREM 3.2.4. *SYLFIND is in NC.*

PROOF. Let $K = O^p(G)$ (see Lemma 2.2.7) and $L = R_{\mathcal{A}}(K)$ (see Lemma 2.2.7). If $L = K$ (tested using Problem 2.1.1) then $R_p(K) = R_{\mathcal{A}}(K)$, which implies $K = 1$, so return G. Otherwise let P = BASECASE2(G, K, L, p) and recurse by returning SYLFIND(P, p).

To analyze the running time of SYLFIND, note that $L = R_{\mathcal{A}}O^p(G)$ so the group P that is passed in the recursive call satisfies $R_{\mathcal{A}}O^p(P) = R_{\mathcal{A}}O^p(L) = (R_{\mathcal{A}}O^p)^2(G)$. Hence the depth of the recursion is logarithmic by Lemma 2.2.5.

Moreover, the procedures BASECASE1 and FRATTINI are in NC. Hence SYLFIND is in NC. □

3.3. SYLCONJ and SYLEMBED. We now describe a procedure SYLCONJ-EMBED which can be used for conjugating Sylow p-subgroups and embedding a p-subgroup into a Sylow p-subgroup. Specifically, we obtain SYLCONJ as a special case of SYLCONJ-EMBED by letting P_1 and P_2 both be Sylow p-subgroups of G. Similarly, SYLEMBED can be implemented by first letting $P_2 = \text{SYLFIND}(G, p)$, then setting $g = \text{SYLCONJ-EMBED}(G, P_1, P_2)$ and returning $P_2^{g^{-1}}$, a Sylow p-subgroup of G containing P_1.

PROBLEM 3.3.1. *SYLCONJ-EMBED*(G, P_1, P_2)

GIVEN: a solvable group $G \leq \text{Sym}(\Omega)$, a p-subgroup P_1 of G, and a Sylow p-subgroup P_2 of G,

FIND: an element $x \in G$ for which $P_1^x \leq P_2$.

THEOREM 3.3.2. *SYLCONJ-EMBED is in NC.*

PROOF. Find $H = O^p(G)$ (see Lemma 2.2.7), so that $P_1H/H \leq G/H = P_2H/H$. Find $K = R_{\mathcal{A}}(H)$ (see Lemma 2.2.7). If $K = H$, then $H = 1$ and $P_1 \leq P_2 = G$. In this case, return $x = 1$.

We may assume $K < H$. We first show that we can find $x \in H$ such that $P_1^x \leq P_2K$. Suppose $P_1 = \langle \mathcal{S} \rangle$. Since $G = P_2H$, for each $s \in \mathcal{S}$, we can factor $s = bh$ with $b \in P_2, h \in H$ (see Problem 2.1.3). For $x \in H$, $s^x \in P_2K$ if and only if $b^{-1}s^x = (x^{-1})^b x h^x \in P_2K$; since $h^x = h[h, x] \in hK$, this happens if and only if $(x^{-1})^b x h \in P_2K \cap H \leq K$. Thus, if $\phi : H \to V$ induces a generalized vector space representation of H/K (see Lemma 2.2.2), then $s^x \in P_2K$ if and only if $\phi(x)$ is a solution to the system of $|\mathcal{S}|$ linear equations

$$\forall s \in \mathcal{S},\ (I - T_{b_s})X + \phi(h_s) = 0,$$

where $s = b_s h_s$ with $b_s \in P_2, h_s \in H$, and where T_{b_s} is defined just before Problem 3.1.1.

Hence, x is obtained by solving this system for $X \in V$ and taking a preimage of X in H (see Lemma 2.2.2).

Let $G^* = \langle P_1^x, P_2 \rangle$ and recursively solve SYLCONJ-EMBED(G^*, P_1^x, P_2) for $y \in G^*$ such that $(P_1^x)^y \leq P_2$. Return the element xy.

To analyze the running time of SYLCONJ-EMBED, note that the group K computed in SYLCONJ-EMBED is equal to $R_{\mathcal{A}}O^p(G)$. Hence the group G^* that is passed in the recursive call satifies $R_{\mathcal{A}}O^p(G^*) = R_{\mathcal{A}}O^p(K) = (R_{\mathcal{A}}O^p)^2(G)$. Thus, the depth of the recursion is logarithmic by Lemma 2.2.5. Therefore, SYLCONJ-EMBED is in NC. □

4. Conclusion

A natural continuation of this work is the search for efficient parallelizations of other group-theoretic sequential algorithms, for example, the problem of finding Sylow normalizers. This problem has been shown to be in polynomial time by Kantor [**5**]. The ability to compute Sylow normalizers may extend the applicability of the Sylow theorems to parallel computation in ways analogous to its use in sequential algorithms [**6**]. These algorithms may also give rise to more efficient sequential algorithms than those initially given by Kantor.

ACKNOWLEDGEMENTS: The author is indebted to William Kantor and Eugene Luks for guidance in his graduate studies.

REFERENCES

1. Babai, L., Luks, E. M., Seress, Á. *Permutation groups in NC*, Proc. 19th ACM STOC, 1987, 409-420.
2. Furst, M. L., Hopcroft, J. and Luks, E. M. *Polynomial time algorithms for permutation groups*, Proc. 21th IEEE FOCS, 1980, 36-41.
3. Kantor, W. M. *Polynomial time algorithms for finding elements of prime order and Sylow subgroups*, J. Algorithms 6, 1985, 478-514.
4. Kantor, W. M. *Sylow's theorem in polynomial time*, J. Comp. Sys. Sci. 30, 1985, 359-394.
5. Kantor, W. M. *Finding Sylow normalizers in polynomial time*, J. Algorithms 11, 1990, 523-563.
6. Kantor, W. M. and Luks, E. M. *Computing in quotient groups*, Proc. 22nd ACM STOC, 1990, 524-534.
7. Kantor, W.M, Luks, E.M., Mark, P.D., In preparation.
8. Luks, E. M., *Isomorphism of graphs of bounded valence can be tested in polynomial time*, J. Comp. Sys. Sci., 25 1982, 42-65.
9. Luks, E. M., *Computing the composition factors of a permutation group in polynomial time*, Combinatorica 7 (1987), 87-99.
10. Luks, E. M., *Parallel algorithms for permutation groups and graph isomorphism*, Proc. 27th IEEE FOCS, 1986, 292-302.
11. Luks, E.M., *Permutation groups and polynomial-time computation*, in these Proceedings.
12. Luks, E. M. and McKenzie, P. *Parallel algorithms solvable permutation groups*, J. Comp. Sys. Sci., 37 1988, 39-62.
13. Mark, P. D. *Sylow's Theorem and Parallel Computation* Ph.D. Dissertation, University of Oregon, 1993.
14. McKenzie, P. and Cook, S. A. *The parallel complexity of the abelian permutation group membership problem*, Proc. 24th IEEE FOCS, 1983, 1-20.
15. Mulmuley, K. *A fast parallel algorithm to compute the rank of a matrix over an arbitrary field*, Combinatorica 7, 1987, 101-104.
16. Pippinger, N. *On simultaneous resource bounds*, Proc. 20th IEEE FOCS 1979, 307-311.
17. Pálfy, P. P., *A polynomial bound for the orders of primitive solvable groups*, J. Algebra 77, 1982, 127-137.
18. Rotman, J. J. *The theory of groups*, 3rd ed.Allyn and Bacon, 1984.
19. Sims, C. C., *Some group-theoretic algorithms*, in Springer Lecture Notes in Math. Vol 697, 1978, 108-124.
20. Wolf, T. R., *Solvable and nilpotent subgroups of $GL(n, q^m)$*, Can. J. Math. 34, 1982, 1097-1111.

DEPARTMENT OF COMPUTER AND INFORMATION SCIENCE, UNIVERSITY OF OREGON, EUGENE, OR 97403

E-mail address: pdm@cs.uoregon.edu

DIMACS Series in Discrete Mathematics
and Theoretical Computer Science
Volume 11, 1993

Computation with Matrix Groups over Finite Fields

CHERYL E. PRAEGER

ABSTRACT. An approach to developing a suite of algorithms for computing with matrix groups over finite fields is discussed. An algorithm is described in outline for deciding whether the subgroup $\langle X \rangle$ of the general linear group $\mathrm{GL}(d,q)$ generated by a given subset $X \subseteq \mathrm{GL}(q)$ contains the special linear group $\mathrm{SL}(d,q)$.

1. Introduction

For problems requiring computation with groups of matrices over finite fields a subgroup G of the general linear group $\mathrm{GL}(d,q)$ of all nonsingular $d \times d$ matrices over the field $F = \mathrm{GF}(q)$ is usually given as $G = \langle X \rangle$, the group generated by a (usually fairly small) subset $X = \{x_1, \dots, x_k\} \subseteq \mathrm{GL}(d,q)$. A few efficient algorithms for computing with such groups have been implemented, most notably Richard Parker's so-called Meataxe algorithm for deciding whether or not G acts irreducibly on the underlying vector space $V = V(d,q)$, see [**7**]. However there are many important and natural questions about G for which no existing algorithms can provide answers. This paper reports on work, completed, in progress, and prospective, of the author and Peter Neumann. Our aim is to produce a collection of efficient general purpose algorithms for recognising the structure of finite matrix groups G which are large in some sense relative to $\mathrm{GL}(d,q)$.

The general class of groups to which we hope our algorithms will apply are subgroups G of $\mathrm{GL}(d,q) = \mathrm{GL}(d,F)$ with, say, $d \leq 60$ and $q \leq 2^{15}$. The groups $G = \langle X \rangle$ will be given by a generating set $X = \{x_1, \dots, x_k\}$ with k fairly small,

1991 *Mathematics Subject Classification.* Primary 20-04, 20G40.

Supported in part by DIMACS grant STC-88-09648. DIMACS is an NSF Science and Technology Center and a cooperative project of Rutgers University, Princeton University, AT&T Bell Labs and Bellcore

This paper is in final form and no version of it will be submitted elsewhere usually $k \leq 10$ and often $k = 2$. Most of our algorithms will be probabilistic in nature, some will rely on fairly deep group theoretic results including the classification of the finite simple groups, and we will try to give complexity analyses as well as proofs of correctness. We hope to produce a collection of algorithms which, together with the Meataxe, would provide a sequence of tests to be performed on a group G to recognise certain of its properties or certain aspects of its structure.

We analyse the complexity of our algorithms in terms of the numbers of field additions or multiplications required. Since in most applications we have in mind the size q of the field is small, this seems reasonable. In the algorithms we have developed so far, the number of field operations required is a polynomial in $\log q$ and d. However, in order to have a practical implementation of these algorithms for which the performance parallels the theoretical estimates, we need a method for making approximately random selections of elements from a group $G = \langle X \rangle \leq \mathrm{GL}(d, q)$ which produces elements which are approximately uniformly distributed over G. Laci Babai [**2**] considered this problem in a more general context. He showed, in [**2**] Proposition 7.2, how to construct, from a given generating set X for a group G of order at most n, a set of $O(\log n)$ elements, at a cost of $O((\log n)^5)$ group multiplications, from which nearly uniformly distributed random elements of G can be obtained at a cost of $O((\log n)^4)$ group multiplications per random element. For matrix group algorithms the group might have order as large as $O(q^{d^2})$, so the cost of producing the set of $O(\log n)$ elements might be as much as $O(d^{10})$, and that for producing each nearly random element $O(d^8)$ (assuming that q is small). Thus, although this work of Babai is on the verge of giving a sound theoretical basis for practical, approximately random selections from groups, it is not quite good enough to provide a practical procedure for use with algorithms like the one in [**5**] which should run at a cost of $O(d^4)$ or $O(d^5)$.

Often the given generating set X for a subgroup $G \leq \mathrm{GL}(d, q)$ will consist of "very non-random" elements. Obtaining a sequence of elements of G with some hope of reaching most of the group elements (let alone having an approximately uniform distribution) might require the formation of very long words in the generating set X. The paper [**5**] suggested choosing words of the form $x_1^{m_1} \dots x_l^{m_l}$, where $l \leq 2d$, the x_i are randomly chosen elements of X, and the m_i are random integers in the interval $[1, q^d]$. Matrices corresponding to such words can be computed efficiently and, if some elements of X have large order, it is plausible that matrices of this sort might be approximately uniformly distributed. However, Derek Holt and Sarah Rees [**4**] found, in their implementation of the algorithm in [**5**], that this procedure was disproportionately slow in comparison with the rest of the algorithm. Holt and Rees focussed on the problem of finding a selection method for which the probability of a selected element having a certain desirable property (namely being irreducible or nearly irreducible, as described in Section 2 below) was almost the same as the proportion of such elements in the whole group. Computer simulations (see [**4**], Section 3) suggested that (for use with this algorithm for $d \leq 60$) it was reasonable to augment X to a set X' by adding a small number of matrices corresponding to randomly chosen words of length $O(d)$ in X, and then to make selections of random words in X' of length 5 to 7.

This problem of obtaining an approximately uniformly distributed sequence of elements from a group $G = \langle X \rangle \leq \mathrm{GL}(d,q)$ is of fundamental importance for the successful implementation of probabilistic algorithms for matrix groups. Sensible criteria for deciding on a practical procedure to use, as well as sound theoretical analyses to justify the choice, are required.

Consider now what tests might be applied to a group $G = \langle X \rangle \leq \mathrm{GL}(d,q)$. The Parker Meataxe algorithm might first be used to test whether G is irreducible on $V = V(d,q)$. This algorithm requires $O(kd^3)$ steps and will correctly verify the irreducibility of G on V with probability $1-\epsilon$, where ϵ is very small but is not easy to estimate. (With high probability an element of the group ring FG with non-zero null space in V or in the dual V^* is found, and non zero vectors of this null space are tested to see if their FG-span is a proper subspace. If the null space is too large for a complete search to be feasible there is a chance that the test may not reveal a proper FG-invariant subspace for a reducible group G, see [**7**].)

Now suppose that, after performing the test for irreducibility with the Meataxe, we are reasonably confident that G is irreducible on V. We might next try to determine whether or not G contains the special linear group $\mathrm{SL}(d,q)$ using the algorithm described in detail in [**5**]. This algorithm requires at worst $O(kd^6)$ steps, though in practice we believe it will run in $O(kd^4)$ steps, and will correctly recognise whether or not G contains $\mathrm{SL}(d,q)$ with high probability. The main ideas behind this algorithm will be outlined in Section 2. So suppose that, with high probability, G is irreducible on V but does not contain $\mathrm{SL}(d,q)$. We next suggest testing whether G preserves a bilinear, sesquilinear or quadratic form on V. Some ideas for such a test will be worked out in detail in [**6**]. If G does preserve such a form we might then test whether G contains the corresponding classical group $\mathrm{Sp}(d,q)$, $\mathrm{SU}(d,q^{1/2})$, or $\Omega^{\epsilon}(d,q)$ where ϵ is $\pm$ or "blank". An algorithm to perform such a test is presently being developed; the general ideas behind it are similar to those used in [**5**].

Where should we go from here? We might wish to test whether the group G preserves on $V(d,q)$ an extension field structure $V(d/b,q^b)$ for some $b>1$ dividing d, and whether $G \geq \mathrm{SL}(d/b,q^b)$. This is a little more difficult than testing whether G is absolutely irreducible on V. One method would involve computing the centraliser of the set X in the ring $M(d,q)$ of all $d \times d$ matrices over $\mathrm{GF}(q)$, and this is certainly feasible for small values of d. Essentially this procedure amounts to solving k systems of simultaneous equations over $\mathrm{GF}(q)$, each a system of d^2 equations in d^2 variables. The cost of this would be almost $O(kd^6)$, and a less expensive method would be preferable.

There are many questions remaining. Which tests are desirable? Which are feasible? How expensive are they, and in what order should they be performed? The reader may be reminded, by the nature of the tests we have suggested, of the list of "geometric" maximal subgroups of $\mathrm{GL}(d,q)$ given by Aschbacher [**1**]. This result of Aschbacher has definitely had a strong influence on our work, both in decisions on which algorithms to work towards and in proving the theoretical results behind the algorithms. However we do not yet have a clear idea of what a complete workable set of algorithms to investigate the structure of the group G might comprise.

There are several problems which we believe may be feasible to solve. We comment on some of them.

1.1. Finding the nonabelian composition factors for G may well be feasible for $d \leq 60$, $q \leq 2^{15}$, $k \leq 10$.

We note that E. Luks is developing algorithms to compute with finite soluble matrix groups, so an answer to this problem may be an important link between the two approaches.

1.2. Finding the non-cyclic chief factors of G could well be feasible for $d \leq 60$, $q \leq 2^{15}$, $k \leq 10$.

1.3. Finding cyclic chief factors of G (for example finding $Z(G)$) is likely to be restricted to $q^d \leq 10^9$.

1.4. Finding a Sylow p-subgroup of G where p is the prime dividing q, will probably be restricted to small d, say $d \leq 10$.

1.5. Finding invariant forms should be feasible for $d \leq 60$, $q \leq 2^{15}$, $k \leq 10$.

1.6. For an irreducible group G testing whether G is imprimitive seems to be difficult.

2. Recognising when G contains $\mathrm{SL}(d,q)$

The basic ideas behind the recognition algorithm [**5**] for the special linear group $\mathrm{SL}(d,q)$ are the following. There is a class $\mathcal{C} = \mathcal{I} \cup \mathcal{N}$ of elements of $\mathrm{GL}(d,q)$ such that, for any group G satisfying $\mathrm{SL}(d,q) \leq G \leq \mathrm{GL}(d,q)$, the proportions of elements of G lying in $\mathcal{I}$ and in $\mathcal{N}$ are relatively large and, for almost all subgroups $G \not\geq \mathrm{SL}(d,q)$, either $G \cap \mathcal{I} = \emptyset$ or $G \cap \mathcal{N} = \emptyset$. Let $\mathcal{L}$ be the list of subgroups $G \not\geq \mathrm{SL}(d,q)$ with $G \cap \mathcal{I} \neq \emptyset$ and $G \cap \mathcal{N} \neq \emptyset$. Several independent random selections of elements of a given subgroup $G \leq \mathrm{GL}(d,q)$ are made. If elements of both $\mathcal{I}$ and $\mathcal{N}$ are found then, after checking that G is not one of the groups in $\mathcal{L}$, we can conclude that $G \geq \mathrm{SL}(d,q)$. On the other hand if, after $5.5d$ independent random selections from G, no element of $\mathcal{I}$ is found, or no element of $\mathcal{N}$ is found, then with probability at least 0.99, G does not contain $\mathrm{SL}(d,q)$ (see [**5**], Lemma 2.7 stated below).

The class $\mathcal{C}$ which we used is the union of two classes $\mathcal{C} = \mathcal{I} \cup \mathcal{N}$, the class $\mathcal{I}$ of irreducible elements and the class $\mathcal{N}$ of nearly irreducible elements. An element x of $\mathrm{GL}(d,q)$ is called *irreducible* if it acts irreducibly on the underlying d-dimensional vector space V over $\mathrm{GF}(q)$ on which $\mathrm{GL}(d,q)$ acts naturally; an element y of $\mathrm{GL}(d,q)$ is said to be *nearly irreducible* if there are y-invariant subspaces V_1 and V_{d-1} of dimensions 1 and $d-1$ respectively such that $V = V_1 \oplus V_{d-1}$ and y is irreducible on V_{d-1}. It was not difficult to show that, for each subgroup G containing $\mathrm{SL}(d,q)$, the proportions of its elements in $\mathcal{I}$ and in $\mathcal{N}$ are sufficiently large that there is a high probability of obtaining elements of G in $\mathcal{I}$ and in $\mathcal{N}$ on repeated random selection from G. However we were not able to get a complete classification of all subgroups G of $\mathrm{GL}(d,q)$ such that both $G \cap \mathcal{I}$ and $G \cap \mathcal{N}$ are non-empty. So we restricted the classes $\mathcal{I}$ and $\mathcal{N}$ a little as follows.

The order $\mathrm{ord}(x)$ of an irreducible element x divides $q^d - 1$ and that of a nearly irreducible element y divides $q^{d-1} - 1$. Moreover it was proved by Zsigmondy [**9**] in 1892 that for a, d positive integers, $a \geq 2$, $d \geq 3$ and $(a,d) \neq (2,6)$, there is a prime dividing $a^d - 1$ which does not divide $a^i - 1$ for any $1 \leq i \leq d-1$. Such a prime is called a *primitive prime divisor* of $a^d - 1$. It is not difficult to see that an element of $\mathrm{GL}(d,q)$ with order divisible by a primitive prime divisor of $q^d - 1$ or $q^{d-1} - 1$ is irreducible or nearly irreducible respectively. However it is possible that elements in $\mathcal{I}$ or $\mathcal{N}$ may have no such primes dividing their orders, for example $\mathrm{GL}(6,2)$ and $\mathrm{GL}(7,2)$ contain elements of order 21 — which are irreducible and nearly irreducible respectively. We shall call an element $x \in \mathcal{I}$ *primitive irreducible* if $\mathrm{ord}(x)$ is divisible by a primitive prime divisor of $q^d - 1$, and similarly we shall call an element $y \in \mathcal{N}$ *primitive nearly irreducible* if $\mathrm{ord}(y)$ is divisible by a primitive prime divisor of $q^{d-1} - 1$. Let $\mathcal{I}_{pr}$ and $\mathcal{N}_{pr}$ denote the sets of all primitive irreducible and primitive nearly irreducible elements of $\mathrm{GL}(d,q)$ respectively. Again it was not difficult to show that, for each subgroup G of $\mathrm{GL}(d,q)$ with $\mathrm{SL}(d,q) \leq G$, the proportions of elements of G in $\mathcal{I}_{pr}$ and in $\mathcal{N}_{pr}$ are large enough to be usable. Also it was possible, using the finite simple group classification, and using Aschbacher's description [**1**] of the subgroups of $\mathrm{GL}(d,q)$, to determine completely the list $\mathcal{L}$ of all subgroups G of $\mathrm{GL}(d,q)$ such that $G \not\geq \mathrm{SL}(d,q)$, and both $G \cap \mathcal{I}_{pr} \neq \emptyset$, and $G \cap \mathcal{N}_{pr} \neq \emptyset$, [**5**] Theorem 1.2. (Christopher Hering (letter dated 25 January 1990), John Walter [**8**] , and Ulrich Dempwolf [**3**] have also looked at this problem.)

For a procedure to check whether $G \geq \mathrm{SL}(d,q)$, given that $G \cap \mathcal{I}_{pr} \neq \emptyset$ and $G \cap \mathcal{N}_{pr} \neq \emptyset$, we therefore needed procedures to check that G was not one of the groups in $\mathcal{L}$. One of the classes of groups in the list $\mathcal{L}$ is the class of groups representable modulo scalars over a subfield, that is groups which are conjugate to a subgroup of $\mathrm{GL}(d,q_0)Z$ for some proper subfield $\mathrm{GF}(q_0)$ of $\mathrm{GF}(q)$, where Z is the subgroup of nonsingular scalar matrices in $\mathrm{GL}(d,q)$. Any special procedures we found to test for such subgroups were awkward and expensive, so we found it best to restrict further the classes $\mathcal{I}$ and $\mathcal{N}$. We say that an irreducible element x or a nearly irreducible element y is *exploitable* if it is primitive and is not definable modulo scalars over a proper subfield; if (d, q) is (6,2) or (7,2) then x or y is said to be exploitable if it is irreducible or nearly irreducible respectively and is not definable modulo scalars over a proper subfield. Let $\mathcal{I}_e$ and $\mathcal{N}_e$ be the sets of exploitable irreducible elements and exploitable nearly irreducible elements respectively. The classification of subgroups of $\mathrm{GL}(d, q)$ meeting $\mathcal{I}_e$ and $\mathcal{N}_e$ nontrivially produces a very short list $\mathcal{L}$ of (conjugacy classes of) "exceptional subgroups". We state the result for $d \geq 4$; the cases $d = 2$ and $d = 3$ are discussed in detail in [**5**] Section 8.

THEOREM. *([**5**] Theorem 3) Suppose that $G \leq \mathrm{GL}(d, q)$, where $d \geq 4$, and that $G \cap \mathcal{I}_e \neq \emptyset$ and $G \cap \mathcal{N}_e \neq \emptyset$. Let Z be the subgroup of scalar matrices contained in G. Then one of the following holds:*

(a) $\mathrm{SL}(d, q) \leq G$;
(b) *d is prime and G is conjugate to a subgroup of $\Gamma L(1, q^d)$;*
(c) *$G \cong \mathrm{PSL}(2, r) \times Z$, where $r = 2d + 1$, d and r are both prime, q is a primitive root modulo d, and q has order d modulo r;*
(d) *$G = M_{23}$ or M_{24} in $\mathrm{GL}(11, 2)$;*
(e) *$G \cong M_{11} \times Z$ in $\mathrm{GL}(5, 3)$.*

Notice that, if $d \geq 5$, then in cases (b)–(e) the dimension d is prime and, for $y \in G \cap \mathcal{N}_e$, $y^d \in Z$. Also in cases (c)–(e) the integer $r = 2d + 1$ is a primitive prime divisor of $q^d - 1$ and, for $x \in G \cap \mathcal{I}_e$, $x^r \in Z$. These provide simple preliminary tests for groups in (b)–(e). Detailed procedures for identifying whether a group G with $G \cap \mathcal{I}_e \neq \emptyset$ and $G \cap \mathcal{N}_e \neq \emptyset$ satisfies one of (b)–(e) are given in [**5**] Section 6. Finally, if $G \geq \mathrm{SL}(d, q)$ then the proportions of the elements of G in $\mathcal{I}_e$ and in $\mathcal{N}_e$ are sufficiently large. The following lemma from [**5**] is a formal statement of the probabilities involved.

LEMMA. *([**5**] Lemma 2.7) Suppose that $\mathrm{SL}(d, q) \leq G \leq \mathrm{GL}(d, q)$ and that n independent random selections are made of elements from G. Let S be the set of elements selected. If $d \geq 4$ then the probability that either $S \cap \mathcal{I}_e = \emptyset$ or $S \cap \mathcal{N}_e = \emptyset$ is at most*

$$\left(1 - \frac{1}{d+1}\right)^n + \left(1 - \frac{1}{d}\right)^n.$$

In particular if $n \geq 5.5d$ then this probability is less than 0.01.

REFERENCES

1. M. Aschbacher, *On the maximal subgroups of the finite classical groups*, Invent. Math. **76** (1984), 469–514.
2. L. Babai, *Local expansion of vertex-transitive graphs and random generation in finite groups*, Proc. 23rd ACM STOC (1991), 164–174.
3. U. Dempwolff, *Linear groups with large cyclic subgroups and translation planes*, Rend. Sem. Mat. Univ. Padova **77** (1987), 69–113.
4. Derek F. Holt and Sarah Rees, *An implementation of the Neumann-Praeger algorithm for the recognition of special linear groups*, (preprint) (1992).

5. Peter M. Neumann and Cheryl E. Praeger, *A recognition algorithm for special linear groups*, Proc. London Math. Soc. (to appear).
6. ———, *Recognition algorithms for classical groups*, (in preparation).
7. R. A. Parker, *The computer calculation of modular characters (the Meat-Axe)*, Computational Group Theory, Proc. of the LMS Symposium on Computational Group Theory (M. D. Atkinson, ed.), Academic Press, London, 1984.
8. J. H. Walter, *Rigid cyclic subgroups in Chevalley groups, I*, J. Algebra **131** (1990), 688–702.
9. K. Zsigmondy, *Zur Theorie der Potenzreste*, Monatsh. für Math. u. Phys. **3** (1892), 265–284.

DEPARTMENT OF MATHEMATICS, UNIVERSITY OF WESTERN AUSTRALIA, NEDLANDS WA 6009, AUSTRALIA

E-mail address: praeger@maths.uwa.oz.au

DIMACS Series in Discrete Mathematics
and Theoretical Computer Science
Volume 11, 1993

Asymptotic results for permutation groups

LÁSZLÓ PYBER

ABSTRACT. In this article we attempt to give a systematic survey of asymptotic properties of classes of permutation groups. Various bounds for the number of generators, orders, base sizes and other invariants are considered. We also comment on enumeration type problems and on the notion of a random permutation group.

0. Introduction

Asymptotic group theory is a very old subject. Indeed, one of the important problems of 19th century group theory was to estimate the order of a primitive permutation group not containing the alternating group in terms of its degree n (see **[PS]** for some historical remarks).

The full solution to this and many similar problems was obtained by using the classification of finite simple groups (from now on to be abbreviated by CFSG). Indeed primitive groups of sufficiently large order have been completely classified **[Cam1]** **[Ka1]**. Moreover CFSG offers some hope to give sharp estimates in cases where obtaining an exact answer does not seem to be possible (such as determining the number of permutation groups of a given degree n **[Py1]**).

Apart from the possibility of obtaining strong general results, the revival of asymptotic group theory is largely due to the fact that asymptotic results often have relevance to theoretical computer science (and may even give hints for those who design practical algorithms).

Somewhat surprisingly the search for better algorithms has led to elementary proofs of quite sharp bounds for the orders of primitive groups **[Ba2]** **[Ba3]** **[Py3]** (note that the results in **[Ba2]** **[Ba3]** were obtained essentially simultaneously with the announcement of CFSG).

1991 *Mathematics Subject Classification.* 20B35.
Supported in part by DIMACS.
This is the final version of this paper.

Elementary proofs offer more insight and the ideas involved often have successful algorithmic applications. It seems therefore important to treat results obtained without CFSG separately.

While our main concern here is with permutation groups, results about linear groups and abstract groups will often be mentioned (see [**Ka2**] [**Ma1**] [**Ma2**] for many others).

Let us say a few words about the basic methods. A typical proof starts with the familiar reduction to primitive groups. In many cases the elementary bounds on the orders of primitive groups (see §2) are sufficient to obtain the desired conclusion. Otherwise the O'Nan-Scott theorem (see [**Cam4**] [**KL3**]) is invoked and usually reduces the problem considered to statements about almost simple and affine primitive groups. (G is *almost simple* if $L \leqslant G \leqslant Aut(L)$ for some simple group L. An *affine primitive* group has a regular elementary abelian p-subgroup, i.e., a $GF(p)$ vector space and the stabiliser of the zero vector is an irreducible linear group.) Sometimes further reduction is necessary in the affine case (see [**Sup**] [**As**]). Finally CFSG is invoked.

The most commonly used consequences of CFSG are that if L is a simple group then $Out(L) = Aut(L)/L$ is solvable (the so-called Schreier conjecture), that $|Out(L)|$ is small (much smaller than $|L|$) and that L can be generated by two elements. Information about the minimal degrees of permutation representations of almost simple groups is often useful (see [**KL3**], Chapter 5).

The above methodology breaks down for example in cases where we want to prove something that is non-trivial even for p-groups. We will try to indicate alternative methods. We will also give short proofs of some statements that do not appear elsewhere.

Several unsolved problems are mentioned in the text. Some of these concern slight improvements of existing results. When trying to prove such results one often stumbles upon something more exciting.

This paper owes much to the excellent surveys by Cameron [**Cam1**] [**Cam3**] [**Cam4**] (indeed the reader is referred to [**Cam4**] both for a modern introduction to the theory of permutation groups and for further reading). However our point of view is different and luckily there is a wealth of new results and applications.

The classical reference for permutation groups is [**Wie2**]; for the classification theorem it is [**Gor**].

Throughout this paper statements depending on the classification theorem will be marked by (CFSG); "log" means logarithm to the base 2.

1. Generators

A useful "subroutine" for dealing with permutation groups is Jerrum's filter [**Jer**]. Given any set of elements of the symmetric group S_n this algorithm quickly outputs a set of at most $n-1$ permutations generating the same group. This of course implies that any subgroup of S_n can be generated by $n-1$ elements.

No algorithmic proof is known for the following result of McIver and Neumann [**MN**] (for a published proof see [**CST**]).

THEOREM 1.1 (CFSG). *Any subgroup of S_n can be generated by at most $max(2, [\frac{n}{2}])$ elements.*

Note that it easily follows from a general reduction theorem of Guralnick and Lucchini [**Gur**][**Luc**] that $[\frac{n}{2}]+1$ elements suffice. According to their result, if every Sylow subgroup of a group G can be generated by r elements then G can be generated by $r+1$ elements (CFSG). But it is not hard to see that 1.1 holds for all p-subgroups of S_n.

The same (folklore) idea works in the case of $GL(n,p)$, p odd. By a result of Patterson [**Pat**], the maximal number of generators of a p-subgroup of $GL(n,p)$ is $\frac{1}{4}n^2$ (for p odd). Using the well-known descriptions of the Sylow r-subgroups ($r \neq p, r$ prime) of $GL(n,p)$ [**CF**] [**We**] and the above reduction theorem we obtain that any subgroup of $GL(n,p)$ ($n \geq 4$) can be generated by $\frac{1}{4}n^2+1$ elements. It would be worthwhile to find the best possible bounds for $GL(n,p^\alpha)$ (especially in the case $p=2$).

The minimal number of generators of any subgroup of a group G is of course bounded by the length of the longest subgroup chain of G. It was proved by Babai [**Ba5**] that any chain of subgroups of S_n has length at most $2n-1$. The best possible bound was obtained by Cameron, Solomon and Turull [**CST**]. They proved that the maximal length of a subgroup chain in S_n is $[\frac{3n-1}{2}]-b$, where b denotes the number of ones in the base 2 expansion of n (CFSG).

For various interesting results on the length of subgroup chains of groups of Lie type, see [**CST**] [**ST1**] [**ST2**] [**SST**] and [**TZ**].

L. G. Kovács (with various co-workers) started investigating general properties of transitive groups. The main question was whether transitivity is a strong restriction (helpful in designing fast algorithms).

It was first proved by Kovács and Praeger [**KP**] that for $G \leq S_n$ we have $|G/G'| \leq 3^{n/3}$ (see [**KP**] for a more detailed statement; the result also appears in [**AG2**]).

Next an example appeared that shattered hopes for a $(\log n)^c$ bound for the number of generators of a transitive group.

EXAMPLE 1.2. For each prime p there is a constant $c_p>0$ such that whenever n is a power of p there is a transitive p-group of degree n which cannot be generated by $c_p n/\sqrt{\log n}$ elements.

For the interesting construction see [**KN**]; similar examples appear in [**AG2**] and [**Au**]. Note that this implies that we cannot have much stronger bounds for $|G/G'|$, $|G/\Phi(G)|$ or $|N_{S_n}(G)/G|$ for transitive groups in general (for a discussion concerning $N_{S_n}(G)/G$ see §5).

Surprisingly, the above example was shown to be the worst case first for nilpotent groups by Kovács and Newman [**KN**] and very recently for solvable groups by Bryant, Kovács and Robinson [**BKR**]. They proved the following.

THEOREM 1.3. *Any solvable transitive subgroup of S_n $(n \geq 2)$ can be generated by $cn/\sqrt{\log n}$ elements for some $c > 0$.*

Kovács and Robinson [**KR1**] also showed that finite irreducible linear groups of dimension n (over an arbitrary field) can be generated by $[\frac{3}{2}n]$ elements (CFSG). In the proof they use a variant of a result of Wiegold [**Wi**]: any direct product of r nonabelian simple groups can be generated by $2 + \log r$ elements (CFSG). Direct analogues of 1.2 and 1.3 are known to hold for finite irreducible linear groups of dimension n over fields having bounded degree over their prime subfields [**DK**] [**BKR**].

Of course the most interesting problem that remains is whether the solvability hypothesis can be removed in the above result 1.3. It would also be of interest to decide whether $|G/G'| < 2^{\frac{cn}{\sqrt{\log n}}}$ holds for transitive subgroups G of S_n for some constant $c > 0$.

A transitive permutation group $G \leq S_n$ is called *minimally transitive* if it has no proper transitive subgroups.

Let $n = \prod_{i=1}^{k} p_i^{\alpha_i}$ be the prime factorization of n, let $\lambda(n) = \sum_{i=1}^{k} \alpha_i$ and $\mu(n) = \max_{i=1}^{k} \alpha_i$ (note that $\log n \geq \lambda(n) \geq \mu(n)$).

The following result was proved by Shepperd and Wiegold in 1963 [**SW**].

THEOREM 1.4. *Suppose that every simple group can be generated by 2 elements. Then any minimally transitive subgroup of S_n can be generated by $\lambda(n)$ elements.*

Of course, by CFSG we know that the hypothesis of the theorem is true. Neumann and Vaughan-Lee [**NV**] gave an elementary proof of the fact that $\log n$ elements suffice; a slightly weaker result was obtained by Babai and T. Sós [**BS2**].

It would be interesting to decide whether in fact $\mu(n) + 1$ elements are always sufficient to generate a minimally transitive group (note that this is not hard to prove for regular or nilpotent groups [**Py6**]).

It is not clear whether any analogue of the above results holds for linear groups.

Call a permutation group G *orbit-minimal* if every proper subgroup H of G has strictly more orbits than G has. Trying to improve an estimate of Babai and T. Sós on the number of vertex transitive graphs (see §5), the present author was led to the (slightly counterintuitive) conjecture that in fact orbit-minimal subgroups of S_n can always be generated by $\log n$ elements, and gave an ugly proof of an $0(\log n \log\log n)$ bound.

Lovász proved the conjecture by an elegant counting argument (see [**Py6**]).

THEOREM 1.5. *Any orbit-minimal subgroup G of S_n can be generated by $\log(n - k + 1)$ elements where k denotes the number of orbits of G.*

Next the following surprising strengthening (which was also proved by a counting argument) appeared in a paper of Babai, Luks and Seress [**BLS2**] on the fast management of permutation groups.

Let $G = \langle \tau_i, \tau_2, \ldots, \tau_\kappa \rangle$. A *random subproduct* of the generators $\tau_i, \ldots, \tau_\kappa$ is an instance of the product $\tau_1^{\epsilon_1} \tau_2^{\epsilon_2} \cdots \tau_\kappa^{\epsilon_\kappa}$ where the ϵ_i are independent, 0-1 valued random variables with $Prob(\epsilon_i = 0) = Prob(\epsilon_i = 1) = 1/2$.

THEOREM 1.6. *Suppose S is a set of generators of the group $G \leq S_n$ and S' is a set of $c \log n$ random subproducts of S. Then with large probability the orbits of G and the subgroup generated by S' agree.*

The above result was used in [**BLS2**] to find $0(\log n)$ generators in Monte Carlo time $0(|S| n \log n)$ for groups that are known to be doubly transitive.

The main idea behind the proof of 1.6 is even more important: under certain conditions products of random subsets of generators of a group G behave like truly random elements. For numerous applications of this idea see [**Ba7**] .

The last class to be considered here is the class of primitive subgroups of S_n. Cameron (personal communication) observed that $0(\log n)$ elements are always sufficient to generate the non-affine primitive groups and suggested the investigation of the general case. That the same holds for affine primitive groups follows from the above result of Kovács and Robinson [**KR1**] on the number of generators of a completely reducible linear group.

Considering the wreath product of A_5 with a transitive 2-group of degree t as in example 1.2, in the product action, we obtain primitive groups of degree $n = 5^t$ (see [**Cam1**]) that cannot be generated by less than $0(\log n/\sqrt{\log\log n})$ elements. It could be quite interesting to find the right estimate.

We end this section with an elegant result of Aschbacher and Guralnick [**AG2**].

THEOREM 1.7 (CFSG). *Let G be a primitive group in S_n. Then $|G/G'| \leq n$.*

2. Orders

We start this section with an old theorem of Landau [**Lan**]

THEOREM 2.1. *The maximal order of a cyclic subgroup of S_n is*

$$e^{(1+o(1))\sqrt{n \ln n}}.$$

This is of course exactly the answer to the following question: how large can be the smallest common multiple of a set of natural numbers the sum of which is n. Theorem 2.1 really belongs to number theory just like many results concerning the typical cycle structure of elements of S_n (see §6).

Dixon [**Dix1**] was the first to prove that solvable subgroups of S_n have order significantly less than $n! = |S_n|$.

THEOREM 2.2. *Let G be a solvable subgroup of S_n. Then $|G| \leq 24^{\frac{n-1}{3}}$.*

It is an easy observation that the automorphism group of a tournament T has odd order (an involution would reverse the direction of some edges). Goldberg and Moon [**GM**] conjectured that $|Aut(T)|$ is always at most $3^{\frac{n-1}{2}}$. Moreover they observed that this is equivalent to the conjecture that the order of an odd-order subgroup of S_n is at most $3^{\frac{n-1}{2}}$. Using the Feit-Thompson theorem Dixon [**Dix2**] proved the latter conjecture. What is more surprising, later Alspach [**Als**] found a beautiful direct proof for the original conjecture thereby obtaining a sharp bound for the order of the odd order subgroups of S_n using hardly any group theory.

Assumptions like nilpotence or transitivity do not imply stronger order bounds on subgroups of S_n. Indeed if n is a power of 2 then it is not hard to find minimally transitive 2-subgroups of S_n of exponential order [**Py6**].

Note, however, the following useful result from [**BCP**]: if G is a subgroup of S_n that has no composition factor isomorphic to an alternating group of degree $> d$ then $|G| < d^{n-1} (d \geq 6)$.

The really important class of permutation groups is that of primitive groups. Despite investigations dating back to Jordan's time (see [**Wie2**]), the first general exponential bound was arrived at by Praeger and Saxl [**PS**].

THEOREM 2.3. *Let G be a primitive subgroup of S_n not containing A_n. Then $|G| \leq 4^n$.*

The proof is based on Wielandt's ideas which he had used to prove that for G uniprimitive the estimate $|G| \leq 24^n$ holds (a permutation group is called *uniprimitive* if it is primitive but not doubly transitive). Wielandt [**Wie3**] used the method of bounding the orders of the Sylow subgroups of G.

Soon after [**PS**], Babai [**Ba2**] used entirely different ideas to obtain the following estimate.

THEOREM 2.4. *Let G be a uniprimitive subgroup of S_n. Then $|G| \leq n^{4\sqrt{n}\ln n}$.*

The group S_m acting on unordered pairs (with $n = \binom{m}{2}$) has order roughly $n^{\sqrt{n}}$, which shows that 2.4 is sharp apart from an $0(\log n)$ factor in the exponent.

In a subsequent paper Babai [**Ba3**] obtained an even better $2^{(2^{c\sqrt{\log n}})}$ bound for the orders of doubly transitive groups not containing A_n. This was improved in [**Py3**].

THEOREM 2.5. *Let G be a doubly transitive subgroup of S_n not containing A_n. Then $|G| < n^{32(\log n)^2}$ for $n \geq 400$.*

Interestingly 2.5 is also sharp apart from an $0(\log n)$ factor in the exponent as shown by $P\Gamma L(d,q)$ in its natural action.

It should be stressed that the proof of 2.4 as simplified by Zemlyachenko [**ZKT**] (see also [**Py4**]), and a somewhat weaker version of 2.5 [**Py3**], could be written up in about 8 pages (and it is independent of CFSG). We will consider some of the ideas behind these proofs in §3.

The Wielandt-Praeger-Saxl and the Babai-Pyber estimates 2.3-2.5 have numerous applications in asymptotic group theory and in computational group theory [**Ba3**][**Ba5**][**BCP**][**BLS2**][**BP1**][**KR2**]. It is a bit distressing that the concise proof of 2.3 in [**PS**] only works for $n > 12000$ (as the authors note, the smaller cases can be settled by a more elaborate version of the proof), and that the asymptotically better bounds in [**Ba2**] [**Ba3**] [**Py3**] are in fact worse for $n \leq 10^7$. To facilitate induction proofs it would be worthwhile to classify primitive groups of order greater than 2^{n-1} (especially if such a classification could be obtained by a reasonably short and elementary argument).

CFSG has had a dramatic effect on this kind of problem. Doubly transitive groups have been completely classified [**Cam1**] [**CKS**] [**Her**]. As a consequence we know that the only 6-transitive groups are A_n and S_n. (Still Jordan's classical $0((\log n)^2/\log\log n)$ bound for the degree of transitivity is used in [**BCFLS**] to obtain CFSG-free superfast Monte Carlo algorithms for handling permutation groups.)

Using results of Kantor [**Ka1**], Cameron [**Cam1**] obtained the following important theorem.

THEOREM 2.6 (CFSG). *There is a constant c such that if G is a primitive subgroup of S_n then one of the following holds:*

i) *G is affine.*
ii) *G is a subgroup of $Aut(T)wrS_\ell$ containing T^ℓ, where T is either an alternating group acting on k-element subsets or a classical group acting on an orbit of subspaces (or in the case $PSL\,(d,q)$, on pairs of subspaces of complementary dimension), and the wreath product has the product action.*
iii) *$|G| \leq n^{c\log\log n}$ for some constant c.*

In particular primitive groups of order greater than $n^{c\log n}$ are of type ii) with T an alternating group. Apart from answering classical questions this result is an indispensable tool in many algorithmic applications[**BKL**][**BLS1**][**BLS2**].

By 2.6 large primitive groups have fairly large simple composition factors. Correspondingly, primitive solvable groups are really small by a result of Pálfy [**Pál**] and Wolf [**Wo**].

THEOREM 2.7. *Let G be a primitive solvable subgroup of S_n. Then $|G| \leq 24^{-\frac{1}{3}}n^{1+c_0}$ where $c_0 = 2.243\ldots$.*

In an equivalent formulation, solvable completely reducible subgroups of $GL(n,p)$ (p prime) have order at most $24^{-\frac{1}{3}}p^{c_0 n}$. Wolf in fact proved the corresponding bound for subgroups of $GL(n,p^\alpha)$ whereas Pálfy gave the best possible bounds for each prime p (for $p \neq 3$ the value of the constant can be improved). These results were unified by Segev [**Se**].

2.7 shows that the derived length of a primitive group is at most roughly $\log n$. For more exact bounds see [**Dix3**] [**Ne**].

The proof of Pálfy uses Suprunenko's results [**Sup**], in particular a structural description of maximal solvable primitive linear groups (for a variant that describes arbitrary solvable primitive linear groups see [**Pá2**]). While in general this approach has been superseded by later developments based on a theorem of Aschbacher [**As**], (see also [**KL3**]), it still provides a particularly easy way to obtain results about solvable linear groups, see [**Pál**] [**Py2**] [**Se**] (and it is a starting point for the recent algorithms of Luks for handling solvable matrix groups [**Lu2**]).

Results like 2.7 have applications in proving asymptotic estimates for abstract groups [**Gl**] [**He**] [**Ma2**] [**Py1**] [**Py2**]. As an illustration we mention a (folklore) consequence of 2.7: if G is a solvable group with Fitting subgroup $F = F(G)$ then $|F|^4 > |G|$ (this is implied by the well-known result of Gaschütz that $G/F(G)$ acts completely reducibly on the elementary abelian group $F(G)/\Phi(G)$ [**Hu, p.279**]).

For any fixed integer d let Γ_d denote the class of groups all of whose non-cyclic composition factors are isomorphic to a subgroup of S_d. The class Γ_d originally arose in the context of testing graph isomorphism. For example the stabilizer of an edge in the automorphism group of a connected graph of maximum degree $d+1$ is in Γ_d. This is used to show that the famous algorithm of Luks [**Lu1**] for testing isomorphism of graphs of bounded valence works in polynomial time. (Various similar algorithms are known for deciding isomorphism of graphs of bounded genus, bounded Hadwiger number, etc. For some of the more recent results in this direction see [**Po1**], [**Po2**].)

Also the groups in Γ_d play a special role in present-day computational group theory (see [**KL2**] for the latest polynomial-time library).

The reason for all this is of course the well-known theorem of Babai, Cameron and Pálfy [**BCP**] (though this result was not used in the first version of the Luks algorithm).

THEOREM 2.8 (CFSG). *Let $G \in \Gamma_d$ be a primitive subgroup of S_n. Then $|G| \leq n^{c(d)}$ for some constant $c(d)$.*

The function $c(d)$ has been further investigated (see [**BKL**]). (Indeed, the present form of the definition of the class Γ_d was developed during these investigations.) A recent result from [**Py5**] settles a problem in [**BKL**].

THEOREM 2.9 (CFSG). *Let $G \in \Gamma_d$ be a primitive permutation group acting on a set Ω of size n. Then either*

i) *$|G| \leq n^{c \log d}$ for some constant c, or*
ii) *the socle N of G is isomorphic to a direct product of t copies of the alternating group A_D, for some t, D (with $|D| \leq d$). Moreover, with this identification there is an N-isomorphism $\Omega \to \binom{D}{k}^t$.*

This result may open up the path towards an $n^{\sqrt{d}}$ d-valent graph isomorphism test (wishful thinking).

Let us also mention another result of a similar flavour [**Py5**].

THEOREM 2.10 (CFSG). *Let G be a primitive subgroup of S_n. Then*

i) *The product of the orders of the abelian composition factors of G is at most $24^{-\frac{1}{3}}n^{1+c_0}$ (where c_0 is the same constant as in 2.6).*
ii) *The number of nonabelian composition factors of G is at most* $\log n$.

This result implies a weaker form of 1.6 and it also shows that primitive groups can be generated by $0(\log n)$ elements. Another consequence of 2.10 is that any chain of normal subgroups in a primitive group has length at most $0(\log n)$, and by a simple induction argument this gives us an $0(\log n)$ bound for the derived length of any subgroup of S_n. These observations improve some lemmas from [**Ba4**] and [**BLS1**].

Part ii) of 2.10 and an earlier result of Cameron [**Cam1**] suggested that the following may be true: the number of nonabelian composition factors of a primitive group of rank r is bounded in r. This conjecture was promptly settled by Cameron during the DIMACS workshop (the proof uses Aschbacher's theorem [**As**]).

We end this section with a problem from (an unpublished version of) [**BCP**]. The following is a simply stated corollary of 2.8 : Let G be a primitive subgroup of S_n such that $|G|$ has no prime divisors larger than p. Then $|G| < n^{c'(p)}$ for some constant $c'(p)$. Prove this without using the classification of finite simple groups!

3. Bases

A *base* for a permutation group $G \leq S_\Omega$ is a sequence $\alpha_1, \alpha_2, \dots, \alpha_b$ of elements of Ω whose pointwise stabilizer is the identity. Bases arose in the algorithms for computing with permutation groups developed by Sims [**Si1**] [**Si2**] [**Si3**].

The order of a group $G \leq S_\Omega$ and the size of its bases are closely related: if G has a non-redundant base of size b then clearly we have $2^b \leq |G| \leq n^b$. This connection is the starting point for Babai's approach to estimating the orders of primitive groups. Another important ingredient is the combinatorial analysis of asymptotic properties of primitive coherent configurations (see below). This in turn was motivated by the graph isomorphism problem (see [**Ba1**] [**Ba2**] [**BL**]).

A *coherent configuration* $\mathcal{C}$ is a partition of the set of pairs $\Omega \times \Omega$ into *color classes* $R_1, R_2, \dots, R_k$ such that, for all i, j,

i) the diagonal $\Delta = \{(\omega, \omega)\}$ is the union of some of the R_i,
ii) R_i^{-1} (the inverse of R_i) is one of the R_j,
iii) the number of paths $(\alpha, \gamma)(\gamma, \beta)$ with $(\alpha, \gamma) \in R_i$ and $(\gamma, \beta) \in R_j$ depends only on the class R_k of (α, β).

$\mathcal{C}$ is called *primitive* if $R_1 = \Delta$ and all the other R_i are connected (as directed graphs). It is called *uniprimitive* if its *rank* k is at least 3.

The classical examples of coherent configurations arise from permutation groups. If $G \leq S_\Omega$ then the orbits R_i of the induced action of G on $\Omega \times \Omega$

determine a coherent configuration $\mathcal{C}(G)$. $\mathcal{C}(G)$ is uniprimitive exactly when G is a uniprimitive subgroup of S_Ω.

We say that a vertex ω *distinguishes* the vertices α, β if (α, ω) and (β, ω) belong to different classes R_i. A set D is called distinguishing for $\mathcal{C}$ if for each pair α, β there is a $\delta \in D$ that distinguishes α and β.

The key result of Babai [**Ba2**] is the following:

LEMMA 3.1. *Let $\mathcal{C}$ be a uniprimitive coherent configuration. Then each pair of vertices is distinguished by at least $\frac{1}{2}(\sqrt{n}-1)$ vertices.*

Applying an Erdős-type probabilistic argument Babai concludes that in fact there is a distinguishing set $D \subset \Omega$ of size $\leq 4\sqrt{n}\, ln\, n$. Clearly D corresponds to a base of $Aut(\mathcal{C})$. Therefore we have the following.

THEOREM 3.2. *Let $\mathcal{C}$ be a uniprimitive coherent configuration.*

i) *$Aut(\mathcal{C})$ has a base of size $\leq 4\sqrt{n}\, ln\, n$.*

ii) *$|Aut(\mathcal{C})| \leq n^{4\sqrt{n}\, ln\, n}$.*

Observe that if G is a uniprimitive group then $G \leq Aut(\mathcal{C}(G))$, and therefore 2.4 follows from 3.2.

It is conjectured [**Ba2**] that the $ln\, n$ factor in in 3.2 (and therefore in 2.4) can be dispensed with.

Another conjecture of Babai [**Ba2**] is that if $\mathcal{C}$ is a primitive strongly regular graph (i.e., the rank of $\mathcal{C}$ is 3) and $Aut(\mathcal{C})$ is of order at least $2^{n^{\frac{1}{2}-\epsilon}}$ for some $0 < \epsilon < 0.1$ (n sufficiently large), then $\mathcal{C}$ is the line graph of a complete graph K_m or a complete bipartite graph $K_{m,m}$.

Next we sketch the proof of the bound 2.5. Throughout the sketch, G will denote a doubly transitive group not containing A_n.

We say that a subset $\Sigma \subseteq \Omega$ is a *halving set* of the group $G \leq S_\Omega$ if the pointwise stabilizer of Σ in G has no orbits of size greater than $\frac{|\Omega|}{2}$. In [**Ba3**] it is proved that if Σ is a halving set of a doubly transitive group G and $g_1, g_2, \ldots, g_\ell$ are random elements of G with $\ell \geq 2\log n$, then the union of the g_i images Σ^{g_i} of Σ is a base of G with positive probability. Another important property of 2-transitive groups proved by Wielandt [**Wie1**] is that such a group G (not containing A_n) does not involve A_k (as a factor group of a subgroup) for $k > 4\log n$ (assuming that $n \geq 400$).

The following lemma is the core of the proof of 2.5 in [**Py3**].

LEMMA 3.3. *Let G be a permutation group of degree n on a set Ω. Suppose that Ω has no subset Δ of size $|\Delta| \geq k$ (where $k \geq 12$) with the setwise stabiliser H of Δ inducing a group H^Δ on Δ that contains A_Δ. Then there is a halving set Σ of G of size at most $4k$. Moreover Σ can be found by the greedy algorithm (that proceeds by successively stabilizing a point of the largest remaining orbit).*

By Wielandt's property and 3.3 if G is doubly transitive then G has a halving set of size at most $16\log n$. Therefore G has a base of size at most $32(\log n)^2$ which in turn implies $|G| \leq n^{32(\log n)^2}$.

It seems that it should be possible to make use of 3.3 in designing new divide and conquer algorithms.

Using CFSG, Liebeck [**Li**] proved the following result.

THEOREM 3.4 (CFSG). *Let G be a primitive subgroup of S_n with no base of size less than* $9 \log n$. *Then G is a subgroup of $S_m wr\, S_r$ containing $(A_m)^r$, where the action of A_m is on k-element subsets and $n = \binom{m}{k}^r$.*

Experience suggests that primitive groups tend to have rather small bases. Let us mention some conjectures in this direction.

Cameron [**Cam4**] conjectures that there is a constant c such that with known exceptions an almost simple primitive group has a base of size at most c (the exceptions correspond to certain permutation representations of the symmetric group, the alternating group and the classical groups as in 2.6 ii).

Babai conjectures that if G is a primitive permutation group in Γ_d then G has a base of size at most $c(d)$ (this would be a qualitative strengthening of 2.8).

It would be interesting to decide whether in fact primitive groups G have bases of size b with $|G| > n^{cb}$ for some (small) constant c. The first interesting case is that of primitive solvable groups.

Apart from theoretical interest such questions are important because special very fast algorithms are known for handling permutation groups with small bases [**BCFS**].

Another classical invariant, much investigated since the last century, is the *minimal degree* of a permutation group G (i.e. the minimal number of points moved by a non-identity element of G). Note that it is easy to show that a transitive subgroup of S_n with minimal degree m and minimal base size b satisfies $bm \geq n$.

The symmetric and alternating groups have minimal degree 2 and 3 respectively. The fact that the minimal degree of a primitive subgroup of S_n not containing A_n tends to infinity with n was proved by Jordan (see [**Wiel**]). As Babai observed, using 3.1 one can prove that in fact the minimal degree of such a G is at least $\frac{1}{2}(\sqrt{n}-1)$ (see [**Py4**]). This bound is best possible up to a (multiplicative) constant. By a really old result of Bochert (see [**Wiel**]) the minimal degree of doubly transitive groups (not containing A_n) is at least $\frac{n}{4} - 1$. This result is used in [**BCFS**].

Of course the CFSG gives much stronger results in this case too. In an important paper Liebeck and Saxl [**LS**] classified all primitive groups with minimal degree at most $\frac{n}{2}$.

We close this section with a (loosely related) deep conjecture of Babai. Let $\mathcal{C}$ be a primitive coherent configuration. The color classes $R_2, R_3 \dots R_k$ are diregular directed graphs say of degree $d_2 \leq d_3 \cdots \leq d_r$ respectively.

Prove that there is a function f such that $d_r \leq f(d_2)$.

The validity of this conjecture in the special case when $\mathcal{C} = \mathcal{C}(G)$ for a primitive group G follows from the so-called Sims conjecture [**CPSS**] (CFSG).

4. Enumeration

The first result of this section serves as a tool for enumerating various kinds of subgroups of S_n [**Py1**].

LEMMA 4.1. *The number of conjugacy classes of maximal solvable subgroups of* S_n *is at most* 2^{17n}.

Let $Mss(G)$ denote the set of maximal solvable subgroups of a group G and let $Sub(G)$ denote the set of all subgroups. As an illustration of how 4.1 can be applied we show the following [**Py1**].

THEOREM 4.2 (CFSG).

$$2^{(\frac{1}{16}+o(1))n^2} \leq |Sub(S_n)| \leq 24^{(\frac{1}{6}+o(1))n^2}.$$

PROOF. By 2.2 the order of a maximal solvable subgroup M of S_n is at most $24^{(n-1)/3}$. According to 1.1 any subgroup of M can be generated by at most $(n+1)/2$ elements. Therefore we have $|Sub(M)| \leq 24^{n^2/6}$.

It follows from 4.1 that $|Mss(S_n)|$, the number of such subgroups M, is at most $n!2^{17n}$.

By a result of Aschbacher and Guralnick [**AG1**] every finite group G can be generated by a solvable subgroup S and an element g of G (CFSG).

This is true in particular for subgroups of S_n, therefore $|Sub(S_n)|$ is bounded by the number of possible pairs S and g. It follows that

$$|Sub(S_n)| \leq |Mss(S_n)|\,|S_n|24^{n^2/6} = 24^{(\frac{1}{6}+o(1))n^2}.$$

To prove the lower bound observe that S_n contains an elementary abelian 2-subgroup H of order $2^{[\frac{n}{2}]}$, and that $|Sub(H)| = 2^{(\frac{1}{16}+o(1))n^2}$ by an easy counting argument. □

Heuristic arguments support the conjecture [**Py1**] that in fact we have $|Sub(S_n)| = 2^{(\frac{1}{16}+o(1))n^2}$. (For some related problems see §6.)

It should be mentioned that an estimate of the form $|Sub(S_n)| \leq 2^{cn^2}$ can be obtained without using CFSG. To see this note that a Sylow p-subgroup P of S_n has order at most $p^{\frac{n}{p-1}}$ and therefore $|Sub(P)| \leq p^{(\frac{n}{p-1})^2}$. The number of such subgroups P is of course less than $n!$. A subgroup H of S_n is determined by a system of Sylow p-subgroups (one for each prime $p \leq n$).

Therefore we have $|Sub(S_n)| \leq (n!)^{\Pi(n)} \prod_{p\leq n} p^{(\frac{n}{p-1})^2}$ (where $\Pi(n)$ denotes the number of primes less than n). By the prime number theorem $\Pi(n) = (1+o(1))\frac{n}{\ln n}$, and $|Sub(S_n)| \leq 2^{cn^2}$ follows by easy computation (for some $c > 0$).

An estimate similar to 4.1 was obtained by Pálfy for the number of conjugacy classes of maximal solvable subgroups of $GL(n,p)$ [**Py1**]. This estimate is an important ingredient in the author's proof [**Py1**](CFSG) of the following theorem: the number of (isomorphism classes of) groups of order n is at most $n^{(\frac{2}{27}+o(1))(\mu(n))^2}$ (see §1 for the definition of $\mu(n)$).

By a very recent result of Borovik and the author [**BP2**] there exists a constant c such that $|Mss(G)| \leq |G|^c$ for every finite group G (CFSG). Combining this with the theorem of Aschbacher and Guralnick [**AG1**] mentioned above, we obtain that $|Sub(G)|$ is at most $|G|^{c+1}$ times the number of subgroups of some solvable subgroup M of G. This seems to be a useful reduction theorem for enumeration type problems.

In the previous sections we saw that transitivity is not a really strong restriction on permutation groups. The next observation shows a reason for that: there are simply too many transitive groups.

PROPOSITION 4.3.

i) *The number of transitive subgroups of* $S_n, n = p^\alpha$ *(*p prime*), is at least* $2^{a_p \frac{n^2}{\log n}}$ *for some constant* a_p.

ii) *The number of solvable transitive subgroups of* S_n *is at most* $2^{\frac{an^2}{\sqrt{\log n}}}$ *for some constant* a.

PROOF. For n a power of p consider a transitive p-group P as in 1.2. By 1.4 P contains a minimally transitive subgroup M generated by $\leq \log n$ elements. Consider the subgroup $Q = \Phi(P)M$ (where $\Phi(P)$ is the Frattini subgroup of P). Clearly Q is a transitive normal subgroup of P and $Q/\Phi(P)$ is a subgroup of the elementary abelian group $P/\Phi(P)$ generated by at most $\log n$ elements. It follows that P/Q is an elementary abelian p-group of order p^t where $t = c_p \frac{n}{\sqrt{\log n}} - \log n$.

By easy counting there are roughly $p^{\frac{t^2}{4}}$ subgroups of P containing Q and all of them are transitive. This proves i).

Part ii) is an obvious consequence of 1.3 and 4.1. □

The number of minimally transitive groups is much smaller [**Py6**].

THEOREM 4.4 (CFSG). *The number of minimally transitive subgroups of* S_n *is at most* $(n!)^{4+o(1)}$.

It is easy to see that S_n has roughly $n!$ regular subgroups, which shows that 4.4 is not far from best possible. However, as the following observation [**BS2**] shows it would really be interesting to give a sharp upper bound for the number $m(n)$ of conjugacy classes of minimally transitive groups: the number of non-isomorphic vertex-transitive graphs on n vertices is at most $m(n)2^{n-1}$.

The number of Cayley graphs of order n is exponential in n and it is easy to construct exponentially many non-conjugate minimally transitive groups when, say n is a power of 2 [**Py6**]. An exponential upper bound on $m(n)$ would therefore be quite sharp and it would also give the order of magnitude of the number of vertex-transitive graphs.

Recently B. McKay and Praeger formulated the following very strong conjecture in this direction: Almost all vertex-transitive graphs are Cayley.

By taking wreath products of A_5 with transitive groups as in 4.3, one can construct roughly $n^{(\log n)^2/\log\log n}$ pairwise non-conjugate primitive groups. This should not be too far from an exact asymptotic estimate.

In a slightly different direction one might consider the number $|Hom(F,S_n)|$ of permutation representations of degree n of the group F. As the following explicit formula of Müller [**Mü**] shows this number is intimately related to $|Sub(F,d)|$, the number of index d subgroups of F.

THEOREM 4.5. *Let F be a finite group of order $m > 1$ and denote by d_2 the greatest divisor of m which is $< m$. Then, as $n \to \infty$, $|Hom(F,S_n)|$ is asymptotically equal to*

$$K_F \exp\{(1-1/m)n\log n - (1-1/m)n + \frac{|Sub(F,d_2)|}{d_2} n^{d_2/m} + \sum_{\substack{d|m \\ d<d_2}} |Sub(F,d)|\left(\frac{1}{d}-\frac{1}{m}\right)n^{d/m}\}$$

where

$$K_F = \begin{cases} m^{-\frac{1}{2}}\exp\left(-\frac{|Sub(F,\frac{m}{2})|^2}{2m}\right), & 2\,|\,m \\ m^{-\frac{1}{2}}, & 2\nmid m. \end{cases}$$

For example Wilf's formula [**Wil**] for the number of solutions of the equation $X^m = 1$ in S_n can be deduced from 4.5.

5. Some additional results

The following estimate was obtained recently by Kovács and Robinson [**KR2**].

THEOREM 5.1. *The number of conjugacy classes (of elements) in any subgroup of S_n is at most 5^{n-1}.*

In the case of solvable groups the bound is improved to $3^{\frac{n-1}{2}}$ $(n > 2)$ [**KR2**]. When n is divisible by four a direct product of $\frac{n}{4}$ copies of S_4 or D_8 shows that we can have $5^{\frac{n}{4}}$ conjugacy classes.

These results are related to the "$k(GV)$ problem": Is it true that if G is a group of linear transformations of the finite-dimensional vector space V over a finite field of characteristic coprime to the order of G, then the number of conjugacy classes of the semidirect product GV is at most $|V|$? (For an account of known partial results see [**KR2**].)

The following appears in [**BP1**].

THEOREM 5.2. *Let G be a subgroup of S_Ω $(|\Omega| = n)$ having no composition factors isomorphic to an alternating group of degree $> d$ $(d \geq 4)$. Then the number of orbits of G on the power set of Ω is at least $2^{cn/d}$ for some constant $c > 0$.*

This result is an important ingredient in the proof of the following theorem **[Py2]** (CFSG): Every finite group of order $n\,(n \geq 4)$ has at least $\frac{\epsilon \log n}{(\log \log n)^8}$ conjugacy classes for some $\varepsilon > 0$.

Another application of 5.2 to the complexity of Boolean functions is to be found in [**BBT**].

For a rich theory of the numbers of orbits of infinite permutation groups on k-sets see [**Cam2**].

The last two problems to be considered here suggest that transitive groups do have some distinctive properties after all.

THEOREM 5.3 (CFSG). *Let G be a transitive subgroup of S_n. Then*

i) $|N_{S_n}(G)| \leq n|G|^{\lambda(n)}$,

ii) *if moreover G is solvable then* $|Aut(G)| \leq |G|^{2\lambda(n)}$.

PROOF. By 1.4, G contains a (minimally) transitive subgroup M generated by $\lambda(n)$ elements. The number of images of M under conjugation by elements of $N_{S_n}(G)$ is at most $|G|^{\lambda(n)}$. The stabilizer of M in $N_{S_n}(G)$ is a subgroup of $C_{S_n}(M)$, and this latter group (the centralizer of a transitive group) is semiregular. It follows that $|N_{S_n}(G)| \leq n|G|^{\lambda(n)}$.

To prove ii) we need a result of Wall [**Wa**]: Let $Max(G)$ denote the set of maximal subgroups of G. If G is solvable then $|Max(G)| < |G|$.

It follows that the number of subgroups of index n of a solvable group G is at most $|G|^{\lambda(n)}$. By a lemma of Gross [**Gr**,Lemma 2.2] if H (resp., R) denotes the stabilizer of a point in G (resp., $N_{S_n}(G)$) then R is isomorphic to the subgroup of $Aut(G)$ consisting of those automorphisms of G which map H onto H. The index of this subgroup in $Aut(G)$ is the number of images of H under $Aut(G)$, and therefore it is at most $|G|^{\lambda(n)}$. The order of this subgroup is at most $|G|^{\lambda(n)}$ by i). The proof is complete. □

Let us comment on the result of Wall used in the above proof. A slightly stronger version appears in [**CWW**], and in both papers it is conjectured that $|Max(G)| < |G|$ holds for an arbitrary finite group G (see [**AG1**] for a related result and conjecture on the number of conjugacy classes of maximal subgroups). Of course the validity of this conjecture would imply that 5.3 ii) extends to arbitrary transitive groups.

The following conjecture is due to Babai: The diameter of any Cayley graph of a transitive subgroup G of S_n, is at most n^c for some constant c. (Note that the assumption of transitivity is an important one as S_n contains large cyclic subgroups.) The conjecture is unproved even for $G = S_n$, however this case seems to be the crucial one (see [**BS1**]).

6. Random elements and subgroups

What does a random element of S_n look like? Such questions belong to "statistical group theory".

Goncharov [**Gon**] proved in 1944 that almost all elements of S_n have roughly $\ln n$ cycles in their cycle-decompositions (here *almost all* means that there can be at most $o(n!)$ exceptions). Erdős and Turán in a series of papers [**ET1**] [**ET2**] [**ET3**] exhibited various other properties of a random element P of the symmetric group. For example they showed that the order $O(P)$ of P is roughly $e^{\frac{1}{2}(\ln n)^2}$, $O(P)$ is divisible by all primes less than $\ln n/\ln\ln n$, and that the size of the largest prime-factor of $O(P)$ is around $\frac{n}{e^{\sqrt{\ln n}}}$.

Further investigations concern the properties of an element of a random conjugacy class of S_n [**ET3**].

Combining some ideas of Erdős and Turán with some classical results on primitive groups, Dixon [**Dix4**] proved the following well-known theorem.

THEOREM 6.1. *A random pair of elements of S_n generates either A_n or S_n with probability at least* $1 - 2/(\ln\ln n)^2$.

By an intermediate result of [**Dix4**] the probability that a random pair generates a primitive group is $1 - 1/n + 0(n^{-2})$. Babai [**Ba6**] proved that in fact the same holds for the probability that a random pair generates A_n or S_n by showing that the number of pairs contained in primitive groups is negligible compared to $n!n^{-2}$ (CFSG).

To illustrate the connection between the "statistical" results and 6.1 we mention an idea of Bovey [**Bov**] which leads to the strongest version of Dixon's theorem obtained by elementary means. Bovey proved the following: Take any α, $0 < \alpha < 1$. Then the minimal degree of the subgroup generated by a random element P is greater than n^α with probability $n^{-\alpha+o(1)}$ as $n \to \infty$. Recall that the minimal degree of a primitive subgroup of S_n is at least $\frac{1}{2}(\sqrt{n}-1)$ (see §3). It follows that a random pair of elements P, Q is contained in a primitive subgroup with probability less than $n^{-1+o(1)}$ and therefore P, Q generates A_n or S_n with probability $1 - n^{-1+o(1)}$.

Very recently Cameron proposed a conjecture which is closely related to Dixon's theorem: A random element Π of S_n lies in no transitive subgroup other than A_n or S_n with probability $1 - n^{-\frac{1}{2}+o(1)}$.

Analogues of 6.1 were proved for the finite simple classical groups by Kantor and Lubotzky [**KL1**] (CFSG). Properties of a random element of $GL(n,q)$ are considered in [**St**].

The following lovely result of Isaacs, Kantor and Spaltenstein [**IKS**] extends statistical investigations in a different direction.

THEOREM 6.2 (CFSG). *Let G be a subgroup of S_n and p a prime dividing $|G|$. Then the order of a random element of G is divisible by p with probability at least* $1/n$.

Let us comment now on the structure of a "typical" subgroup of S_n. The results of section 4 together with some heuristic arguments suggest that almost all subgroups of S_n have structure similar to an elementary abelian 2-group with about $n/2$ orbits. For example Kantor conjectures that almost all subgroups of

S_n are nilpotent. Another plausible conjecture is that almost all subgroups of S_n are not transitive (by 4.3 this is true at least for almost all solvable subgroups of S_n).

All of the preceding conjectures should also be true for elements of almost all conjugacy classes of subgroups of S_n (note that the size of any conjugacy class of subgroups is negligible compared to $|Sub(S_n)|$).

During the conference Kantor suggested the following as "everybody's conjecture": Almost all subgroups of S_n have very large base size.

Confirming this we give a CFSG-free proof of the following.

THEOREM 6.3. *There exists a constant $\beta > 0$ such that almost all subgroups of S_n have minimal base size at least βn.*

In the proof we are going to use the following consequence of the estimates 2.4 and 2.5 for the orders of primitive groups: primitive groups of degree d can be generated by $c\sqrt{d}\,(\log d)^2$ elements for some constant c. It will be convenient to use the fact that any subgroup of S_d can be generated by $d-1$ elements (see section 1).

LEMMA 6.4. *There exists a constant m_0 such that if $G \leq S_\Omega$ ($|\Omega| = n$) and $m \geq m_0$ then G is generated by a set of elements R, $|R| \leq \frac{n}{(\log m)^2}$ and a subgroup H with no orbits of length greater than m.*

PROOF. We will define a series of subgroups $G = G_0 \geq G_1 \geq \ldots \geq G_t = H$ inductively. We will associate to G_i a set $\mathcal{S}_i$ of disjoint subsets of Ω of size at least $\sqrt{m}$ each of which is the union of some orbits of G_i. Furthermore we will associate to each G_i two sets of elements P_i and Q_i such that G_i, P_i and Q_i together generate G, $|P_i| \leq |\mathcal{S}_i|$ and

$$|Q_i| \leq \frac{i + t_i}{\left(\frac{4m^{\frac{1}{4}}}{c(\log m)^2}\right)}$$

where t_i is the number of orbits of G_i.

Let P_0, Q_0 and $\mathcal{S}_0$ be empty.

Suppose we have defined $G_0, \ldots, G_i$ and G_i has an orbit Δ of size greater than m. Consider a decomposition $B_1 \cup \cdots \cup B_d\, (d \geq 2)$ of Δ into blocks of imprimitivity with d as small as possible (we allow $d = |\Delta|$). Define G_{i+1} as the intersection of the setwise stabilizers of the blocks B_j. Clearly G_{i+1} is normal in G_i and G_i/G_{i+1} acts (faithfully) as a primitive group on the B_j.

Let R_{i+1} be a minimal set of generators for G_i/G_{i+1}.

If $d > \sqrt{m}$ then we take $P_{i+1} = P_i$, $Q_{i+1} = Q_i \cup R_{i+1}$ and $\mathcal{S}_{i+1} = \mathcal{S}_i$. We have

$$|R_{i+1}| \leq c\sqrt{d}\,(\log d)^2 = \frac{d}{\left(\frac{\sqrt{d}}{c(\log d)^2}\right)} \leq \frac{d}{\left(\frac{4m^{\frac{1}{4}}}{c(\log m)^2}\right)}.$$

Now $(i+1)+t_{i+1}=i+t_i+d$ and therefore we obtain

$$|Q_{i+1}| \le \frac{(i+1)+t_{i+1}}{\left(\frac{4m^{\frac{1}{4}}}{c(\log m)^2}\right)}$$

by induction.

If $d \le \sqrt{m}$ then we take $P_{i+1}=P_i \cup R_{i+1}$ and $Q_{i+1}=Q_i$. Also in this case we obtain $\mathcal{S}_{i+1}$ from $\mathcal{S}_i$ by adding the sets $B_1,\dots,B_d$ and deleting the element of $\mathcal{S}_i$ that contains Δ (if there is such as element). Note that the B_j have size $\frac{|\Delta|}{d} > \sqrt{m}$. Now $|P_{i+1}| \le |P_i|+|R_{i+1}| \le |P_i|+d-1 \le |\mathcal{S}_i|+d-1 \le |\mathcal{S}_{i+1}|$ (by induction).

If G_i has no orbits of size greater than m then we set $G_i=H$ and $R=P_i \cup Q_i$.

It is straightforward to see that $|R| \le \frac{n}{\sqrt{m}} + \frac{2n}{\left(\frac{4m^{\frac{1}{4}}}{c(\log m)^2}\right)}$, which is less than $\frac{n}{(\log m)^2}$ for m sufficiently large.

The proof of the lemma is complete. □

The maximal subgroups of S_n with given orbits are called Young subgroups. We will need some simple facts about such subgroups. It is easy to see that there are at most $n!$ such subgroups. If Y is a Young subgroup of S_n with no orbits of size greater than m then $|Y| \le m^n$.

The proof of Theorem 6.3:

We may suppose that in Lemma 6.4 we have $m_0 \ge 2^{(2^7)}$. Define t by $t = [\log\log n - \log\log m_0]$. Applying the lemma inductively we associate to each subgroup G of S_n a series of subgroups $G=G_0 \ge G_1 \ge \dots \ge G_t$ such that G_i has no orbits of size greater than $n_i = n^{\frac{1}{2^i}}$ and G_{i-1} is generated by G_i and a set R_i of size at most $\frac{n}{(\log n_i)^2}$. Note that we have $m_0^2 \ge n_t \ge m_0$ by the choice of t.

Claim. For almost all subgroups G of S_n, $|G_t| \ge 2^{\epsilon n}$ where $\epsilon = \frac{1}{128 \log m_0}$.

To confirm the claim we estimate the number of groups G with $|G_t| < 2^{\epsilon n}$. Clearly G_t can be generated by $[\epsilon n]$ elements. G_t is a subgroup of at least one of the Young subgroups Y of S_n with all orbits smaller than m_0^2. In each such group Y we can choose $[\epsilon n]$ elements in at most $|Y|^{\epsilon n} \le \left((m_0^2)^n\right)^{\epsilon n}$ ways.

There are less than $n!$ possible groups Y, therefore the number of choices for G_t is at most $n!\, m_0^{2\epsilon n^2} = n!\, 2^{\frac{n^2}{64}}$.

We can apply the same argument to estimate the number of choices for G_{i-1} with a given G_i. G_{i-1} is contained in a Young subgroup Y with all orbits of size less than m_{i-1}. Given G_i and Y the number of choices for the set R_i is at most $(n_{i-1}^n)^{|R_i|}$.

Using the definition of the G_i we obtain that

$$(n_{i-1}^n)^{|R_i|} = n_i^{2n|R_i|} \le 2^{n^2\left(\frac{2}{\log n_i}\right)} = 2^{\frac{2^{i+1} n^2}{\log n}}.$$

It follows that the total number of choices for G is at most

$$(n!)^{t+1} \cdot 2^{\frac{n^2}{64}} \cdot 2^{\frac{n^2}{\log n}(2^2+\cdots+2^{t+1})} \leq (n!)^{\log\log n} 2^{n^2(\frac{1}{64}+\frac{2^{t+2}}{\log n})}.$$

We have

$$2^{t+2} \leq 2^{\log\log n - \log\log m_0 + 2} \leq \frac{4\log n}{\log m_0} \leq \frac{4\log n}{2^7} = \frac{\log n}{32}.$$

Therefore the number of choices for G is at most $(n!)^{\log\log n} \cdot 2^{\frac{3n^2}{64}}$ and this is negligible compared to $|Sub(S_n)|$ by 4.2. This proves the claim.

Suppose now that $|G_t| \geq 2^{\epsilon n}$ for some G. If B is a base of G then it is also a base of G_t. G_t has no orbits of size greater than m_0^2, and it follows by an easy argument that $|G_t| \leq (m_0^2)^{|B|}$. Therefore we have $|B| \geq (\frac{\epsilon}{2\log m_0})n$. The proof of the theorem is complete. □

The results and conjectures of this section suggest that if we want to develop a meaningful theory of random subgroups of S_n we have to avoid the most obvious definitions of a random permutation group.

References

[Als] B. Alspach, *A combinatorial proof of a conjecture of Goldberg and Moon*, Canad. Math. Bull. **11** (1968), 655–661.

[As] M. Aschbacher, *On the maximal subgroups of the finite classical groups*, Invent. Math. **76** (1984), 469–514.

[Au] M. S. Audu, *Transitive permutation groups of prime-power order,*, Ph.D. thesis, Oxford, 1983.

[AG1] M. Aschbacher and R. M. Guralnick, *Solvable generation of groups and Sylow subgroups of the lower central series*, J. Algebra **77** (1982), 189–201.

[AG2] M. Aschbacher and R. M. Guralnick, *On abelian quotients of primitive groups*, Proc. Amer. Math. Soc. **107** (1989), 89–95.

[Ba1] L. Babai, *On the complexity of canonical labeling of strongly regular graphs*, SIAM J. Comp. **9** (1980), 212–216.

[Ba2] ———, *On the order of uniprimitive permutation groups*, Annals of Math. **113** (1981), 553–568.

[Ba3] ———, *On the order of doubly transitive permutation groups*, Invent. Math. **65** (1982), 473–484.

[Ba4] ———, *A Las Vegas-NC algorithm for isomorphism of graphs with bounded multiplicities of eigenvalues*, Proc. 27th IEEE FOCS (1986), 303–312.

[Ba5] ———, *On the length of chains of subgroups in the symmetric group*, Comm. in Algebra **14** (1986), 1729–1736.

[Ba6] ———, *The probability of generating the symmetric group*, J. Comb. Theory (A) **52** (1989), 148–153.

[Ba7] ———, *Computational complexity in finite groups*, Proc. Int. Congress of Math., Kyoto, Japan, 1990, Springer, Tokyo, 1991, pp. 1479–1489.

[BBT] L. Babai, R. Beals and P. Takácsi-Nagy, *Symmetry and complexity*, Proc. 24th ACM STOC (1992), 438–449.

[BCFLS] L. Babai, G. Cooperman, L. Finkelstein, E. M. Luks and Á. Seress, *Fast Monte Carlo algorithms for permutation groups*, Proc. 23rd ACM STOC (1991), 90–100.

[BCFS] L. Babai, G. Cooperman, L. Finkelstein and Á. Seress, *Nearly linear time algorithms for permutation groups with a small base*, Proc. ISSAC'91, (Int. Symp. on Symbolic and Algebraic Computation), Bonn (1991), 200–209.

[BCP] L. Babai, P. J. Cameron and P. P. Pálfy, *On the orders of primitive groups with restricted nonabelian composition factors*, J. Algebra **79** (1982), 161–168.

[BKL] L. Babai, W. M. Kantor and E. M. Luks, *Computational complexity and the classification of finite simple groups*, Proc. 24th IEEE FOCS (1983), 162–171.

[BKR] R. M. Bryant, L. G. Kovács and G. R. Robinson, *personal communication.*

[BL] L. Babai and E. M. Luks, *Canonical labeling of graphs*, Proc. 15th ACM STOC (1983), 171–183.

[BLS1] L. Babai, E. M. Luks and Á. Seress, *Permutation groups in NC*, Proc. 19th ACM STOC (1987), 409–420.

[BLS2] L. Babai, E. M. Luks and Á. Seress, *Fast management of permutation groups*, 28th IEEE FOCS (1988), 272–282.

[Bov] J. D. Bovey, *The probability that some power of a permutation has small degree*, Bull. London Math. Soc. **12** (1980), 47–51.

[BP1] L. Babai and L. Pyber, *Permutation groups without exponentially many orbits on the power set*, J. Comb. Theory (A) (to appear).

[BP2] A. Borovik and L. Pyber, *On the number of maximal solvable subgroups of a finite group*, in preparation.

[BS1] L. Babai and Á. Seress, *On the diameter of Cayley graphs of symmetric groups*, J. Comb. Theory (A) **49** (1988), 175–179.

[BS2] L. Babai and V. T. Sós, *Sidon sets in groups and induced subgraphs of Cayley graphs*, Europ. J. Comb. **6** (1985), 101–114.

[Cam1] P. J. Cameron, *Finite permutation groups and finite simple groups*, Bull. London Math. Soc. **13** (1981), 1–22.

[Cam2] ———, *Oligomorphic permutation groups*, LMS Lecture Note Series 152, Cambridge Univ. Press, 1990.

[Cam3] ———, *Some open problems on permutation groups*, Proc. Durham Symp. Groups and Combinatorics (1990) (to appear).

[Cam4] ———, *Permutation groups*, Handbook of Combinatorics (to appear).

[CF] R. Carter and P. Fong, *The Sylow 2-subgroups of the classical groups*, J. Algebra (1964), 139–151.

[CKS] C. W. Curtis, W. M. Kantor and G. M. Seitz, *The 2-transitive representations of the finite Chevalley groups*, Trans. Amer. Math. Soc. **218** (1976), 1–57.

[CPSS] P. J. Cameron, C. E. Praeger, J. Saxl and G. M. Seitz, *On the Sims conjecture and distance transitive graphs*, Bull. London Math. Soc. **15** (1983), 499–506.

[CST] P. J. Cameron, R. Solomon and A. Turull, *Chains of subgroups in symmetric groups*, J. Algebra **127** (1989), 340–352.

[CWW] R. J. Cook, J. Wiegold and A. G. Williamson, *Enumerating subgroups*, J. Austral. Math. Soc. (A) **43** (1987), 220–223.

[Dix1] J. D. Dixon, *The Fitting subgroup of a linear solvable group*, J. Austral. Math. Soc. **7** (1967), 417–424.

[Dix2] ———, *The maximum order of the group of a tournament*, Canad. Math. Bull. **10** (1967), 503–505.

[Dix3] ———, *The solvable length of a solvable linear group*, Math. Z. **107** (1968), 151–158.

[Dix4] ———, *The probability of generating the symmetric group*, Math. Z. **110** (1969), 199–205.

[DK] J. D. Dixon and L. G. Kovács, *Generating finite nilpotent irreducible linear groups*, preprint.

[ET1] P. Turán and P. Erdős, *On some problems of statistical group theory,* I, Zeitschr. für Warscheinlichkeitstheorie und verw. Gegiete **4** (1965), 175–182.

[ET2] ———, *On some problems of statistical group theory,* II, Acta. Math. Hung. **18** (1967), 151–163.

[ET3] ———, *On some problems of statistical group theory,* VII, Periodica Math. Hung. **2** (1972), 149–163.

[Gl] D. Gluck, *On the $k(GV)$ problem*, J. Algebra **89** (1984), 46–55.

[GM] M. Goldberg and J. W. Moon, *On the maximum order of the group of a tournament*, Canad. Math. Bull. **9** (1966), 563–569.

[Gon] V. L. Goncharov, *On the field of Combinatory Analysis*, Isvestija Akad. Nauk., SSSR, Ser. mat. 8 (1944), 3–48 (Russian); English transl.: Transl. AMS, Ser. 2 **19** (1962), 1–46.

[Gor] D. Gorenstein, *Finite simple groups: an introduction to their Classification*, Plenum, New York, 1982.

[Gr] F. Gross, *Automorphisms of induced extensions*, J. Algebra **117** (1988), 457–471.

[Gur] R. M. Guralnick, *On the number of generators of a finite group*, Arch. Math. **53** (1989), 521–523.

[He] H. Heineken, *Nilpotent subgroups of finite soluble groups*, Arch. Math. **56** (1991), 417–423.

[Her] C. Hering, *Transitive linear groups and linear groups which contain irreducible subgroups of prime order,* II, J. Algebra **93** (1985), 191–198.

[Hi] D. G. Higman, *Invariant relations, coherent configurations and generalized polygons*, Combinatorics (M. Hall,Jr. and J. H. van Lint, eds.), D. Reidel, Dordrecht, 1975, pp. 347–363.

[Hu] B. Huppert, *Endliche Gruppen*, Springer, Berlin, 1967.

[IKS] I. M. Isaacs, W. M. Kantor and N. Spaltenstein, *On the probability that a group element is p-singular*, J. Algebra (to appear).

[Jer] M. Jerrum, *A compact representation for permutation groups*, J. Algorithms **7** (1986), 36–41.

[Ka1] W. M. Kantor, *Permutation representations of the finite classical groups of small degree or rank*, J. Algebra **60** (1979), 158–168.

[Ka2] ______, *Some topics in asymptotic group theory*, Proc. Durham Symp. Groups and Combinatorics (1990) (to appear).

[KL1] W. M. Kantor and A. Lubotzky, *The probability of generating a finite classical group*, Geom. Dedicata **36** (1990), 67–87.

[KL2] W. M. Kantor and E. M. Luks, *Computing in quotient groups*, Proc. 22nd ACM STOC (1990), 524–534.

[KL3] P. B. Kleidman and M. W. Liebeck, *The subgroup structure of the finite classical groups*, LMS Lecture Note Series 129 (1990), Cambridge Univ. Press.

[KN] L. G. Kovács and M. F. Newman, *Generating transitive permutation groups*, Quarterly J. Math. Oxford (2) **39** (1988), 361–372.

[KP] L. G. Kovács and C. E. Praeger, *Finite permutation groups with large abelian quotients*, Pacific J. Math. **136** (1989), 283–292.

[KR1] L. G. Kovács and G. R. Robinson, *Generating finite completely reducible linear groups*, Proc. Amer. Math. Soc. **112** (1991), 357–364.

[KR2] ______, *On the number of conjugacy classes of a finite group*, preprint.

[Lan] E. Landau, *Handbuch der Lehre van der Verteilung der Primzahlen, Vol.* I, Teubner, Leipzig, 1909, pp. 222–229.

[Li] M. W. Liebeck, *On the minimal degrees and base sizes of primitive groups*, Arch. Math. **43** (1984), 11–15.

[LS] M. W. Liebeck and J. Saxl, *Minimal degrees of primitive permutation groups with an application to monodromy groups of covers of Riemann surfaces*, Proc. London Math. Soc. (3) **63** (1991), 266–314.

[Luc] A. Lucchini, *A bound on the number of generators of a finite group*, Arch. Math. **53** (1989), 313–317.

[Lu1] E. M. Luks, *Isomorphism of graphs of bounded valence can be tested in polynomial time*, J. Comp. Sys. Sci. **25** (1982), 42–65.

[Lu2] ______, *Computing in solvable matrix groups*, preprint.

[Ma1] A. Mann, *Generators of p-groups*, Proc. Groups-St. Andrews 1985 (1986), Cambridge, 273–281.

[Ma2] ______, *Some applications of powerful p-groups*, Proc. Groups-St. Andrews 1989, Cambridge, 1991, pp. 370–385.

[MN] A. McIver and P. M. Neumann, *Enumerating finite groups*, Quart. J. Math. Oxford (2) **38** (1987), 473–488.

[Mü] T. Müller, *Counting subgroups of finite index in free products of finite groups*, preprint.

[Ne] M. F. Newman, *The soluble length of soluble linear groups*, Math. Z. **126** (1972), 59–70.

[NV] P. M. Neumann and M. R. Vaughan-Lee, *An essay on BFC groups*, Proc. London Math. Soc. (3) **35** (1977), 213–237.

[Pat] A. R. Patterson, *The minimal number of generators for p-subgroups of $GL(n,p)$*, J. Algebra **32** (1974), 132–140.

[Pá1] P. P. Pálfy, *A polynomial bound for the orders of primitive solvable groups*, J. Algebra **77** (1982), 127–137.

[Pá2] ———, *Proc. Second Int. Group Theory Conf.*, Bressanone/Brixen 1989, Suppl. Rendiconti Circ. Mat. Palermo, vol. 23, 1990, pp. 253–263.

[Po1] I. N. Ponomarenko, *The isomorphism problem for classes of graphs*, Dokl. Akad. Nauk. SSSR 304 (1989),552-556 (Russian); English translation in Soviet Math. Dokl **39** (1989), 119–122.

[Po2] ———, *The isomorphism problem for classes of graphs closed under contraction*, Zap. Nauchn. Sem. LOMI **174** (1988), 144–177 (Russian); English translation in J. Sov. Math. **55** (1991), 1621–1643.

[Py1] L. Pyber, *Enumerating finite groups of given order*, Annals of Math (to appear).

[Py2] ———, *Finite groups have many conjugacy classes*, J. London Math. Soc. (2) (to appear).

[Py3] ———, *On the orders of doubly transitive permutation groups, elementary estimates*, J. Comb. Theory (A) (to appear).

[Py4] ———, *The minimal degree of primitive permutation groups*, Handbook of Combinatorics (to appear).

[Py5] ———, *Pálfy–Wolf type theorems for completely reducible subgroups of $GL(n,p^\alpha)$*, in preparation.

[Py6] ———, *Transitive graphs and minimally transitive groups*, in preparation.

[PS] C. E. Praeger and J. Saxl, *On the order of primitive permutation groups*, Bull. London Math. Soc. **12** (1980), 303–308.

[Se] Y. Segev, *On completely reducible subgroups of $GL(n,\Delta)$*, Israel J. Math. **51** (1985), 163–176.

[SST] G. M. Seitz, R. Solomon and A. Turull, *Chains of subgroups in groups of Lie type,* II, J. London. Math. Soc. (2) **42** (1990), 93–100.

[Si1] C. C. Sims, *Determining the conjugacy classes of a permutation group*, Computers in Algebra and Number Theory, SIAM–AMS Proc. 4, 1970, pp. 191–195.

[Si2] ———, *Computational Problems in Abstract Algebra* (J. Leech, ed.), Pergamon Press, New York, 1970, pp. 169–183.

[Si3] ———, *Proc. Second Symp. Symbolic and Algebraic Manipulation*, ACM, New York, 1971, pp. 23–28.

[ST1] R. Solomon and A. Turull, *Chains of subgroups in groups of Lie type,* I, J. Algebra **132** (1990), 174–184.

[ST2] ———, *Chains of subgroups in groups of Lie type,* III, J. London Math. Soc. (2) **44** (1991), 437–444.

[St] R. Stong, *Some asymptotic results on finite vector spaces*, Adv. in Appl. Math. **9** (1988), 167–199.

[Sup] D. A. Suprunenko, *Matrix groups*, AMS, Providence, 1976.

[SW] J. A. M. Shepperd and J. Wiegold, *Transitive groups and groups with finite derived groups*, Math. Z. **81** (1963), 279–285.

[TZ] A. Turull and A. Zame, *Number of prime divisors and subgroup chains*, Arch. Math. **55** (1990), 333–341.

[Wa] G. E. Wall, *Some applications of the eulerian functions of a finite group*, J. Austral. Math. Soc. 2 (1961), 35–59.

[We] A. Weir, *Sylow p-subgroups of the classical groups over finite fields with characteristic prime to p*, Proc. Amer. Math. Soc. **6** (1955), 529–533.

[Wi] J. Wiegold, *Growth sequences of finite groups* III, J. Austral. Math. Soc. (A) **25** (1978), 142–144.

[Wie1] H. Wielandt, *Abschatzungen für den Grad einer Permutationsgruppe von vorgeschriebenen Transitivitatsgrad*, Dissertation, Berlin, 1934, Schriften des Math. Seminars und des Instituts für angew, Math. der Universitat, Vol. 2, Berlin, 1934, pp. 151–174.

[Wie2] ———, *Finite permutation groups*, Acad. Press, New York, 1964.

[Wie3] ———, *Permutation groups through invariant relations and invariant functions*, Lecture Notes, Ohio State University, Columbus, Ohio, 1969.

[Wil] H. S. Wilf, *The asymptotics of* $e^{P(z)}$ *and the number of elements of each order in* S_n, Bull. Amer. Math. Soc. **15** (1986), 228–232.

[Wo] T. R. Wolf, *Solvable and nilpotent subgroups of* $GL(n, q^m)$, Canad. J. Math. **34** (1982), 1097-1111.

[ZKT] V. M. Zemlyachenko, N. M. Kornienko and R. I. Tyshkevich, *Graph isomorphism problem*, Zap. Nauchn. Sem. LOMI **118** (1982), 83–158 (Russian); English translation in J. Sov. Math. **29** (1985), 1426–1481.

MATHEMATICAL INSTITUTE OF THE HUNGARIAN ACADEMY OF SCIENCES, BUDAPEST, P.O.B. 127, 1364 HUNGARY

E-mail address: H1130Pyb@huella.bitnet

DIMACS Series in Discrete Mathematics
and Theoretical Computer Science
Volume **11**, 1993

Computations in Associative Algebras

LAJOS RÓNYAI

ABSTRACT. We give an overview of algorithmic results on computational problems related to structural invariants of associative algebras over fields which support exact (symbolic) computations, in most cases algebras over finite fields or algebraic number fields. Such problems are the computation of the radical, the simple components, direct decompositions of modules, etc. We are interested in algorithms which are efficient from the perspective of complexity theory. For this reason, for most of the paper, polynomial time algorithms (both deterministic and randomized) are considered. We discuss applications of algorithms for associative algebras to various computational problems related to group representations.

1. Introduction

Our aim is to provide a survey on algorithmic results related to associative algebras. In most cases we will be interested in polynomial time methods (deterministic and/or randomized) for computing various invariants, structural ingredients, decompositions, attached structures, etc. First we give some basic definitions on algebras. For more details the reader is referred to Pierce [**57**], Herstein [**35**], Kertész [**44**], and Reiner [**62**]. A linear space $\mathcal{A}$ over the field F is an *associative algebra* over F if it is equipped with a binary, F-bilinear operation (called multiplication). This multiplication is associative:

$$x(yz) = (xy)z \textit{ holds for every } x, y, z \in \mathcal{A}.$$

We shall assume throughout that $\mathcal{A}$ is finite dimensional over F, i.e. $dim_F \mathcal{A} = n < \infty$. Recall that $\mathcal{A}$ is a *commutative algebra* if $xy = yx$ holds for every $x, y \in \mathcal{A}$. An F-subspace $\mathcal{B}$ of $\mathcal{A}$ is a *subalgebra* of $\mathcal{A}$ if $\mathcal{B}$ is closed under multiplication. An important example is the *center* $C(\mathcal{A}) = \{x \in \mathcal{A};\ xy = yx \textit{ for every } y \in \mathcal{A}\}$.

1991 *Mathematics Subject Classification.* Primary 68Q40, 11Y40; Secondary 68Q25.

Research partially supported by Hungarian National Foundation for Scientific Research (OTKA), Grant 2581.

This paper is in final form and no version of it will be submitted for publication elsewhere.

Clearly we have $F \leq C(\mathcal{A})$, provided that $\mathcal{A}$ has an identity element $1 = 1_{\mathcal{A}}$. In this case $\mathcal{A}$ is *central* (over F) if $F = C(\mathcal{A})$. An F-subspace $\mathcal{I}$ of $\mathcal{A}$ is a *left ideal* of $\mathcal{A}$ if $yx \in \mathcal{I}$ holds whenever $x \in \mathcal{I}$ and $y \in \mathcal{A}$. A *right ideal* is defined in an analogous fashion. An F-subspace $\mathcal{I}$ of $\mathcal{A}$ is an *ideal* of $\mathcal{A}$ if $\mathcal{I}$ is both left and right ideal of $\mathcal{A}$. If $\mathcal{I}$ is an ideal of $\mathcal{A}$ then we can form the *factor algebra* $\mathcal{A}/\mathcal{I}$. The notions of *homomorphism* and *$\mathcal{A}$-module* are used in the standard way. A pair of nonzero elements $x, y \in \mathcal{A}$ is a pair of *zero divisors* in $\mathcal{A}$ if $xy = 0$. A nonzero element $e \in \mathcal{A}$ is an *idempotent* if $e^2 = e$. An idempotent e is a *primitive idempotent* of $\mathcal{A}$ if there is no idempotent $f \in \mathcal{A}$ such that $f \neq e$ and $fe = f$.

An algebra $\mathcal{A}$ is *simple* if it has only trivial ideals (i.e., (0) and $\mathcal{A}$) and $\mathcal{A}\mathcal{A} \neq (0)$. We say that $\mathcal{A}$ is the direct sum of its (left) ideals $\mathcal{A}_1, \ldots, \mathcal{A}_k$ (written as $\mathcal{A}_1 \oplus \cdots \oplus \mathcal{A}_k$) if $\mathcal{A}$ is the direct sum of these linear subspaces.

Examples.

a) If the field K is a finite algebraic extension of the field F then K is a finite dimensional simple and commutative algebra over F.

b) $M_n(F)$, the algebra of all n by n matrices over the field F. $M_n(F)$ is a simple algebra over F and $dim_F M_n(F) = n^2$.

c) Subalgebras of $M_n(F)$, i.e. linear subspaces of $M_n(F)$ closed under matrix multiplication.

The latter examples are typical. If $\mathcal{A}$ has an identity element, then $\mathcal{A}$ can be efficiently embedded as a subalgebra of $M_n(F)$. This is easily seen using the *regular representation* . For $x \in \mathcal{A}$ we define the linear map $R_x : \mathcal{A} \to \mathcal{A}$ as $R_x(y) = xy$ for every $y \in \mathcal{A}$. It is straightforward that R is an algebra homomorphism of $\mathcal{A}$ to the algebra of linear transformations of the linear space $\mathcal{A}$. Moreover, if $\mathcal{A}$ has an identity element then R is an injective map. If $\mathcal{A}$ has no identity element then first we can adjoin one using the Dorroh extension ([**44**] p 43). This increases the dimension by one and now the regular representation gives an embedding of $\mathcal{A}$ into $M_{n+1}(F)$.

An element $x \in \mathcal{A}$ is *nilpotent* if $x^m = 0$ for some positive integer exponent m. Similarly $x \in \mathcal{A}$ is *strongly nilpotent* if xy is nilpotent for every $y \in \mathcal{A}$. The *radical* $Rad(\mathcal{A})$ of $\mathcal{A}$ is the set of strongly nilpotent elements of $\mathcal{A}$. It can be seen that $Rad(\mathcal{A})$ is an ideal of $\mathcal{A}$ and the factor algebra $\mathcal{A}/Rad(\mathcal{A})$ has no strongly nilpotent elements, except 0. An algebra $\mathcal{A}$ with $Rad(\mathcal{A}) = (0)$ is a *semisimple algebra* . There is a very strong and useful structure theorem for semisimple algebras, due to Wedderburn.

WEDDERBURN'S THEOREM. *Let $\mathcal{A}$ be a finite dimensional semisimple algebra over the field F. Then $\mathcal{A}$ is a direct sum of simple algebras*

$$\mathcal{A} = \mathcal{A}_1 \oplus \mathcal{A}_2 \oplus \cdots \oplus \mathcal{A}_k,$$

where the $\mathcal{A}_i$ are the only minimal nontrivial ideals of $\mathcal{A}$. Moreover, each A_i is isomorphic to a full matrix algebra $M_{n_i}(F_i)$, where F_i is a not necessarily commutative extension field of F.

If F is finite, then the theorem of Wedderburn on finite division rings (cf. [**35**], p 71) implies that the fields F_i are commutative.

We are interested in exact (symbolic) computations. For this reason we restrict our attention to ground fields admitting efficient exact arithmetic. In most cases F will be either a finite field or an algebraic number field. Even when we consider algebras over the reals $\mathbf{R}$ or the complex numbers $\mathbf{C}$, we shall work in an appropriate algebraic number field.

Now we set some conventions about the input of the algorithmic questions considered. To obtain sufficiently general results, we consider algebras to be given by a collection of *structure constants.* If $\mathcal{A}$ is an algebra over the field F and $a_1, a_2, \ldots, a_n$ is a linear basis of $\mathcal{A}$ over F then multiplication can be described by representing the products $a_i a_j$ as linear combinations of the a_i

$$a_i a_j = \gamma_{ij1} a_1 + \cdots + \gamma_{ijn} a_n.$$

The coefficients $\gamma_{ijk} \in F$ are called structure constants. In this paper algebras are considered to be given as an array of structure constants. We shall also work with rings whose additive groups are finitely generated free Abelian groups. In this case the structure constants $\gamma_{ijk} \in \mathbf{Z}$ are given with respect to a basis over the integers $\mathbf{Z}$. Substructures (such as subalgebras, ideals, subrings, subspaces) are represented by bases whose elements are given as linear combinations of basis elements of a larger structure (algebra or ring).

If F is either a finite field, or an algebraic number field, then F can be considered as an algebra over its prime field P, therefore F can also be inputted with structure constants from P. (If F is finite then $P = F_p$ for some prime p, if F is a number field then $P = \mathbf{Q}$). In these cases F can be, and often is, specified by giving the (monic) minimal polynomial f of a single generating element α over the prime field P. Notice that this representation is a special case of the representation with structure constants. The coefficients of f give the structure constants with respect to the basis $1, \alpha, \alpha^2, \ldots, \alpha^{n-1}$ of F where $n = dim_P F$. If F is a number field, then we require that $f(x) \in \mathbf{Z}[x]$, i.e. that α is integral over the integers $\mathbf{Z}$.

Another important way to represent an algebra is in the form of a matrix algebra. In this case it suffices to specify a set of matrices which generates the algebra. The results described here are applicable in this setting as well, since from this representation one can efficiently find a basis of the algebra and then structure constants with respect to this basis. In the opposite direction, the regular representation gives an efficient method to obtain a matrix representation from structure constants.

Our main aim is to discuss algorithms which run in time polynomial in terms of the *input size* (cf.[**36**], Chapters 12-13). The size of an integer is the number of its binary digits. The size of a rational number p/q expressed in lowest terms is size(p)+size(q). Modulo p residue classes have size $\lceil \log_2(p+1) \rceil$. The size of compound objects (polynomials, vectors, matrices, etc.) is the sum of the sizes of their components.

We now briefly comment on the most important algorithmic ingredients of the methods to be described later on. There are deterministic polynomial time algorithms for the basic seminumerical problems (such as arithmetical operations in F and polynomial arithmetic over F) if F is a finite field or an algebraic number field. The reader is referred to Collins, Mignotte, and Winkler [**19**], Knuth [**45**] for more details. Eberly [**24**], in Section 1.2 discusses the parallel complexity of these problems. The basic algorithmic tasks of linear algebra (such as testing linear dependence, computing ranks, determinants, and solving systems of linear equations) can be accomplished in deterministic polynomial time over the fields considered. If F is finite then the well-known textbook methods demonstrate this point. In the number field case it will be sufficient to solve linear algebra problems over the field of rationals $\mathbf{Q}$. Polynomial time methods are available to solve systems of linear equations over $\mathbf{Q}$ and over $\mathbf{Z}$ ([**5**], [**27**], [**43**], [**53**]).

A very important algorithmic question in this context is the problem of factoring polynomials over F. Lenstra, Lenstra, and Lovász [**51**] proposed the first deterministic polynomial time procedure to factor polynomials over $\mathbf{Q}$. This result was extended to arbitrary number fields by Chistov, Grigoryev [**18**], Landau [**47**], A. K. Lenstra [**49**]. For factoring polynomials over finite fields a deterministic method was given by Berlekamp [**6**], [**52**]. The time complexity of this algorithm is a polynomial in the parameters p, s and $deg(f)$ where $f \in F_q[x]$ is the polynomial to be factored and $q = p^s$, p prime. Note that the input size in this case is $O((1 + deg(f)) \log q)$, consequently the running time of the method is not polynomial in the input size. The problem can be solved in polynomial time if we allow randomization. The first such method was given by Berlekamp [**7**]. This method belongs to a special kind of randomized algorithms. For an arbitrary input they either give correct solution or, with small probability, they report failure (i.e. never give incorrect answer). Such methods are called *Las Vegas algorithms* . The term was introduced by Babai [**1**].

Some of the algorithms discussed in the paper rely on solutions of subproblems not known to have deterministic polynomial time algorithms. These are the problem of finding the prime factorization of an integer and the problem of factoring polynomials over a finite (prime) field. We intend to make clear the dependence of the methods on these procedures. To this end we define a deterministic algorithm to be an *ff-algorithm* if it is allowed to call oracles for these two problems. Similarly an *f-algorithm* is a deterministic method which is allowed to call an oracle for factoring polynomials over finite fields. In both cases the cost of a call is the size of the input of the call.

A polynomial time f-algorithm can be considered as practical, at least in a theoretical sense, because there are Las Vegas polynomial time algorithms for factoring polynomials over finite fields. In contrast, ff-algorithms are impractical as there are no polynomial time methods for factoring integers.

The reader interested in recent developments in the above algebraic and number theoretical methods is referred to the excellent survey papers of von zur Gathen [**32**], Strassen [**68**], Lenstra and Lenstra [**50**], Bach [**4**].

The organization of the paper is as follows. In Section 2 we consider algorithms for computing the radical of an associative algebra. Section 3 covers algorithms for computing the simple direct summands of a semisimple algebra (Wedderburn decomposition). In Section 4 we review algorithms (and some negative results) for the problem of expressing a simple algebra as a direct sum of minimal left ideals. While for most of the paper the ground field is either a finite field or a number field, Section 5 contains material on algorithms for algebras over the "large" fields **R** and **C**. Finally, in Section 6 we discuss algorithms for linear representations of groups.

Acknowledgements. The author is indebted to L. Babai and W. M. Kantor for their help making possible for me to attend the Workshop.

2. Computing the radical

The algorithmic problem considered here is the following. Given is an associative algebra $\mathcal{A}$ as a collection of structure constants over the ground field F. Our objective is to find a basis over F of $Rad(\mathcal{A})$. The elements of this basis are expressed as F-linear combinations of the input basis $a_1, a_2, \dots, a_n$ of $\mathcal{A}$ (with respect to which the the structure constants are interpreted). If F is an algebraic number field, then a nice solution is provided by a classic theorem of Dickson [**21**], p 106, see also Friedl and Rónyai [**31**].

DICKSON'S THEOREM. *Let F be a field, $charF = 0$ and let $\mathcal{A} \leq M_k(F)$ be a subalgebra of a full matrix algebra over F. Then*

$$Rad(\mathcal{A}) = \{x \in \mathcal{A};\ Tr(xy) = 0 \ for\ every\ y \in \mathcal{A}\}.$$

Using the fact that Tr is an F-linear function we obtain

COROLLARY 2.1. *Let $\mathcal{A}$ be an algebra of matrices over the field F, $charF = 0$. Let $a_1, a_2, \dots, a_n$ be a basis of $\mathcal{A}$ over F. Then*

$$Rad(\mathcal{A}) = \{x \in \mathcal{A};\ Tr(xa_i) = 0 \ for\ i = 1, \dots, n\}.$$

Corollary 2.1 shows that the radical can be described by a system of linear equations over F. It suffices to express x as a linear combination (with unknown coefficients) of $a_1, a_2, \dots, a_n$. If $\mathcal{A}$ is inputted with structure constants, then we first embed it into a matrix algebra (regular representation). The procedure outlined runs in deterministic time polynomial in the input size.

We mention here that Drazin [**23**] discusses further connections between algebraic properties and the trace form in matrix algebras over fields of characteristic zero.

We turn now to the case when F is finite. Given a finite dimensional associative algebra $\mathcal{A}$ over the field F_q by a collection of structure constants, we want to find a basis of $Rad(\mathcal{A})$, in time polynomial in the input size, i.e. in the parameters $n = dim_{F_q}\mathcal{A}$ and $\log q$.

To illustrate the difficulties in this case, consider the algebra $\mathcal{S}$ of all p by p scalar matrices over F_p. Clearly $\mathcal{S} \cong F_p$ is semisimple and $\mathcal{S} \leq M_p(F_p)$. On the other hand, the trace form is identically zero on $\mathcal{S}$, demonstrating that we can not use directly a Dickson-type trace condition.

It suffices to solve the problem for algebras over the prime field F_p, because $\mathcal{A}$ can be considered as an algebra over F_p, and the radical, the set of strongly nilpotent elements of $\mathcal{A}$, does not depend on the ground field. Also, this change does not affect the input size. From an F_p-basis of $Rad(\mathcal{A})$ a basis over F_q is easily obtained.

In [**30**] and [**63**] a sequence of ideals $I_{-1}, I_0, \ldots, I_l$ of $\mathcal{A}$ and a sequence of functions $g_i : I_{i-1} \rightarrow F_p$, $i = 0, 1, \ldots, l$ ($l \leq \log_p n$, where $n = dim_{F_p}\mathcal{A}$) is defined with the following properties.

1. $I_{-1} = \mathcal{A}$ and $I_l = Rad(\mathcal{A})$.
2. g_i is a F_p-linear function for $i = 0, \ldots, l$.
3. $I_i = \{\ x \in I_{i-1}\ ;\ g_i(xy) = 0\ for\ every\ y \in \mathcal{B}\ \}$. (Here $\mathcal{B}$ is the Dorroh extension of $\mathcal{A}$).
4. $g_i(x)$ can be computed for any $x \in \mathcal{A}$ in time polynomial in n and $\log p$.

These properties imply that from a basis of I_{i-1} one can obtain a basis of I_i by solving a system of linear equations over F_p. By the linearity of the functions g_i and by property 4 the coefficients of this system can be obtained in polynomial time. If we start with $I_{-1} = \mathcal{A}$ then we obtain $Rad(\mathcal{A})$ using $l + 1 = O(\log n)$ iterations, therefore (a basis of) $Rad(\mathcal{A})$ over F_p is computed at the cost of $(n + \log p)^{O(1)}$ bit operations.

The theoretical basis of the radical-algorithm is a characterization of the radical with trace-like functions [**63**]. This result was extended by Eberly [**24**] and recently in [**40**], Theorem 3.1. We state here the latter result.

THEOREM 2.2. *Let n be a positive integer, R an integral domain of characteristic zero and p be a rational prime such that the principal ideal pR generated by the element $p1_R$ is a maximal ideal of R. Let F denote the residue class field $F = R/pR$. For nonnegative integers j consider the following trace functions $\phi_j : M_n(R) \rightarrow R/p^{j+1}R$ defined as*

$$\phi_j : X \mapsto Tr(X^{p^j}) + p^{j+1}R.$$

We have the following.

(i) *The functions ϕ_j can be considered as functions on*

$$M_n(F) \cong M_n(R)/pM_n(R),$$

i.e. for all nonnegative integers j and for all $X, Y \in M_n(R)$,

$$X \equiv Y \pmod p \Rightarrow \phi_j(X) = \phi_j(Y).$$

(ii) *The function ϕ_j is p^j-linear on subrings where ϕ_i vanishes for $0 \leq i < j$, i.e. if $\mathcal{A}$ is a multiplicatively closed subset of $M_n(R)$ such that $\phi_i(Z) = 0$ for all $Z \in \mathcal{A}$ and for all nonnegative integers $i < j$, then for all scalars $a, b \in R$ and matrices $X, Y \in \mathcal{A}$ we have*

$$\phi_j(aX + bY) = a^{p^j}\phi_j(X) + b^{p^j}\phi_j(Y).$$

(iii) *For a matrix $X \in M_n(R)$, the matrix $X + pM_n(R) \in M_n(F)$ is nilpotent if and only if $\phi_j(X^k) = 0$ holds for all integers $0 \leq j$, $0 < k$; this is in turn equivalent to $\phi_j(X^k) = 0$ for all $1 \leq k \leq n$ and $0 \leq j$ such that $p^j \leq n$.*

(iv) *If $\mathcal{A}$ is a subalgebra of $M_n(F)$, then the subsets I_j defined by $I_{-1} := \mathcal{A}$ and*

$$I_j := \{X \in I_{j-1} \mid \forall Y \in \mathcal{A} \cup \{1_{M_n(F)}\}\ \phi_j(XY) = 0\}$$

for $j \geq 0$ are two-sided ideals of $\mathcal{A}$. For an element $X \in I_{j-1}$ ($j \geq 0$) let $\tilde{X} \in M_n(R)$ be an arbitrary representative of X. Then

$$Tr_j(X) := \frac{1}{p^j}Tr(\tilde{X}^{p^j}) + pR \in F$$

is independent from the choice of $\tilde{X}$. Moreover $Tr_j : I_{j-1} \rightarrow F$ is a p^j-linear function (I_{j-1} is considered here as an F-module).

(v) *We have*

$$I_j = \{X \in I_{j-1} \mid \forall Y \in \mathcal{A} \cup \{1_{M_n(F)}\}\ Tr_j(XY) = 0\}.$$

The sequence of ideals I_j converges to the Jacobson radical of $\mathcal{A}$ in the sense that:

$$I_j = Rad(\mathcal{A}) \text{ if } j \geq \lfloor \log_p n \rfloor.$$

This result can be applied to finite algebras if we set $R = \mathbf{Z}$. It can be applied also to algebras over the function field $F = F_q(X_1, \ldots, X_m)$. Let $g(X) \in \mathbf{Z}[X]$ be a polynomial whose reduction modulo p is the minimal polynomial of a generating element α of F_q over F_p. For the local ring $R = \mathbf{Z}[X][X_1, \ldots, X_m]_{(p)}/(g(X))$ we have the isomorphism $R/pR \cong F$. The method runs in deterministic polynomial time if m is bounded [**40**].

Finally we mention that "associative" radical algorithms can be used to compute radicals of Lie algebras over finite fields or algebraic number fields. There are deterministic polynomial time methods to compute the nilradical of a Lie algebra [**31**], [**63**]. For computing the solvable radical, Beck, Kolman and Stewart [**8**] gave a direct method based on the Killing form in characteristic zero. In the finite case, the problem can be efficiently reduced to the problem of computing the nilradical [**31**], [**63**].

3. The Wedderburn decomposition

In this section we consider algorithms for computing the Wedderburn decomposition (the minimal nontrivial ideals) of a semisimple algebra $\mathcal{A}$ over F. The input algebra $\mathcal{A}$ is given with structure constants.

From Wedderburn's Theorem we have

$$\mathcal{A} = \mathcal{A}_1 \oplus \mathcal{A}_2 \oplus \cdots \oplus \mathcal{A}_k,$$

where $\mathcal{A}_1, \mathcal{A}_2, \ldots, \mathcal{A}_k$ are the minimal ideals of $\mathcal{A}$. The task of computing (bases over F of) the ideals $\mathcal{A}_i$ can be reduced in deterministic polynomial time to the case when $\mathcal{A}$ is commutative. One can work with the center $C(\mathcal{A})$ of $\mathcal{A}$. A basis of the F-algebra $C(\mathcal{A})$ can be obtained by solving a system of linear equations over F. $C(\mathcal{A})$ is a semisimple associative algebra and its Wedderburn decomposition relates nicely to that of $\mathcal{A}$:

$$C(\mathcal{A}) = C(\mathcal{A}_1) \oplus C(\mathcal{A}_2) \oplus \cdots \oplus C(\mathcal{A}_k).$$

Having the subalgebras $C(\mathcal{A}_i)$, we can easily compute the ideals $\mathcal{A}_i$ because of the relations $\mathcal{A}_i = C(\mathcal{A}_i)\mathcal{A}$ (which in fact show that $\mathcal{A}_i$ is the ideal of $\mathcal{A}$ generated by $C(\mathcal{A}_i)$).

We can therefore assume that $\mathcal{A}$ is a commutative semisimple algebra over F. The first efficient decomposition procedure was given by Friedl [**30**], [**31**]. It is an iteration which processes sequentially a basis $a_1, \ldots, a_n$ of $\mathcal{A}$. The result is either a decomposition of $\mathcal{A}$ into a direct sum of two smaller ideals $\mathcal{A} = \mathcal{I} \oplus \mathcal{J}$ or a proof that $\mathcal{A}$ is a field, therefore indecomposable. In the former case this cutting procedure can be applied to the ideals $\mathcal{I}$ and $\mathcal{J}$ and so on, since $\mathcal{I}$ and $\mathcal{J}$ are semisimple commutative algebras and their ideals are also ideals of $\mathcal{A}$. If F is a number field, then, to control sizes, one has to "reduce" the bases resulted from the calls of the cutting algorithm.

The outline of the cutting procedure follows. A basis $a_1, \ldots, a_n$ of $\mathcal{A}$ is considered. Suppose that after processing the elements $a_1, \ldots, a_i$, the algebra $\mathcal{F}_i$ generated by $a_1, \ldots, a_i$ over F is a field ($\mathcal{F}_0 = F$). If $i = n$ then we conclude that $\mathcal{A}$ is a field, therefore it has no proper ideals. If $i < n$ then the minimal polynomial f of $b = a_{i+1}$ over the field $\mathcal{F}_i$ is computed. This involves finding the first linear dependence over $\mathcal{F}_i$ of the elements $1_{\mathcal{A}}, b, b^2, \ldots, b^n$. Next we factor f into irreducible factors over $\mathcal{F}_i$. If f is irreducible, then $\mathcal{F}_i(b)$ is a field, we set $\mathcal{F}_{i+1} = \mathcal{F}_i(b)$ and the i-th step is finished. If f is reducible then we have $f = gh$ where g and h are nonconstant relatively prime polynomials because $\mathcal{A}$ is a direct sum of fields. Now one can put $\mathcal{I} = \mathcal{A}g(b)$ and $\mathcal{J} = \mathcal{A}h(b)$. It is not hard to verify that $\mathcal{A} = \mathcal{I} \oplus \mathcal{J}$ is a proper decomposition.

The cutting procedure and therefore the method for finding the Wedderburn decomposition runs in deterministic polynomial time if F is a number field. If $F = F_q$, $q = p^k$, where p is a prime, then the algorithmic bottleneck is factoring polynomials over $\mathcal{F}_i$. We have nevertheless a polynomial time f-algorithm. If we substitute a Las Vegas (deterministic, resp.) procedure into the factoring oracle, then we obtain Las Vegas time polynomial in $dim_{F_q}\mathcal{A}$ and $\log q$ (deterministic time polynomial in $dim_{F_q}\mathcal{A}$, p and k, resp.) for computing the Wedderburn decomposition of $\mathcal{A}$.

Eberly in **[25]** presents an efficient Las Vegas algorithm which avoids iteration over the basis elements of $\mathcal{A}$, and seems to be promising from a practical perspective as well. The key idea is in the use of splitting elements (first introduced in **[24]** in an algorithmic context). Here we use a slightly different definition. An element $x \in \mathcal{A}$ is a *splitting element* of $\mathcal{A}$ if the minimal polynomial of x over F is squarefree and has maximal degree among the elements of $\mathcal{A}$. The Schwartz lemma **[67]** implies that splitting elements can be found in randomized polynomial time. If F is a sufficiently large perfect field, and $\mathcal{A}$ is a semisimple commutative algebra over F, then $\mathcal{A}$ contains a splitting element with minimal polynomial of degree $n = dim_F\mathcal{A}$. Eberly shows that the factorization of the minimal polynomial of a splitting element in this case gives the complete Wedderburn decomposition, rather than a partial decomposition. If F is a small finite field, then one can work with an extension E of F with $[E : F] < 4n$.

In **[25]** Eberly provides also efficient parallel algorithms for computing the Wedderburn decomposition of algebras over finite fields and over algebraic number fields.

In **[33]**, Gianni, Miller and Trager consider a problem more general than the computation of the Wedderburn decomposition. The structure theorem of (commutative, unital) Artinian rings implies, that a finite commutative algebra $\mathcal{A}$ is isomorphic to a product of finite local algebras. In **[33]** an algorithm is presented to find this decomposition. It is shown that it suffices to find the primitive idempotents in the Berlekamp subalgebra

$$B(\mathcal{A}) = \{x \in \mathcal{A};\ x^q = x\}$$

(here $\mathcal{A}$ is a commutative algebra over F_q, $q = p^k$). The methods for finding the primitive idempotents in $B(\mathcal{A})$ (and hence the Wedderburn decomposition of $B(\mathcal{A})$) are more direct and efficient than those suggested in **[63]**, Section 3. Their algorithms run in Las Vegas time polynomial in $dim_{F_q}\mathcal{A}$ and $\log q$ and deterministic time polynomial in $dim_{F_q}\mathcal{A}$, p and k.

They also outline an algorithm to find the Wedderburn decomposition of a semisimple commutative algebra $\mathcal{A}$ over $\mathbf{Q}$. First a $\mathbf{Z}$-order $\Gamma \leq \mathcal{A}$ is selected. (Orders are defined and discussed in Section 4.3.) Next, for an appropriate prime p the primitive idempotents of the (finite) algebra $\Gamma/p\Gamma$ are computed. Then a simple, elegant, and fast lifting method is applied to lift the primitive idempotents into idempotents of $\Gamma/p^N\Gamma$, where the exponent N is sufficiently large to allow the reconstruction of the idempotents of $\mathcal{A}$. The reconstruction requires lattice basis reduction **[51]**. The method can be implemented to run in deterministic polynomial time. The algorithm was adapted in **[40]** to algebras over function fields over finite fields. As a sample result, we mention that the Wedderburn decomposition of a semisimple algebra over the field $F_q(X_1, \ldots, X_m)$

can be computed in randomized polynomial time, provided that m is fixed.

4. The decomposition of simple algebras

Suppose that $\mathcal{A}$ is a simple algebra over F. Then, by Wedderburn's Theorem $\mathcal{A}$ is isomorphic to a full matrix algebra $M_k(D)$, where k is a positive integer and D is a division ring, with $[D:F] < \infty$ and $F \leq C(D)$. The *explicit isomorphism problem* (*EI* for short) is formulated as follows: suppose that $\mathcal{A}$ is a simple algebra over F, given by structure constants; find k and (a basis over F of) D, and an F-algebra isomorphism $\phi : \mathcal{A} \to M_k(D)$ (i.e. represent $\mathcal{A}$ as the algebra of all linear transformations of a k-dimensional linear space over D).

A closely related algorithmic question is the *problem of zero divisors* (*ZD* for short): an associative algebra $\mathcal{A}$ over F is given by structure constants; find a pair of zero divisors in $\mathcal{A}$ (or verify that $\mathcal{A}$ contains no zero divisors). From the results discussed in Sections 2-3 it is clear, that the problem admits a polynomial time solution (deterministic if F is a number field, randomized if F is finite) if $\mathcal{A}$ is not simple. It is also clear that from a solution of *EI* for a simple algebra $\mathcal{A}$ we can easily obtain a solution of *ZD* for $\mathcal{A}$ as well.

4.1. Simple algebras over finite fields.

If F is finite then the problems *EI* and *ZD* turn out to be equivalent. In [**63**], Section 4, a polynomial time f-algorithm is described to solve *ZD*. As we have remarked, the interesting case is when $\mathcal{A}$ is simple. The solution of this case is based on an almost constructive proof of Wedderburn's theorem on finite division algebras given in [**35**] pp. 71-72. One possible way to phrase Wedderburn's theorem is to state that a finite noncommutative algebra $\mathcal{A}$ must contain zero divisors, and indeed, Herstein's argument exhibits a pair of zero divisors in $\mathcal{A}$. From this theorem we know, that if $\mathcal{A}$ is simple, then $\mathcal{A} \cong M_k(D)$, where D is a *field*. In [**63**] Herstein's proof is adapted to full matrix algebras, with the purely existential steps replaced by constructive ones.

Next we discuss some applications of the *ZD*-algorithm, the first one being a solution to *EI*. From results considered so far one can easily see the existence of an efficient f-algorithm to decide if a given finite algebra $\mathcal{A}$ is isomorphic to a full matrix algebra. It suffices to test if $Rad(\mathcal{A}) = (0)$ and if $\mathcal{A}$ is directly indecomposable. In case of an affirmative answer, say if $\mathcal{A} \cong M_l(D)$, then we can also find l and D. The problem remains to establish an explicit isomorphism from $\mathcal{A}$ to $M_l(D)$. To this end, it suffices to find an idempotent $e \in \mathcal{A}$ such that $rank(e) = 1$ (here e is viewed as an element of $M_l(D)$). We put $V = M_l(D)e$. It is well-known that $dim_D V = l$ and $M_l(D)$ acts nontrivially therefore faithfully on V via multiplication from the left. We can use the easy fact that if $e \in M_l(D)$ is an idempotent such that $rank(e) = m$, then $eM_l(D)e \cong M_m(D)$. We infer, that it is enough to give a method to find a *singular* idempotent in $\mathcal{A}$: if $e \in M_l(D)$, $rank(e) = m < l$, then we reduced the problem to the smaller instance $e\mathcal{A}e \cong M_m(D)$. If f is a rank one idempotent of $e\mathcal{A}e$, then f is a rank one idempotent in $\mathcal{A}$ as well, because e is the identity element of $e\mathcal{A}e$ and this implies that $D \cong fe\mathcal{A}ef = f\mathcal{A}f$.

To obtain a singular idempotent, we call the *ZD*-algorithm. This produces a zero divisor, provided that $l > 1$. Now the right identity element e of the left ideal $\mathcal{A}x$ is obviously a singular idempotent and can be found by solving a system of linear equations (describing that e is the identity element of $\mathcal{A}x$), once x is given. We have obtained a polynomial time f-algorithm for *EI*.

An explicit isomorphism is useful because the usual representation of $M_l(D)$ (as the algebra of all l by l matrices over D) is easy to handle. For example it can be conveniently decomposed into a direct sum of minimal left ideals. If e_{ii} denotes the matrix having 1 in position (i, i) and zeroes elsewhere then we have

$$M_l(D) = M_l(D)e_{11} \oplus M_l(D)e_{22} \oplus \cdots \oplus M_l(D)e_{ll}.$$

It is easy to check that $M_l(D)e_{ii}$ is a minimal left ideal for $i = 1, \ldots, l$. If we have an explicit isomorphism from $\mathcal{A}$ to $M_l(D)$, then we readily obtain a decomposition of $\mathcal{A}$. Using results of Section 3, this observation can be generalized to an arbitrary finite semisimple algebra $\mathcal{B}$. $\mathcal{B}$ can be decomposed into a direct sum of minimal left ideals by a polynomial time f-algorithm.

With the help of the *ZD*-algorithm we can provide a polynomial time f-algorithm to find *common invariant subspaces* of linear transformations over a finite ground field. Here we just formulate the problem, for details we refer to [**63**].

Let $X_1, \ldots X_k \in M_l(D)$ and consider their action on the linear space V of l by 1 column vectors over D. The objective is to find a subspace $(0) < U < V$ such that $X_iU \leq U$ for $i = 1, \ldots, k$, if such subspace U exists.

The method for computing common invariant subspaces can be extended to obtain an efficient method to solve a fundamental problem of representation theory: to decompose finite modules over finite semisimple algebras into a direct sum of simple submodules. In fact simple submodules can be found by the method for common invariant subspaces. From simple submodules one can find a direct decomposition using standard methods (cf. Herstein [**35**], pp. 97-98).

Finally we outline an application to permutation groups. Let $G \leq S_n$ and let $K < H$ normal subgroups of G such that H/K is an elementary Abelian p-group for some prime p. Suppose that G, H, K are given by generating sets. The problem is to find a minimal normal subgroup L of G such that $K < L \leq H$.

Here the factor group $V := H/K$ can be considered as a linear space over F_p. The elements of G act on V as F_p-linear transformations via conjugation, and the problem is equivalent to finding a minimal G-invariant subspace of V. This can be done with the common invariant subspace algorithm. As p is small, the method runs in deterministic polynomial time.

4.2. Quaternion algebras over Q.

Over algebraic number fields the problems *ZD* and *EI* seem to be much more difficult. To illustrate this point, we discuss first the smallest nontrivial case, namely, when $\mathcal{A}$ is a simple noncommutative algebra over $\mathbf{Q}$ and $dim_{\mathbf{Q}}\mathcal{A} = 4$. This condition implies that either $\mathcal{A}$ is a division ring or $\mathcal{A}$ is isomorphic to $M_2(\mathbf{Q})$. In both cases $\mathcal{A}$ is central over $\mathbf{Q}$. Following [**64**], we consider the algorithmic problem *zd4*: the input is a central simple algebra $\mathcal{A}$ over $\mathbf{Q}$, $dim_{\mathbf{Q}}\mathcal{A} = 4$, given by structure constants. The problem is to decide if $\mathcal{A}$ contains zero divisors (or, equivalently, if $\mathcal{A} \cong M_2(\mathbf{Q})$). (Note that in contrast to *ZD*, *zd4* is a decision problem.)

It turns out that *zd4* is equivalent (in the sense of Karp reductions) to a decision problem related to ternary quadratic forms over the rationals, which we call *3qf*. The input of *3qf* is a triplet a, b, c of nonzero integers. The objective is to decide if there exist integers x, y, z, not all zero, such that $ax^2 + by^2 + cz^2 = 0$.

The proof of the equivalence is based on properties of *quaternion algebras* (see O'Meara [**54**], Section 57). For nonzero integers a and b the associative algebra $H(a, b)$ over Q is defined by generators and relations as follows.

$$H(a,b) = \langle\, 1,\ u,\ v;\ 1u = u1 = u;\ 1v = v1 = v;\ u^2 = a;\ v^2 = b;\ uv = -vu\rangle.$$

$H(a, b)$ is called a quaternion algebra. It is not difficult to see that $H(a, b)$ is central simple over $\mathbf{Q}$ and $dim_{\mathbf{Q}}H(a, b) = 4$. Every central simple algebra $\mathcal{A}$ over $\mathbf{Q}$, $dim_{\mathbf{Q}}\mathcal{A} = 4$ can be efficiently represented as a quaternion algebra. Moreover $c1 + du + ev$ (c, d, e, are rational numbers) is a zero divisor in $H(a, b)$ if and only if $c^2 - d^2a - e^2b = 0$; and if $H(a, b)$ contains zero divisors then so does the subspace spanned by the elements 1, u, v.

Based on the equivalence, one can obtain algorithmic results on *zd4* by working with *3qf*, making use of the fact that ternary quadratic forms are extensively studied objects of number theory. A classical theorem of Legendre (cf. Cassels [**17**], Chapter 6, Niven, Zuckerman [**56**], Section 5.12) implies that *3qf*, and therefore *zd4* belongs to the complexity class $NP \cap coNP$. Similarly, both problems can be solved in ff-polynomial time (cf. [**64**], and Pollard, Schnorr [**59**]). The proof in [**64**] uses the method of Kannan [**42**] for finding short vectors in low dimensional lattices. In [**64**] evidence is presented that *3qf* and hence *zd4* admits no polynomial time algorithms, by linking it to number theoretic problems believed to be difficult. One such question is the *quadratic residuosity problem*. Let $p_1, \ldots, p_k$ be different odd primes, $n = p_1 \ldots p_k$, and m an integer, $0 < m < n$, $gcd(m, n) = 1$, such that the Jacobi symbol $\left(\frac{m}{n}\right) = 1$. The problem is to decide if the congruence $x^2 \equiv m(mod\ n)$ is solvable. The problem is formulated in Goldwasser and Micali [**34**] for $k = 2$.

A related stronger question is to find a solution of the congruence $x^2 \equiv m(mod\ n)$ if a solution exists. This problem is at least as hard as finding the prime factors of n in the sense that if we can efficiently solve the above quadratic congruence then we can factor n in random polynomial time (cf. Rabin [**61**], Goldwasser and Micali [**34**]).

In [**64**] a Las Vegas polynomial time Karp reduction from quadratic residuosity to *3qf* is given. The result is conditional on the Generalized Riemann Hypothesis (GRH for short). In particular, it follows that finding zero divisors (using randomization) is at least as hard as factoring squarefree integers. The reduction is based on a strong version of the Chebotarev Density Theorem obtained by Lagarias and Odlyzko [**46**] with the assumption of GRH. For generalizations of the Riemann Hypothesis and their significance in computational number theory the reader is referred to Bach [**3**].

4.3. Simple algebras over number fields.

Recall that if $\mathcal{A}$ is a central simple algebra over a number field F, then $\mathcal{A}$ is isomorphic to a full matrix algebra $M_k(D)$, where D is a central skewfield over F. We have $dim_F D = d^2$ for some positive integer d and this d is called the *index* of $\mathcal{A}$. Clearly $\mathcal{A}$ contains zero divisors iff $d < \sqrt{n}$. The input of the algorithmic problem *INDEX* is a central simple algebra $\mathcal{A}$ over an algebraic number field F and a positive integer $d \leq \sqrt{dim_K \mathcal{A}}$. $\mathcal{A}$ is given as a collection of structure constants. The problem is to decide if the index of $\mathcal{A}$ is d.

The main result of [**65**] states that the value of the index of $\mathcal{A}$ has a short and efficiently verifiable certificate. In technical terms, we have ([**65**], Theorem 1.1)

THEOREM 4.1. *INDEX*$\in NP \cap coNP$.

Two special cases of *INDEX* deserve special attention:

The problem *MATRIX* obtained by restricting the value of d to be 1. This is the problem of recognizing full matrix algebras over number fields. A small basis of matrix units (i.e. elements e_{ij} such that $e_{ij}e_{kl} = \delta_{jk}e_{il}$) would be a quite natural candidate for a certificate. However, the existence of such a basis (or even the existence of small zero divisors) is still open.

The problem *SKEWFIELD* obtained by setting the value of d to be $\sqrt{dim_F \mathcal{A}}$. This is the problem of recognizing division rings over F. The result for *INDEX* implies that *MATRIX*$\in NP \cap coNP$, and *SKEWFIELD*$\in NP \cap coNP$.

The proof uses noncommutative algebraic number theory. It turns out, that $\mathcal{A}$ has some subrings, called *maximal orders*, which behave much like the ring of algebraic integers A does in F. To illustrate the ideas involved, we present here some facts on orders and outline the certificate and the verification process. For details we refer to Reiner [**62**], and Rónyai [**65**].

Let $R \leq F$, $R \neq F$ be a Dedekind domain (a commutative, integrally closed Noetherian domain in which all nonzero prime ideals are maximal). For basic algebraic number theory the reader is referred to Ireland and Rosen [**38**], and Lang [**48**]. Let $\mathcal{A}$ be a semisimple algebra over F. An *R-order* in $\mathcal{A}$ is a subring $\Lambda \leq \mathcal{A}$ with the following properties: Λ is a finitely generated module over R; Λ has an identity element (this is necessarily the same as the identity element of $\mathcal{A}$ and R); Λ generates $\mathcal{A}$ as a linear space over F.

When R is clear from the context, we simply use the term *order*. Important examples are:

(1) If $\mathcal{A} = M_n(F)$, then $\Lambda = M_n(R)$ is an order in $\mathcal{A}$.

(2) If $R = \mathbf{Z}$, $F = \mathbf{Q}$ and $\mathcal{A}$ is an algebraic number field, then the full ring of algebraic integers A of $\mathcal{A}$ is an order in $\mathcal{A}$.

An R-order Λ in $\mathcal{A}$ is a *maximal R-order* if it is not a proper subring of any other R-order of $\mathcal{A}$. The orders of examples (1), (2) are maximal orders.

A convenient way to construct orders in $\mathcal{A}$ is as follows. Let M be a finitely generated R-submodule of $\mathcal{A}$ such that M generates $\mathcal{A}$ as an F-space. For example, if $a_1, \dots, a_n$ is a basis of $\mathcal{A}$ over F then the R-submodule generated by $a_1, \dots, a_n$ can serve as M. The *left order* $O_l(M)$ of M is defined as

$$O_l(M) = \{x \in \mathcal{A};\ xM \subseteq M\}.$$

It is easy to see that $O_l(M)$ is an R-order in $\mathcal{A}$. The *discriminant* $d(\Lambda) = d(\Lambda/R)$ of an R-order Λ can be defined in a way similar to the discriminant of a ring of algebraic integers in a number field. The reader is referred to [**62**], Section 10.

Assume that Λ is a maximal R-order in the central simple algebra $\mathcal{A}$ over F, $dim_F\mathcal{A} = m^2 = n$. The integer m_∞ is defined to be 2 if there exists an embedding of F to $\mathbf{R}$, the field of real numbers, such that via this embedding $\mathcal{A} \otimes_F \mathbf{R}$ is isomorphic to a full matrix algebra over the skewfield $\mathbf{H}$ of Hamiltonian quaternions; the value of m_∞ is 1 otherwise. Thus, $m_\infty = 1$ iff for every real embedding of F, the algebra $\mathcal{A} \otimes_F \mathbf{R}$ is isomorphic to a full matrix algebra over $\mathbf{R}$.

For the proof of the following theorem we refer to [**62**], Theorems 23.6, 25.3, 32.1, 32.19.

THEOREM 4.2. *The prime ideals $\mathcal{P}$ of Λ and the prime ideals P of R are in one-to-one correspondence, given by*

$$P = R \cap \mathcal{P},\ \mathcal{P} \supseteq P\Lambda.$$

(i) *The ideals of Λ are exactly the products of prime ideals.*
(ii) *For a prime ideal P of R there exists a unique natural number m_P such that $P\Lambda = \mathcal{P}^{m_P}$. The numbers m_P are divisors of n.*
(iii) *The discriminant $d(\Lambda)$ of Λ is independent from the choice of the maximal order Λ. Moreover $m_P > 1$ implies that $P|d(\Lambda)$.*
(iv) *If $R = A$ is the ring of algebraic integers of F, then the index of $\mathcal{A}$ is the* lcm *of the numbers m_∞ and m_P, where P runs through the prime ideals of A.*

As for the proof of Theorem 4.1, it is shown that $\mathcal{A}$ contains a maximal A-order Λ admitting a polynomial size description. Once a $\mathbf{Z}$-basis of Λ is given, the maximality is checked by a noncommutative variant of a criterion of Zassenhaus [**70**], [**58**] [**60**] for the local maximality of orders in number fields. As a by-product, we obtain all the local indices m_P, such that $m_P > 1$. Here we use an auxiliary algorithm for finding the maximal ideals over a rational prime p of a ring S given by structure constants over $\mathbf{Z}$. (The result generalizes an algorithm of Huang [**36**] for the case $S = A$). It will be indicated in the next section, that m_∞ can be computed efficiently. Finally we check if the least common multiple of m_∞ and the local indices m_P equals to d.

In [**65**] evidence is given that the task of *constructing* a maximal order is at least as difficult as factoring integers and factoring polynomials over finite fields. In the opposite direction the following results are proved in [**39**].

THEOREM 4.3. *Let $\mathcal{A}$ be a finite dimensional semisimple algebra over $\mathbf{Q}$ given by structure constants. Then a maximal $\mathbf{Z}$-order Λ can be constructed by a polynomial time ff-algorithm.*

COROLLARY 4.4. *Let F be an algebraic number field, $\mathcal{A}$ a finite dimensional central simple algebra over F. Let A denote the ring of algebraic integers of F. Suppose that $\mathcal{A}$ is given by structure constants over F. Then a maximal A-order Λ in $\mathcal{A}$ can be constructed by a polynomial time ff-algorithm.*

The algorithms are based on Jacobinski's [**41**] theory of hereditary orders. For a rational prime p let $\mathbf{Z}_p$ denote the ring

$$\mathbf{Z}_p = \{r/s \in \mathbf{Q};\ r, s \in \mathbf{Z},\ gcd(p, s) = 1\}.$$

$\mathbf{Z}_p$ is a discrete valuation ring with unique maximal ideal $p\mathbf{Z}_p$. Let $\mathcal{A}$ be a semisimple algebra over a number field F. For $\mathbf{Z}_p$-orders in $\mathcal{A}$ we introduce the following partial ordering: Γ *radically contains* Λ if and only if $\Gamma \geq \Lambda$ and $Rad\Gamma \geq Rad\Lambda$. (The radical of an order Λ is the intersection of the maximal left ideals of Λ.) The orders maximal with respect to this partial ordering are called *extremal.* Maximal orders are obviously extremal. In the construction of a maximal order the following statement serves as a theoretical tool for extending orders. It is a simplified version of [**41**], Proposition 2.

PROPOSITION 4.5. *Let $\Lambda \leq \Gamma$ be $\mathbf{Z}_p$-orders in $\mathcal{A}$. Suppose that Λ is extremal and Γ is minimal among the $\mathbf{Z}_p$-orders properly containing Λ. Then there exists an ideal $\mathcal{I}$ of Λ minimal among those containing $Rad\Lambda$ such that $O_l(\mathcal{I}) \geq \Gamma$.*

Based on the theory of orders, a deterministic polynomial time method is given in [**66**] to construct splitting elements in central simple algebras over number fields.

THEOREM 4.6. *Let $\mathcal{A}$ be a central simple algebra over a number field F, $dim_F\mathcal{A} = m^2$, given with structure constants. We can find in deterministic polynomial time an element $a \in \mathcal{A}$ such that the minimal polynomial f of a over F is squarefree and $\deg f = m$. It is possible to achieve, that f is irreducible over F. (In this case $F(a)$ is a maximal subfield of $\mathcal{A}$.)*

Applications of Theorem 4.6 are given in Sections 5 and 6. The element a is found in a $\mathbf{Z}$-order of $\mathcal{A}$. A modular argument reduces the search to finding splitting elements in finite full matrix algebras. More precisely the following holds ([**66**], Theorem 2.3):

THEOREM 4.7. *Let Γ be a $\mathbf{Z}$-order in $\mathcal{A}$, and p be a rational prime such that for the discriminant $d(\Gamma/\mathbf{Z})$ we have $gcd(p, d(\Gamma/\mathbf{Z})) = 1$. Let $\mathcal{Q}$ be a maximal ideal of Γ containing $p\Gamma$ and write $\mathcal{B} = \Gamma/\mathcal{Q}$. Then $\mathcal{B}$ is isomorphic to a finite full matrix ring. Moreover, if $a \in \Gamma$ is an element such that $a + \mathcal{Q}$ is a splitting element of $\mathcal{B}$ over $C(\mathcal{B})$, then a is a splitting element of $\mathcal{A}$.*

5. Decomposition of algebras over R and C

Suppose that we have an associative algebra $\mathcal{A}$ specified by structure constants over a number field F. One can quite naturally consider the algebra $\mathcal{B} = \mathcal{A} \otimes_F K$, obtained by allowing coefficients from an extension field K of F. If K is a number field (i.e., if $[K : F]$ is finite) then the algorithmic results discussed so far can be applied without difficulty. But what happens if K is a "big" field, for example, if $K = \mathbf{R}$ or $K = \mathbf{C}$? We can pose the same algorithmic questions as we did before, with the hope that the result (e.g. the radical, simple direct summands, etc.) over K can be represented over some number field $E \geq F$ admitting a description of modest size. For example we have $Rad(\mathcal{A} \otimes_F K) = Rad(\mathcal{A}) \otimes_F K$ if $K = \mathbf{R}$ or $K = \mathbf{C}$, hence a basis over F of the radical of $\mathcal{A}$ is a basis of the radical of the K-algebra defined by $\mathcal{A}$. This implies that the radical algorithm for F will work over K as well.

We turn now to decomposition problems of semisimple algebras. Most of the results discussed here are from Eberly [**24**], [**25**]. Let $\mathcal{A}$ be a semisimple algebra over F. Eberly calls a number field $K > E \geq F$ a *semi-splitting field* for $\mathcal{A}$ over F and K if the simple components of $\mathcal{A} \otimes_F E$ remain simple algebras in $(\mathcal{A} \otimes_F E) \otimes_E K \cong \mathcal{A} \otimes_F K$. Unfortunately the degree of E over F can be as large as $n!$, where $n = dim_F\mathcal{A}$ (cf. [**25**], Lemmas 3.2, 3.4). It is possible to give a deterministic exponential time method for computing the Wedderburn decomposition of $\mathcal{A} \otimes_F K$ along this line.

If we allow that different simple components of $\mathcal{A} \otimes_F K$ are represented over different number fields, then more efficient methods can be obtained (see [**25**], Theorem 3.5).

THEOREM 5.1. *Suppose that $\mathcal{A}$ is given by structure constants over F. Then the number of simple components k over K, number fields $F \leq E_i < K$ and simple direct summands $\mathcal{A}_i$ of $\mathcal{A} \otimes_F E_i$ such that the simple direct summands of $\mathcal{A} \otimes_F K$ are exactly $\mathcal{A}_i \otimes_{E_i} K$ $(1 \leq i \leq k)$ can be computed in deterministic polynomial time. For the degrees of the extension fields E_i produced by the algorithm we have $[E_i : F] \leq n$, if $K = \mathbf{C}$ and $[E_i : F] \leq \binom{n}{2}$, if $K = \mathbf{R}$.*

Eberly in [**26**], Theorems 4.2 and 4.8 considered the explicit isomorphism problem for simple algebras $\mathcal{A}$ over F, such that $\mathcal{A} \otimes_F K$ remains simple if $K = \mathbf{C}$ or $K = \mathbf{R}$. We sketch the approach taken for the case $K = \mathbf{C}$. The requirement that $\mathcal{A} \otimes_F \mathbf{C}$ is simple means that $\mathcal{A}$ is central over F. Suppose that $dim_F\mathcal{A} = m^2$. Eberly selects a splitting element $a \in \mathcal{A}$. Let f be the minimal polynomial of a over F, and let $E \geq F$ be an extension of F generated by a root of f. Now for an efficiently computable polynomial $p(x) \in E[x]$ the rank of $p(a)$ is 1 (here $p(a)$ is viewed as an element of $M_m(\mathbf{C})$), and consequently the left ideal $(\mathcal{A} \otimes_F E)p(a)$ of $\mathcal{A} \otimes_F E$ has dimension m over E. As in the finite case, we immediately obtain an explicit isomorphism $\mathcal{A} \otimes_F E \to M_m(E)$.

THEOREM 5.2. *Given $\mathcal{A}$ over F, a specification of a field extension $E \geq F$ with degree at most m over F and an idempotent $e \in \mathcal{A} \otimes_F E$ that remains primitive in $(\mathcal{A} \otimes_F E) \otimes_E \mathbf{C} \cong \mathcal{A} \otimes_F \mathbf{C}$ can be computed in Las Vegas polynomial time.*

If $K = \mathbf{R}$, then the argument is similar, but slightly more complicated. Note that in this case $\mathcal{A}$ need not be central over F. We have however $dim_F C(\mathcal{A}) =$ 1 or 2. In both cases a splitting element is required over $C(\mathcal{A})$. Note also, that in this case $\mathcal{A} \otimes_F \mathbf{R}$ is isomorphic to a full matrix algebra over $\mathbf{R}$, or $\mathbf{C}$, or over the skewfield of Hamiltonian quaternions $\mathbf{H}$.

THEOREM 5.3. *Suppose that we are given a simple algebra $\mathcal{A}$ over an algebraic number field $F \leq \mathbf{R}$, $dim_F \mathcal{A} = n$, such that $\mathcal{A} \otimes_F \mathbf{R}$ remains simple over $\mathbf{R}$. Then a specification of a real extension $E \geq F$ with degree at most n over F and an idempotent $e \in \mathcal{A} \otimes_F E$ that remains primitive in $(\mathcal{A} \otimes_F E) \otimes_E \mathbf{R} \cong \mathcal{A} \otimes_F \mathbf{R}$ can be computed in Las Vegas polynomial time.*

An analogous result is given by Babai and Rónyai [**2**], Theorem 1.4 for the decomposition of modules over $\mathbf{C}$. The difficulty arising from the possibly too large degree of a semi-splitting field is circumvented by an approach similar to the one in Theorem 5.1. The irreducible submodules are described over different number fields.

THEOREM 5.4. *Suppose that we are given a semisimple algebra $\mathcal{A}$ over an algebraic number field F and an $\mathcal{A}$-module M. Assume that the center of each minimal ideal of $\mathcal{A}$ is a Galois extension of F. Then $M \otimes_F \mathbf{C}$ can be decomposed into a direct sum of irreducible $\mathcal{A} \otimes_F \mathbf{C}$-submodules in Las Vegas polynomial time.*

In Theorems 5.2, 5.3, and 5.4 randomization is used only for selecting splitting elements from central simple algebras over number fields. By Theorem 4.6 this can be done by a deterministic procedure as well. As a consequence, the adjective "Las Vegas" in the statement of the theorems can be replaced by "deterministic".

6. Computations for group representations

Computations with linear representations of groups is probably the oldest and most important area of machine computations involving algebraic structures. Numerous algorithms have been put forward to compute characters, representations and solve decomposition problems. We refer to the surveys of Butler [**13**], Cannon [**15**], Neubüser [**55**], and the bibliography of Felsch [**29**]. Much of the literature on computational representation theory, however, focuses on observed practical performance of the methods proposed. Little is known about the efficiency of the algorithms in terms of (asymptotic) bit complexity. Here we discuss some computational tasks involving group representations primarily on the basis of Eberly [**24**] and Babai, Rónyai [**2**] where exact (symbolic) computations and polynomial time algorithms are considered. For the basic concepts of representation theory we refer to Curtis, Reiner [**20**].

Eberly in [**24**], Section 3.2 considers finitely generated matrix groups $G \leq GL(m, F)$ over a field F. G is given by a set S of generators. Recall, that $env(G) \leq M_m(F)$ denotes the (associative) algebra of matrices generated by the elements of G, and $com(G) \leq M_m(F)$ stands for the algebra of matrices from $M_m(F)$ which commute with every $g \in G$. Several algorithmic problems related to G are shown to be efficiently reducible to algorithmic problems for $env(G)$ and $com(G)$. Here "efficiently" means arithmetic-Boolean circuits of polynomial size and polylogarithmic depth. The result (cf. [**24**], Theorems 3.2.8, 3.2.17) is formulated below.

THEOREM 6.1. *We have the following efficient reductions over an arbitrary field F:*

(a) *The problem of deciding if G is completely reducible to the problem of deciding if an algebra is semisimple.*
(b) *The problem of computing the isotypic components of a completely reducible G to the problem of computing the Wedderburn decomposition of a semisimple algebra.*
(c) *The problem of computing the irreducible components of an isotypic G to the problem of expressing a simple algebra as a direct sum of minimal left ideals.*
(d) *If F is infinite, then in (b) and (c) there are efficient reductions in the opposite direction as well.*

The reductions are correct for Boolean computations over number fields and finite fields (in (*d*) the field F is required to be sufficiently large), hence results for algebras are applicable to matrix groups. For example, the equivalence of the problems in (*c*) over a number field implies that the problem of finding the irreducible components of G is at least as difficult as factoring squarefree integers.

In [**24**], Section 3.3 algorithms for computing the character table of a finite group G are considered. Here, as input, G is given by a multiplication table. The famous methods of Burnside [**12**] and Dixon [**22**] are discussed and analyzed. It is shown that Burnside's method can be implemented in deterministic polynomial time if the number field F is a splitting field for G. A polynomial bound on the complexity of Dixon's algorithm is established. The bound involves a small exponent if Linnik's constant for the first prime in arithmetic progressions is small (this follows from GRH, cf. Titchmarsh [**69**]). Some modifications of Dixon's algorithm are suggested and an efficient probabilistic variant is proposed.

In Section 3.4 of [**24**] representations of symmetric groups and full linear groups are considered. Some basic algorithms (such as computation of characters) and algorithmic problems are discussed. Eberly points out that (with the usual representation of the input objects) the interesting problems seem to be difficult from the perspective of polynomial time computations. On the other hand, most of the methods admit implementations in polynomial space.

The main result of **[2]** is a method for computing a complete set of inequivalent complex irreducible representations, i.e. a *dual object* of a finite group.

THEOREM 6.2. *Given a finite group G by its multiplication table, one can find, in deterministic polynomial time, a complete set of inequivalent complex irreducible representations of G. The matrix elements will belong to the cyclotomic field $\mathbf{Q}_e$, where e is the exponent of the group G.*

Here $\mathbf{Q}_e = \mathbf{Q}(\xi_e)$, where ξ_e is a primitive $e-th$ root of unity over $\mathbf{Q}$. The *exponent* of a group G is the smallest positive e such that $g^e = 1$ for every $g \in G$.

It was conjectured by Maschke that a dual object with matrix elements from $\mathbf{Q}_e$ exists, and proved some 40 years later by R. Brauer. It was not known, however, that a solution (whether over $\mathbf{Q}_e$ or not) admitting a short description existed at all. The *existence* of a dual object of *polynomial size* is a consequence of the algorithm. From Theorem 6.2 routine computations with modules give the following.

THEOREM 6.3. *Given a representation $\mathcal{V}$ of the finite group G over the algebraic number field F by the list of matrices corresponding to each group element, one can decompose $\mathcal{V}$, in polynomial time, into complex irreducible components. The output will be given over the field $F\mathbf{Q}_e$ (the smallest field containing F and $\mathbf{Q}_e$).*

Next we briefly outline the main ideas involved in the algorithm of Theorem 6.2. A subgroup $H \leq G$ is *elementary* if it is a direct product of a cyclic group and a p-group for some prime p. A subgroup $H \leq G$ is an *E-subgroup* if it is of the form $H =< g, P >$, where $g \in G$ and P is a Sylow p-subgroup of the centralizer $C_G(g)$ for some prime p. We shall make use of the following theorem of R. Brauer (**[10]**, **[11]**, **[29]**, (15.1)).

BRAUER'S THEOREM. *Every character χ of G can be represented in the form*

$$\chi = \sum a_i \psi_i^G$$

where ψ_i is an irreducible character of some E-subgroup $H_i \leq G$ and the coefficients a_i are integers.

Actually, this result is formulated in **[11]** with elementary subgroups in the place of E-subgroups. This version, however, immediately follows by the transitivity of induction and the observation that every elementary subgroup is contained in an E-subgroup. The purpose of this modification is to reduce (below a polynomial bound) the number of elementary subgroups one has to consider. In the algorithm the irreducible characters $\psi_1, \psi_2, \ldots, \psi_s$ of the E-subgroups of G

are computed. We have $s < n^3 \log_2 n$, where $n = |G|$. Using a result of Blichfeldt (cf. [**29**], Theorem (10.2)), we can efficiently construct representations $\mathcal{U}_i$ of G over $\mathbf{Q}_e$ such that ψ_i^G is the character afforded by $\mathcal{U}_i$, for $i = 1, \dots, s$.

Now let χ be a given nonlinear irreducible character of G, and $k = \chi(1)$ be the degree of χ. Let $\mathcal{I} \leq \mathbf{Q}_e[G]$ be the minimal ideal corresponding to χ and ρ be a minimal left ideal of $\mathcal{I}$. Since $\mathbf{Q}_e$ is a splitting field of G, we see that $\mathcal{I} \cong M_k(\mathbf{Q}_e)$, as algebras over $\mathbf{Q}_e$.

By Brauer's theorem from the representations $\mathcal{U}_i$ we can construct (pasting step) an $\mathcal{I}$-module M which is a direct sum of $t = kl + 1$ copies of ρ, where $t < n \log_2 n$. In M we compute a submodule N (cutting step) which is isomorphic to kl copies of ρ. Then $N' = M/N$ will be the desired irreducible $\mathbf{Q}_e[G]$-module with character χ. The construction of N is based on an efficient algorithm to find large free submodules in modules over $M_k(F)$ (F is a number field).

In Theorem 1.3 of [**2**], and Theorem 4.1 of [**66**] the problem of decomposing a given representation is considered also in a sense stronger than the one in Theorem 6.2. In the next result we allow a more concise representation of the input. The statement is, however, weaker in that the constituents may be given over different fields. The theorem is an immediate consequence of Theorem 5.4 and the subsequent remark.

THEOREM 6.4. *Let $S \subset G$ be a set of generators of the finite group G. Assume a complex representation $\mathcal{V}$ of G with matrix elements from an algebraic number field F is given by the list $\{\mathcal{V}(g); g \in S\}$ of matrices corresponding to the generators. Then $\mathcal{V}$ can be decomposed to complex irreducible subrepresentations in deterministic polynomial time.*

Finally we mention an algorithmic result from [**39**] related to group representations over number fields (which are not necessarily splitting fields for the group). The result seems to be new even in the case $F = \mathbf{Q}$. The method uses orders and the algorithm of Theorem 4.6.

THEOREM 6.5. *Let F be a number field and G be a finite group given by its multiplication table. Then the degrees of the irreducible representations of G over F can be computed in deterministic time polynomial in the input size and in the discriminant $|d(f)|$, where f is the defining polynomial of F given in the input.*

REFERENCES

1. L. Babai, *Monte-Carlo algorithms in graph isomorphism testing*, Tech. Rep. 79-10, Dep. Math. Stat., Universite de Montreal (1979).
2. L. Babai and L. Rónyai, *Computing irreducible representations of finite groups*, Mathematics of Computation **192** (1990), 705–722.
3. E. Bach, *Fast algorithms under the extended Riemann Hypothesis: a concrete estimate*, Proc 14th ACM STOC, San Francisco, California, 1982, pp. 290–295.
4. ———, *Number theoretic algorithms*, Annual Review of Computer Science, vol. 4, 1990, pp. 119–172.
5. E. H. Bareiss, *Sylvester's identity and multistep integer-preserving Gaussian elimination*, Mathematics of Computation **103** (1968), 565–578.

6. E. R. Berlekamp, *Algebraic coding theory*, McGraw-Hill, 1968.
7. ———, *Factoring polynomials over large finite fields*, Mathematics of Computation **24** (1970), 713–715.
8. R. E. Beck, B. Kolman, and I. N. Stewart, *Computing the structure of a Lie algebra*, Computers in nonassociative rings and algebras, Academic Press, 1977, pp. 167–188.
9. M. Ben-Or, *Probabilistic algorithms over finite fields*, Proc. of the 22th IEEE FOCS, 1981, pp. 394–398.
10. R. Brauer, *Applications of induced characters*, American J. of Mathematics **69** (1947), 709–716.
11. R. Brauer and J. Tate, *On the characters of finite groups*, Annals of Mathematics **62** (1955), 1–7.
12. W. Burnside, *Theory of groups of finite order*, 2nd ed., Dover, 1955.
13. G. Butler, *Computational group theory*, Technical Report 276, Dept. of Computer Science, University of Sydney (1986).
14. P. Camion, *A deterministic algorithm for factorizing polynomials of $F_q[x]$*, Annals of Discrete Math **17** (1983), 149–157.
15. J. J. Cannon, *Computers in group theory: A survey*, Communications of the ACM **12** (1969), 3–12.
16. D. G. Cantor and H. Zassenhaus, *A new algorithm for factoring polynomials over finite fields*, Mathematics of Computation **36** (1981), 587–592.
17. J. W. S. Cassels, *Rational quadratic forms*, Academic Press, 1978.
18. A. L. Chistov and D. Yu. Grigoryev, *Polynomial time factoring of the multivariable polynomials over a global field*, LOMI preprint E-5-82, Leningrad (1982).
19. G. E. Collins, M. Mignotte, and F. Winkler, *Arithmetic in basic algebraic domains*, Computer algebra. Symbolic and algebraic computation, 2nd ed., Springer-Verlag, 1983, pp. 189–220.
20. C. W. Curtis and I. Reiner, *Representation theory of finite groups and associative algebras*, John Wiley and Sons, 1966.
21. L. E. Dickson, *Algebras and their arithmetics*, University of Chicago, 1923.
22. J. D. Dixon,, *Computing irreducible representations of groups*, Mathematics of Computation **24** (1970), 707–712.
23. M. P. Drazin, *Structure matrices of algebras*, Journal of Algebra **87** (1984), 247–260.
24. W. M. Eberly, , *Computations for algebras and group representations*, Ph. D. Thesis, Dept. of Computer Science, University of Toronto, 1989.
25. ———, *Decomposition of algebras over finite fields and number fields*, Computational Complexity **1** (1991), 179–206.
26. ———, *Decompositions of algebras over* **R** *and* **C**, Computational Complexity **1** (1991), 207 230.
27. J. Edmonds, *System of distinct representatives and linear algebra*, Journal of Research of the National Bureau of Standards **718, 4** (1967), 241–245.
28. W. Feit, *Characters of finite groups*, W. A. Benjamin, Inc., 1967.
29. V. Felsch,, *A bibliography on the use of computers in group theory and related topics: algorithms, implementations, and applications*, SIGSAM Bulletin **12** (1987), 23–86.
30. K. Friedl, *Algorithms in algebra*, Diploma Thesis, Eötvös University, Budapest, 1983. (Hungarian)
31. K. Friedl and L. Rónyai, *Polynomial time solutions of some problems in computational algebra*, Proc. 17th ACM STOC, Providence, Rhode Island, 1985, pp. 153-162.
32. J. von zur Gathen, *Algebraic Complexity Theory*, Annual Review of Computer Science, vol. 3, 1988, pp. 317-347.
33. P. Gianni, V. Miller, and B. Trager, *Decomposition of algebras*, Lecture Notes in Computer Science 358 (G. Goos and J. Hartmanis, eds.), Springer-Verlag, 1988, pp. 300-308.
34. S. Goldwasser and S. Micali, *Probabilistic encryption and how to play mental poker keeping secret all partial information*, Proc. 14th ACM STOC, San Francisco, California, 1982, pp. 365-378.
35. I. N. Herstein, *Noncommutative rings*, Mathematical Association of America, 1968.

36. J. E. Hopcroft and J. D. Ullman, *Introduction to automata theory, languages and computation*, Addison-Wesley, 1979.
37. M. A. Huang, *Factorization of polynomials over finite fields and factorization of primes in algebraic number fields*, Proc. 16th ACM STOC, Washington, D.C., 1984, pp. 175–182.
38. K. Ireland and M. Rosen, *A classical introduction to modern number theory*, Springer-Verlag, 1982.
39. G. Ivanyos and L. Rónyai, *On the complexity of finding maximal orders in semisimple algebras over* **Q**, manuscript, 1992.
40. G. Ivanyos, L. Rónyai, and Á. Szántó, *Decomposition of algebras over* $F_q(X_1, \dots, X_m)$, manuscript, 1992.
41. H. Jacobinski, *Two remarks about hereditary orders*, Proc. of the Amer. Math. Soc. **28** (1971), 1–8.
42. R. Kannan, *Improved algorithms for integer programming and related lattice problems*, Proc. 15th ACM STOC, Boston, Massachusetts, 1983, pp. 193–206.
43. R. Kannan and A. Bachem, *Polynomial algorithms for computing the Smith and Hermite normal forms of an integer matrix*, SIAM J. on Computing **4** (1979), 499–507.
44. A. Kertész, *Lectures on Artinian rings*, Akadémiai Kiadó, Budapest, 1987.
45. D. E. Knuth, *The art of computer programming*, Vol. 2, Seminumerical algorithms, Addison-Wesley, 1981.
46. J. C. Lagarias and A. M. Odlyzko, *Effective versions of the Chebotarev density theorem*, Algebraic number fields (L-functions and Galois theory) (A. Fröhlich, ed.), Academic Press, 1977, pp. 409-464.
47. S. Landau, *Factoring polynomials over algebraic number fields*, SIAM J. on Computing **14** (1985), 184-195.
48. S. Lang, *Algebraic number theory*, Springer-Verlag, 1986.
49. A. K. Lenstra, *Factoring polynomials over algebraic number fields*, Lecture Notes of Computer Science 162 (Proc. of EUROCAL), Springer-Verlag, 1983, pp. 245–254.
50. A. K. Lenstra and H. W. Lenstra Jr., *Algorithms in number theory*, Handbook of Theoretical Computer Science (Jan van Leeuven, ed.), Elsevier, 1990, pp. 673–715.
51. A. K. Lenstra, H. W. Lenstra Jr., and L. Lovász, *Factoring polynomials with rational coefficients*, Math. Annalen **261** (1982), 515–534.
52. R. Lidl and H. Niederreiter, *Finite fields*, Addison-Wesley, 1983.
53. L. Lovász, *Some algorithmic problems on lattices*, Theory of Algorithms, Proc. Colloq. Math. Soc. J. Bolyai 1984, North Holland, 1985, pp. 323–338.
54. O. T. O'Meara, *Introduction to quadratic forms*, Springer-Verlag, 1963.
55. J. Neubüser, *Computing with groups and their character tables*, Computer algebra. Symbolic and algebraic computation, 2nd ed., Springer-Verlag, 1983, pp. 45–56.
56. I. Niven and H. S. Zuckerman, *An introduction to the theory of numbers*, 3rd ed., John Wiley and Sons, 1971.
57. R. S. Pierce, *Associative algebras*, Springer-Verlag, 1982.
58. M. E. Pohst, *Three principal tasks of computational algebraic number theory*, Number theory and applications, Proc. NATO Advanced Study Institute, Kluwer Academic Publishers, 1989, pp. 123–133.
59. J. M. Pollard and C. P. Schnorr, *An efficient solution of the congruence* $x^2 + ky^2 \equiv m(n)$, IEEE Trans. on Information Theory **33** (1987), 702–709.
60. M. E. Pohst and H. Zassenhaus, *Algorithmic algebraic number theory*, Cambridge University Press, 1989.
61. M. O. Rabin, *Digitalized signatures and public-key functions as intractable as factorization*, MIT/LCS TR-212, Technical Memo MIT (1979).
62. I. Reiner, *Maximal orders*, Academic Press, 1975.
63. L. Rónyai, *Computing the structure of finite algebras*, Journal of Symbolic Computation **9** (1990), 355–373.
64. ———, *Zero Divisors in Quaternion Algebras*, Journal of Algorithms **9** (1988), 494–506.
65. ———, *Algorithmic properties of maximal orders in simple algebras over* **Q**, Computational Complexity (to appear).

66. ———, *A deterministic method for computing splitting elements in simple algebras over* **Q**, manuscript, 1992..
67. J. T. Schwartz, *Fast probabilistic algorithms for verification of polynomial identities*, Journal of the ACM **27** (1980), 701–717.
68. V. Strassen, *Algebraic complexity theory*, Handbook of Theoretical Computer Science (Jan van Leeuven, ed.), Elsevier, 1990, pp. 635–672.
69. E. C. Titchmarsh, *A divisor problem*, Rendiconti del Circolo Matematico di Palermo **54** (1930), 414–429.
70. H. Zassenhaus, *Ein Algorithmus zur Berechnung einer Minimalbasis über gegebener Ordnung*, Funktional-analysis, Birkhäuser, 1967, pp. 90-103.

Computer and Automation Institute, Hungarian Academy of Sciences, Budapest, Victor Hugo u. 18-22, H-1132 Hungary

E-mail address: h631ron@ella.hu

DIMACS Series in Discrete Mathematics
and Theoretical Computer Science
Volume **11**, 1993

Cayley Graphs and Direct-Product Graphs

ARNOLD L. ROSENBERG

ABSTRACT. Every Cayley graph whose underlying group has nonprime order can be embedded with dilation 2 *onto* a direct-product graph. When the underlying group has "nice" structure, the embedding becomes even more desirable computationally, in that the factor graphs in the direct product have small node-degrees. These results, which were obtained jointly with Marc Baumslag, have applications for the problem of efficiently emulating large interconnection networks on small ones. Complete proofs appear in [**3**].

1. Introduction

1.1. Framework. We are concerned with a genre of parallel architecture that comprises a set of identical processors that communicate via an interconnection network. It is convenient for us to model such *processor networks* to a greater level of detail than is common in the literature, using arc-labeled directed (multi-)graphs whose nodes represent the processors of the network and whose arcs represent interprocessor adjacencies. Both faithful modeling and mathematical tractability recommend that we view the arc labels in our networks as being induced by transformations of the set of nodes. Formally, we model a processor network by an *operator graph.*

DEFINITION 1.1. *An* operator graph *(*opg, *for short) is given by a set* V *of* nodes *and a set* Π *of permutations of* V, *that yield* labelled arcs:
For each $v \in V$ *and each* $\pi \in \Pi$ *there is a link labelled* π *from node* v *to node* $v\pi$.
We denote the opg by $\mathcal{G} = (V; \Pi)$.

1991 *Mathematics Subject Classification.* 94C15, 68C25, 05C99, 68E10.
This research was supported in part by NSF Grant #CCR-90-13184.
This paper is based on "Processor-time tradeoffs for Cayley graph interconnection networks" by Marc Baumslag and Arnold L. Rosenberg, which appeared in *6th Distributed Memory Computing Conf.*, Portland, Oregon, April 28 - May 1, 1991, pp. 630-636. ©1991 IEEE.

Most popular interconnection networks — the de Bruijn [**5, 16**] and shuffle-exchange [**17**] networks, to name just two — are opgs.

When we want to represent networks that have nonuniform node-degrees, such as one observes in a tree (at the root) or in a rectangular mesh (along the boundaries), we must relax the definition of Pi, allowing it to contain (even partial) transformations of V. In this paper, we restrict attention to opgs.

The ultimate in structural uniformity (from our vantage point) is enjoyed by such popular networks as the toroidal mesh [**9**], the butterfly [**13**], and the hypercube [**6**]:

DEFINITION 1.2. *A* Cayley graph *(*cag, *for short) is an opg $\mathcal{G} = (V; \Pi)$, where V is the group* $\mathbf{Gr}(\Pi)$ *generated by Π under functional composition.*
Each $\pi \in \Pi$ acts on $\mathbf{Gr}(\Pi)$ *by right multiplication: $\pi \in \Pi$ "leads" node $g \in$* $\mathbf{Gr}(\Pi)$ *to node $g \cdot \pi$.*

1.2. Direct-Product Networks. The principal objects of our study are opgs and cags that "almost" enjoy the following direct-product structure.

The *direct product* $\mathcal{G}_1 \times \mathcal{G}_2$ of opgs $\mathcal{G}_1 = (V_1; \Pi_1)$ and $\mathcal{G}_2 = (V_2; \Pi_2)$ is the opg $\mathcal{G}_{1,2} = (V_1 \times V_2; \Pi_{1,2})$, whose arcs are induced by the following permutations:

- For each permutation $\pi \in \Pi_1$, there is a permutation $\pi^{(1)} \in \Pi_{1,2}$ such that, for each node $\langle v_1, v_2 \rangle \in V_1 \times V_2$, $\langle v_1, v_2 \rangle \pi^{(1)} = \langle v_1 \pi, v_2 \rangle$.
- For each permutation $\pi \in \Pi_2$, there is a permutation $\pi^{(2)} \in \Pi_{1,2})$ such that, for each node $\langle v_1, v_2 \rangle \in V_1 \times V_2$, $\langle v_1, v_2 \rangle \pi^{(2)} = \langle v_1 v_2 \pi \rangle$.

The computational benefits that accrue when one's network enjoys a direct-product structure are well documented in the literature.

(i) Direct-product networks admit schemes for deterministic off-line permutation routing — generally a hard problem — that are roughly as efficient as the schemes that their factor networks admit [**2**].

(ii) A direct-product network admits an efficient partition into a network of subnetworks that are uniform in both size and structure. This partitionability can be exploited to impart to an architecture flexibility in apparent structure that would be prohibitively expensive to provide in hardware [**11**].

(iii) Direct-product networks can be efficiently emulated by their factor networks, via emulations that are balanced in all important computational measures [**3, 7, 8, 15**].

These benefits of the direct-product structure would be of much less computational significance if a network literally had to *be* a direct-product network to enjoy the benefits. However, it is well documented in the cited sources that it suffices for a network "almost" to be a direct-product network. In the case of the first two of the cited benefits, the demands of the word "almost" are satisfied if the network in question *can efficiently emulate* a direct-product network; this is transparently true for the strong notion of emulation studied in [**1, 4, 15**], but it remains true for many applications using even the weaker notion of emulation studied in [**8**]. In the case of the third cited benefit, the demands of the word "almost" are satisfied if the network in question *can efficiently be emulated by* a direct-product network. It is this third application that motivates the work described here.

1.3. *Networks That Are "Almost" Direct Products.*

DEFINITION 1.3. *The* square *of the digraph* $\mathcal{G} = (V; E)$ *is the digraph* $\mathcal{G}^{(2)} = (V; E^{(2)})$, *where*

$$E^{(2)} = E \cup \{(u, w) \in V \times V \mid \exists v \in V,\ \{(u, v), (v, w)\} \subseteq E\}.$$

DEFINITION 1.4. *The digraph* $\mathcal{G}$ *is* almost *the digraph* $\mathcal{H}$ *just when* $\mathcal{G}$ *is a spanning subgraph of* $\mathcal{H}^{(2)}$.[1]

In the terminology of [**1, 3, 4**], $\mathcal{G}$ is *almost* $\mathcal{H}$ just when one can *embed* $\mathcal{G}$ *onto* $\mathcal{H}$ *with dilation* 2.

The question of interest here is:

When is a digraph $\mathcal{G}$ almost a (nontrivial) direct-product digraph?

One example of such a digraph is the order-n de Bruijn graph $\mathcal{D}_n$. One shows easily (cf. [**15**]) that, for all m and n, $\mathcal{D}_{m+n}$ is almost $\mathcal{D}_m \times \mathcal{D}_n$.

We find that the answer to our question is, "almost always," whenever the digraph $\mathcal{G}$ is a cag. The next section makes this informal statement precise.

2. Cags that are Almost Direct Products

Let us focus on a cag $\mathcal{G} = (G; \Pi)$, where $G = \mathbf{Gr}(\Pi)$. We find three conditions that guarantee that $\mathcal{G}$ is almost a direct-product graph. Before announcing these conditions, though, we must recall some group-theoretic notions.

2.1. Group-Theoretic Notions.

Let H be a subgroup of $G = \mathbf{Gr}(\Pi)$.

DEFINITION 2.1. ***(a)*** *The* right coset *of* H *by* $g \in G$ *is the set* $Hg = \{h \cdot g \mid h \in H\}$.
(b) *The* quotient *of* G *by* H *(denoted* G/H*) is the collection of all right cosets of* H *in* G.
(c) *A* transversal *for* H *in* G *is a subset of* G *that contains precisely one element from each right coset of* H *in* G.

DEFINITION 2.2. *The* coset graph *of* G *with respect to* H *and* Π *is the opg* $(G/H; \Pi)$:

$\pi \in \Pi$ *acts on* $Hg \in G/H$ *via right multiplication:* $(Hg)\pi = H(g \cdot \pi)$.

2.2. The Main Results.

[1]That is, $\mathcal{G}$ is a connected subgraph of $\mathcal{H}^{(2)}$ that contains all nodes of $\mathcal{H}^{(2)}$.

2.2.1. *The Schreier Generator Condition.* The first — and most general — condition we present builds on a result of O. Schreier that constructs a generating set for *any* nontrivial subgroup H of G, in terms of *any* transversal for H in G.

LEMMA 2.1. [**14**] *For any nontrivial subgroup H of G and any transversal T for H in G, the set*

$$\Sigma(H;T,\Pi) =_{\text{def}} \{t\pi[t\pi]^{-1} \mid t \in T, \pi \in \Pi\}$$

is a generating set for H (called the Schreier generators for H with respect to T and Π*).*

THEOREM 2.1. *Every subgroup H of G has a set of generators Π_H such that the cag $\mathcal{G} = (G;\Pi)$ is almost the direct-product graph $\mathcal{K} = (G/H;\Pi) \times (H;\Pi_H)$.*

Proof Sketch. Let T be any transversal for H in G. Using the fact that each $g \in G$ is uniquely expressable as a product $g = h_g \cdot [g]$, where $h_g \in H$ and $[g] \in T$, one verifies as follows that the set $\Pi_H = \Sigma(H;T,\Pi)$ can serve as the generating set for H that proves the theorem.

The association that proves the theorem maps node $g \in G$ to the pair

$$\langle Hg, h_g \rangle \in G/H \times H.$$

One verifies that, for each $\pi \in \Pi$, there exists an $h_{[g],\pi} \in \Sigma(H;T,\Pi)$ such that $g \cdot \pi = h_{[g],\pi} \cdot [g \cdot \pi]$. (This property embodies the importance of the Schreier generators here.) Therefore, under this association, the arc $(g, g \cdot \pi)$ of $\mathcal{G}$ is "routed" along a length-2 path in $\mathcal{K}$:[2]

$$\langle Hg, h_g \rangle \;\rightarrow\; \langle Hg, h_g \cdot h_{[g],\pi} \rangle \;\rightarrow\; \langle H(g \cdot \pi), h_g \cdot h_{[g],\pi} \rangle.$$

□

2.2.2. *The Complementary Subgroup Condition.* The condition we present now is somewhat surprising, in that it seems to depend only on the cardinalities of subgroups of G, rather than on their structures.

DEFINITION 2.3. *Subgroups H and K of G are* complementary *just when*[3]

- *H and K are* orthogonal *in the sense of having trivial intersection; i.e., $H \cap K = \{e_G\}$;*
- *H and K have complementary sizes; i.e., $|H| \cdot |K| = |G|$.*

THEOREM 2.2. *If G has complementary subgroups H and K, then the cag $\mathcal{G} = (G;\Pi)$ is almost the direct-product graph $\mathcal{L} = (G/H;\Pi) \times (G/K;\Pi)$.*

---

[2]For ease of reading, arcs in the path are denoted by arrows.

[3]e_G denotes the identity of G; $|G|$ denotes the size of G.

Proof Sketch. The correspondence that witnesses the claimed relationship assigns node g of $\mathcal{G}$ to node $\langle Hg, Kg\rangle$ of $(G/H;\Pi) \times (G/K;\Pi)$. This assignment is bijective because of the cardinalities of the groups H and K, together with the fact that the groups are orthogonal. One verifies that, under this association, the arc $(g, g\cdot\pi)$ of $\mathcal{G}$ is "routed" along the following length-2 path in $\mathcal{L}$.

$$\langle Hg, Kg\rangle \to \langle Hg\cdot\pi, Kg\rangle \to \langle Hg\cdot\pi, Kg\cdot\pi\rangle.$$

□

2.2.3. *The Group-Independent Generator Condition.* The condition we present now differs from the other two we have discussed, in that it depends on the particular set of elements we use to generate the group G, rather than on the inherent nature of G (i.e., its structure or subgroups).

DEFINITION 2.4. *The subset $\Pi' \subset \Pi$ is* independent *in G just when the following function $\phi : \Pi \to \Pi' \cup \{e_G\}$ extends to a homomorphism of G.*

$$\phi(\pi) = \begin{cases} \pi & \text{if } \pi \in \Pi' \\ e_G & \text{if } \pi \in \Pi - \Pi' \end{cases}$$

REMARK. One can test the independence of a set $\Pi' \subset \Pi$ in time linear in the size of the cag $\mathcal{G} = (G;\Pi)$.

THEOREM 2.3. *If Π has a subset $\Pi' \subset \Pi$ that is independent in G, then the cag $\mathcal{G} = (G;\Pi)$ is almost the direct-product graph $\mathcal{H} = (\mathbf{Gr}(\Pi');\Pi')\times(G/\mathbf{Gr}(\Pi');\Pi)$.*

Proof Sketch. The algebraic underpinnings of the condition of generator independence are not transparent, but they are genuine, as we see now. These underpinnings yield an elegant proof of the Theorem.

DEFINITION 2.5. *For any subgroup H of G:*
(a) *Let N be a normal subgroup of G that is orthogonal to H. We say that G is* a semidirect product *of H and N, written $G = H \otimes_{sd} N$, just when each $g \in G$ can be written in the form $g = h\cdot n$ for some $h \subset H$ and $n \subset N$.*[4]
(b) $\mathbf{Ncl}_G(H)$ *is the smallest normal subgroup of G that contains H.*

LEMMA 2.2. *The subset $\Pi' \subset \Pi$ is independent in G iff G is the semidirect product*

$$G = \mathbf{Gr}(\Pi') \otimes_{sd} \mathbf{Ncl}_G(\mathbf{Gr}(\Pi - \Pi')).$$

Proof Sketch. The proof builds on three facts.
For brevity, let $N_G = \mathbf{Ncl}_G(\mathbf{Gr}(\Pi - \Pi'))$.

(i) *G is the complex product $G = \mathbf{Ncl}_G(\mathbf{Gr}(\Pi')\cdot N_G$.*[5]
(ii) *If $\mathbf{Gr}(\Pi')$ is orthogonal to N_G, then Π' is independent in Π.*
(iii) *If Π' is independent in Π, then $\mathbf{Gr}(\Pi')$ is orthogonal to N_G.*

[4] Equivalently, G has complementary subgroups H and N, with N normal in G.
[5] This claim is true for *any* group G, *any* set of generators Π, and *any* $\Pi' \subset \Pi$.

Facts 1 and 2 establish the necessity of the semidirect-product condition for independence; Facts 1 and 3 establish the sufficiency. □-Lemma

The proof of the Theorem can now be completed in two steps.

First, one shows that, when $\Pi' \subset \Pi$ is independent in G, then the following subgroups of G are complementary. The first subgroup is $\mathbf{Gr}(\Pi')$; the second is the normal subgroup $N =_{\text{def}} \mathbf{Ncl}_G(\mathbf{Gr}(\Pi - \Pi'))$.

It is not clear that we get the desired (i.e., claimed) factors in the direct product, so we next note that N, being normal in G, is "pseudo-commutative" in G, in the sense that, for all $n \in N$ and all $g \in G$, there is an $n' \in N$ such that $n \cdot g = g \cdot n'$. This property allows us to write any element of G as the product of an element of $\mathbf{Gr}(\Pi')$ by an element of $\mathbf{Ncl}_G(\mathbf{Gr}(\Pi - \Pi'))$. □

3. Consequences

We illustrate our main results by looking at two closely related families of cags that are important in the world of parallel architectures, the *butterfly* and *cube-connected cycles* (CCC) [**12**] networks. For brevity we define these families via their underlying groups and generator sets. Graph-theoretic definitions of these networks, and proofs that the group- and graph-theoretic definitions coincide appear in [**1**].

A single group underlies both the order-n butterfly network $\mathcal{B}_n$ and the order-n CCC network $\mathcal{C}_n$, namely, the *(standard) wreath product* $G_n =_{\text{def}} Z_2 \wr Z_n$ [**14**], which we view as a group of permutations of the set $Z_2^n =_{\text{def}} \{0,1\}^n$. Each element of G_n is a pair $\langle \alpha; \vec{\beta} \rangle$ ($\alpha \in Z_n$, $\vec{\beta} \in Z_2^n$) which acts on $\vec{\delta} \in Z_2^n$ via a modulo-2 vector addition[6] of $\vec{\beta}$ to $\vec{\delta}$, followed by a sequence of α left-cyclic shifts. Multiplication in G_n is composition of permutations.

Three elements of G_n are of particular interest:

$$\begin{aligned} \sigma_1^{(n)} &= \langle 1; 0, 0, \dots, 0, 0 \rangle \\ \sigma_2^{(n)} &= \langle 1; 0, 0, \dots, 0, 1 \rangle \\ \sigma_3^{(n)} &= \langle 0; 0, 0, \dots, 0, 1 \rangle \end{aligned}$$

It is shown in [**1**] that $\mathcal{B}_n$ is the cag $(G_n; S_{\text{but}}^{(n)})$, and $\mathcal{C}_n$ is the cag $(G_n; S_{\text{ccc}}^{(n)})$, where $S_{\text{but}}^{(n)} =_{\text{def}} \{\sigma_1^{(n)}, \sigma_2^{(n)}\}$, and $S_{\text{ccc}}^{(n)} =_{\text{def}} \{\sigma_1^{(n)}, \sigma_3^{(n)}\}$.

The results in the previous section yield the following information about the cags $\mathcal{B}_n$ and $\mathcal{C}_n$.

(i) Using the Schreier generator condition, one shows that both $\mathcal{B}_n$ and $\mathcal{C}_n$ are almost the direct-product graph $\mathcal{Q}_n \times \mathcal{R}_n$, where $\mathcal{Q}_n$ is the n-dimensional hypercube, and $\mathcal{R}_n$ is the n-node cycle.

(ii) Using the Complementary subgroup condition, one shows that
- $\mathcal{B}_n$ is almost the direct-product graph $\mathcal{R}_n \times \mathcal{D}_n$, where $\mathcal{D}_n$ is the order-n de Bruijn graph.

[6] We write the product in cyclic groups *additively*.

- $\mathcal{C}_n$ is almost the direct-product graph $\mathcal{R}_n \times \mathcal{S}_n$, where $\mathcal{S}_n$ is the order-n shuffle-exchange graph.

(iii) Using the Group-independent generator condition, one shows that $\mathcal{C}_n$ is almost the direct-product graph $\mathcal{R}_n \times \mathcal{S}_n$. The technique does not apply to $\mathcal{B}_n$.

REFERENCES

1. F.S. Annexstein, M. Baumslag, A.L. Rosenberg (1990): Group action graphs and parallel architectures. *SIAM J. Comput. 19*, 544-569.
2. M. Baumslag and F.S. Annexstein (1991): A unified approach to global permutation routing on parallel networks. *Math. Syst. Th. 24*, 233-251.
3. M. Baumslag and A.L. Rosenberg (1991): Processor-time tradeoffs for Cayley graph interconnection networks. *6th Distributed Memory Computing Conf.*, 630-636.
4. S.N. Bhatt, F.R.K. Chung, J.-W. Hong, F.T. Leighton, B. Obrenić, A.L. Rosenberg, E.J. Schwabe (1991): Optimal emulations by butterfly-like networks. *J. ACM*, to appear.
5. N.G. de Bruijn (1946): A combinatorial problem. *Proc. Koninklijke Nederlandische Akademe van Wetenschappen (A) 49*, Part 2, 758-764.
6. S.L. Johnsson (1987): Communication efficient basic linear algebra computations on hypercube architectures. *J. Parallel Distr. Comput. 4*, 133-172.
7. R. Klasing, R. Lüling, B. Monien (1990): Compressing cube-connected cycles and butterfly networks. *2nd IEEE Symp. on Parallel and Distributed Processing.*
8. R. Koch, F.T. Leighton, B. Maggs, S. Rao, A.L. Rosenberg, E.J. Schwabe (1990): Work-preserving emulations of fixed-connection networks. Submitted for publication; see also, *21st ACM Symp. on Theory of Computing*, 227-240.
9. M. Kunde (1988): Routing and sorting on mesh-connected arrays. *VLSI Algorithms and Architectures* (Aegean Wkshp. on Computing) *Lecture Notes in Computer Science 319*, Springer-Verlag, N.Y., 423-433.
10. B. Obrenić (1991): An approach to emulating separable graphs. *3rd ACM Symp. on Parallel Algorithms and Architectures*, 159-167.
11. B. Obrenić, M.C. Herbordt, A.L. Rosenberg, C.C. Weems, F.S. Annexstein, M. Baumslag (1991): Using emulations to construct high-performance virtual parallel architectures. Tech. Rpt. 91-40, Univ. Massachusetts. See also, Achieving multigauge behavior in bit-serial SIMD architectures via emulation. *3rd IEEE Symp. on Frontiers of Massively Parallel Computation* (1990), 186-195.
12. F.P. Preparata and J.E. Vuillemin (1981): The cube-connected cycles: a versatile network for parallel computation. *C. ACM 24*, 300-309.
13. A.G. Ranade (1991): Optimal speedup for backtrack search on a butterfly network. *3rd ACM Symp. on Parallel Algorithms and Architectures*, 40-48.
14. D.J.S. Robinson (1980): *A Course in the Theory of Groups.* Springer-Verlag, New York.
15. A.L. Rosenberg (1992): Product-shuffle networks: toward reconciling shuffles and butterflies. *Discr. Appl. Math. 37/38*, 465-488.
16. M.R. Samatham and D.K. Pradhan (1989): The de Bruijn multiprocessor network: a versatile parallel processing and sorting network for VLSI. *IEEE Trans. Comp. 38*, 567-581.
17. H. Stone (1971): Parallel processing with the perfect shuffle. *IEEE Trans. Comp., C-20*, 153-161.

DEPARTMENT OF COMPUTER SCIENCE, UNIVERSITY OF MASSACHUSETTS, AMHERST, MASSACHUSETTS 01003

E-mail address: rosenberg @cs.umass.edu

DIMACS Series in Discrete Mathematics
and Theoretical Computer Science
Volume **11**, 1993

Group Membership for Groups with Primitive Orbits

NAMITA SARAWAGI, GENE COOPERMAN, AND

LARRY FINKELSTEIN

ABSTRACT. This paper considers a permutation group $G = \langle S \rangle$ of degree n with t orbits such that the action on each orbit is primitive. It presents a $O(tn^2 \log^c(n))$ time Monte Carlo group membership algorithm for some constant c. The algorithm is notable for its use of a new theorem showing how to find $O(t \log^2 n)$ generators in $O\tilde{}(|S|n)$ time under a more general form of the above hypotheses. The algorithm relies on new combinatorial methods for computing with groups [**CF92**] and previous work of Babai, Luks and Seress [**BLS88**]. In addition, it makes extensive use of a structure theorem for primitive groups by Cameron [**Cam81**], which can be derived from results of Kantor [**Kan79**] and the classification of finite simple groups.

1. Introduction

New combinatorial methods for computing with permutation groups have recently been developed which have led to Monte Carlo algorithms for solving fundamental problems that have superior worst case asymptotic performance [**BCF$^+$91, BCF^{++}91, CF92**]. The main objective of this paper is to explore the power of these methods when additional assumptions are made concerning the nature of the action on the underlying point set. In particular, we prove the following result. We use the notation $O\tilde{}(f(n))$ to mean $O(f(n) \log^c n)$ for some constant c. This work is based on the thesis of Sarawagi [**Sar92**].

THEOREM 1.1. *Let $G = \langle S \rangle$ be a group acting on Ω, such that $S \subseteq \mathrm{Sym}(\Omega)$, $|\Omega| = n$, G has t orbits, and the action of G on each orbit is primitive. Then*

1991 *Mathematics Subject Classification.* Primary 20C40, 20P05.
The authors were supported in part by NSF Grants # CCR-8903952 and CCR-9204469.
This paper is in final form and no version of it will be submitted for publication elsewhere.

a strong generating set for G can be determined in $O^{\sim}(n|S| + tn^2)$ Monte Carlo time.

The proof of Theorem 1.1 is based on the procedure `SGS_Primitive_Orbits` in section 2 and given in Theorem 2.1.

The algorithm used to prove Theorem 1.1 relies on a classification of the orbits of a permutation group (Theorem 3.8). This classification of the primitive orbits makes extensive use of the classification of finite simple groups through a result of Cameron [**Cam81**], which in turn relies on work of Kantor [**Kan79**]. For our purposes, if G is a permutation group on an n-element set Ω and O is an orbit of G with G^O primitive, then O is a *small base* orbit if $|G^O| < |O|^{5\log|O|}$ (Definition 3.13). The *Cameron* orbits (Definition 3.7) include all orbits that are not small base. In the latter case the structure of G^O is well determined by Theorem 3.9. In the special case where G^O contains $\mathrm{Alt}(O)$ we use an elegant algorithm for finding a 3-cycle due to Babai, Luks and Seress [**BLS88**]. If such an orbit is not small base, it is called a *giant* (Definition 3.13), as in [**BLS87**]. Note that if O is a giant orbit, then $|O| \geq 35$ (since otherwise $|G^O| < |O|^{5\log|O|}$).

The algorithm proceeds in two phases. As a pre-processing step, the giant orbits are recognized and the points are re-ordered so that the points of the giant orbits are last in the ordering.

The first phase, described in section 3, uses a typical Sims-type control structure [**Sim**]. One finds a strong generating set for the action of the group on a non-giant orbit O, and generators for the pointwise set stabilizer of O. The action of the stabilizer subgroup on the next non-giant orbit is then considered. The key to this phase is controlling the number of Schreier generators for each point stabilizer subgroup as we proceed through a sequence of base points in O. Two techniques are used to achieve this control. First, it is noted that if O is a Cameron orbit, then the action on O, of the point stabilizer subgroup, is faithful on its second smallest orbit. Further, the length of that orbit is at most $3\sqrt{|O|}$. Second, a new theorem on reduction of generators is proved (Theorem 3.15), which shows that if we are given $O^{\sim}(t)$ generators for the current point stabilizer subgroup, then one can efficiently construct a generating set of size $O^{\sim}(t)$ for the next point stabilizer subgroup which guarantees that the total time to "process" the orbit is $O^{\sim}(nt|O|)$ Monte Carlo time. This reduction is accomplished using a generalization of combinatorial techniques first introduced in [**BCF$^+$91**] (see also [**CF92**] in this volume) rather than sifting. An interesting consequence of these techniques is the following corollary to Theorem 3.15 (for $H = G$).

COROLLARY 1.2. *Let $G = \langle S \rangle$ be a group acting on Ω with $|\Omega| = n$, and suppose G has t orbits, and the action of G on each orbit is primitive. Then a generating set for G containing $O(t\log^2 n)$ elements can be constructed in $O^{\sim}(n|S|)$ Monte Carlo time.*

In the second phase, described in section 5, one begins with generators for a normal subgroup N of G which is the pointwise set stabilizer in G of all non-giant orbits. Initially one works with the socle of N, denoted $\mathrm{Soc}(N)$. Because each giant orbit has size at least 35, the projection of $\mathrm{Soc}(N)$ on a remaining orbit is either trivial or a giant. The key to finding a strong generating set for $\mathrm{Soc}(N)$ is the "fast-giant" technique developed by Babai, Luks and Seress [**BLS88**], which allows the construction of a 3-cycle from a generating set for a giant in $O^{\sim}(n^2)$ Las Vegas time. Also needed is a fast normal closure algorithm described in [**BCF$^+$91**] The full strong generating set can then be constructed by viewing $N/\mathrm{Soc}(N)$ as a subgroup of an elementary abelian 2-group, and employing techniques of linear algebra.

The algorithms we present in this paper are all Monte Carlo. A Monte Carlo algorithm is a randomized algorithm whose reliability (probability of success) can be increased at the cost of additional time. The Monte Carlo nature of the main algorithm arises through multiple invocations of Monte Carlo subroutines. The subroutines have some small probability of returning an incorrect answer, and multiple invocations of the algorithms can lead to a large overall probability of error. However, all of the subroutines satisfy the property that if they run in time c for some input, and if they are allowed to run for an additional factor of time, t, then the probability of error will be ce^{-t}. Under these circumstances, one can always argue that if such an algorithm is invoked k times, then allowing an additional factor of $t = \log k$ time for each invocation will cause the overall probability of error to be bounded by c. The details of the argument are contained in Theorem A.2 of the appendix of [**CF92**], and formal definitions and related theorems are contained in the same appendix. Thus, some additional number of $\log n$ factors suffice to retain a reasonably small probability of error and for this reason, we may omit explicit proofs of reliability.

2. Overview of the Main Algorithm

In this section, we provide an overview of our main algorithm and the supporting subroutines. In order to highlight the underlying algorithmic ideas, we defer the formal timing analyses until later sections.

The ordering of Ω ultimately is determined by the algorithm as it processes the orbits in a top down fashion. We will denote by O_j, the jth orbit which has been processed and set $O^{(i)} = \cup_{j=1}^{i-1} O_j$ for $1 \le i \le t+1$.

Procedure `SGS_Primitive_Orbits`
Input: (S, Ω) where $\langle S \rangle = G$ acts on Ω and has t primitive orbits
Output: A set U which is an SGS for G

Replace S by S' such that $G = \langle S' \rangle$ and $|S'| = O^{\sim}(t)$
`Test_Giants`(S, Ω)
Reorder $\Omega = \cup_{i=1}^{t} O_i$ such that giant orbits are last

Initialize $S'' \leftarrow S'$, $i \leftarrow 1$ and $U \leftarrow \emptyset$
While $i \leq t$ and O_i is a non-giant orbit do
  $(U', S'') \leftarrow$ `SGS_Non-Giant_&_Orbit_Stabilizer`(S', S'', O_i, Ω)
    [U' is a SGS for $G_{O(i)}^{O_i}$ and $G_{O(i+1)} = \langle S'' \rangle$]
  Add U' to U
  Let A be the union of non-trivial orbits of $\langle S'' \rangle$
Add `SGS_Giant_Orbits`(S'', A) to U
Return U

THEOREM 2.1. *The procedure SGS_Primitive_Orbits is correct and runs in* $O\tilde{}(n|S| + tn^2)$ *Monte Carlo time.*

PROOF. The first step is to ensure that $|S|$ is not too large. This is accomplished using Theorem 3.3, which shows how to replace S by a generating set S' so that $|S'| = O\tilde{}(t)$ in $O\tilde{}(n|S|)$ Monte Carlo time. The algorithm then classifies the orbits into types giant and non-giant. This can be done in $O\tilde{}(tn^2)$ time (see Section 4). Using this classification, the orbits are reordered so that the non-giant orbits appear before the giant orbits. Let t' be the number of non-giant orbits.

For each i, $1 \leq i \leq t'$, `SGS_Non-Giant_&_Orbit_Stabilizer`(S', S'', O_i, Ω) is invoked. This routine outputs a SGS, U', for $G^{O^{(t'+1)}}$ and generators S'' for $G_{O(t'+1)}$ in $O\tilde{}(tn|O_i| + n^2)$ Monte Carlo time by Theorem 3.1. Furthermore, the size of S'' is $O^{\sim}(t)$.

Continuing with i, $t' < i \leq t$, either $G_{O(t'+1)}^{O_i}$ is a giant or acts trivially. If the action is that of a a giant, the algorithm invokes the procedure `SGS_Giant_Orbits` using the generators for $G_{O(t'+1)}$. It is shown in Theorem 5.2 that this phase is correct and takes $O\tilde{}(tn^2)$ Monte Carlo time. Hence `SGS_Primitive_Orbits` is correct and runs in the stated time. □

3. Non-Giant Primitive Orbits

The iterative step in the main algorithm for a non-giant orbit can be formulated as follows. We are given a generating set S of size $O^{\sim}(t)$ for the normal subgroup $N = G_{O(i)}$ of G. We must produce a set U of strong generators for $N^{O^{(i)}}$ and a generating set S' for $N_{O(i)} = G_{O(i+1)}$ of size $O^{\sim}(t)$. This is accomplished by the next procedure.

Procedure `SGS_Non-Giant_&_Orbit_Stabilizer`
  Input: (S, S', O, Ω) where $S, S' \subset \mathrm{Sym}(\Omega)$, $|S| = O\tilde{}(t)$, $|S'| = O\tilde{}(t)$,
    $N = \langle S' \rangle \trianglelefteq G = \langle S \rangle$, G has t primitive orbits and
    O is a non-giant G orbit.
  Output: A set $U \subset N$ which is an SGS for N^O
    and a generating set S'' for N_O of size $O\tilde{}(t)$.

Let x_1 be an arbitrary point of O

Let O'_{x_1} be the smallest orbit of G_{x_1} acting on $O \setminus \{x_1\}$
Reorder the points of O such that x_1 is first followed by O'_{x_1}
Set $S'' \leftarrow S'$, $U \leftarrow \emptyset$
For each consecutive base point x_i $(i \geq 1)$ in O do
  Let T be $O^{\sim}(1)$ elements of $\langle S'' \rangle$ such that $x_i^{\langle T \rangle} = x_i^{N_{x_1,\ldots,x_{i-1}}}$
    [T is constructed using Lemma 3.2(ii).]
  Add T to U
  Let C be a transversal for $N_{x_1,\ldots,x_i}$ in $N_{x_1,\ldots,x_{i-1}}$ built from T
  Let S_i be a set of Schreier generators for $N_{x_1,\ldots,x_i}$ built from S'' and C
  Replace S'' by a generating set for $N_{x_1,\ldots,x_i}$ of size $O^{\sim}(t)$ built from S_i
    [S'' is constructed using the argument in Theorem 3.15]
Output U and S''

THEOREM 3.1. *The procedure SGS_Non-Giant_&_Orbit_Stabilizer runs in Monte Carlo time $O^{\sim}(n|O|t)$ and outputs a generating set of size $O^{\sim}(t)$ for the point stabilizer of O.*

PROOF. The proof can be understood by quoting results from Sections 3.2 and 3.3 and deferring formal definitions and proofs of these results until those sections.

We analyze the time to construct S'' through each iteration of the loop, since this will dominate the running time for the procedure. First observe that by Lemma 3.17 the hypotheses of Theorem 3.15 are satisfied with regard to the subgroup $N_{x_1,\ldots,x_i}$.

If $\log |G^O| \leq 5 \log^2 n$ (O is a small base orbit), then there are $O^{\sim}(t|O|)$ Schreier generators, for $N_{x_1,\ldots,x_i}$. By Theorem 3.15, this generating set can be reduced to one of size $O^{\sim}(t)$ in Monte Carlo time $O^{\sim}(t|O|n)$. Since there are at most $O(\log^2 n)$ base points on O, the total time spent in a small base orbit O is $O^{\sim}(t|O|n)$.

If $\log |G^O| > 5 \log^2 n$, then O is a Cameron type orbit for which the smallest orbit of G_{x_1} on $O \setminus \{x_1\}$ has length at most $3\sqrt{|O|}$, by Lemma 3.12. There are $O^{\sim}(t|O|)$ Schreier generators for N_{x_1} and so the time to find $O^{\sim}(t)$ generators for N_{x_1} is $O^{\sim}(t|O|n)$, as for the small base case. For successive point stabilizers with $i > 1$, there are at most $O^{\sim}(t\sqrt{|O|})$ Schreier generators. One can replace them with $O^{\sim}(t)$ Schreier generators in $O^{\sim}(t\sqrt{|O|}n)$ Monte Carlo time by Theorem 3.15. Since there are at most $3\sqrt{|O|}$ base points on this orbit, that bounds the number of iterations, and the total time spent on a Cameron type orbit is also $O^{\sim}(t|O|n)$. □

3.1. Random Subproducts and Subgroup Chain Lengths. The algorithms presented in this paper make heavy use of combinatorial methods for computing with groups based on the notion of random subproducts. This has led to a rich supply of new randomized tools for computing with groups. In this section, we will briefly review the specific ideas we require for this paper, and refer the reader to [**CF92**] in this volume for a more through treatment, including proofs.

A *random subproduct* of a sequence of group elements $(g_1, g_2, \dots, g_k)$ is a product of the form ${g_1}^{e_1}{g_2}^{e_2}\cdots{g_k}^{e_k}$ where $e_i \in \{0,1\}$ are selected independently from the uniform distribution over $\{0,1\}$. Given a set, S, of generators of a group, G, and a transversal T of a point stabilizer subgroup G_x in G, a *random Schreier subproduct* is formed as follows. Form a random subproduct, g, of the sequence of group elements in S (for $h \in G$, let $\overline{h}$ be the unique element in T such that $h\overline{h} \in G_x$), form all Schreier generators $tg(\overline{tg})^{-1}$ for $t \in T$, and then form a random subproduct on this set of Schreier generators.

Random subproducts and random Schreier subproducts, though not purely random elements of G and G_x, respectively, have useful properties. The following lemmas describe these properties.

LEMMA 3.2. *Let $G = \langle S\rangle$ be an arbitrary permutation group acting on Ω with $|\Omega| = n$.*

(i) *Let $H < G$ be a proper subgroup. Then the probability that a random subproduct r (formed using S) is not in H is at least $1/2$.*

(ii) *There exists a constant $c > 0$ such that for arbitrary $d \geq 1$, if S' is a set of $cd\log n$ random subproducts on S, then with probability at least $1 - 1/n^d$, $\langle S'\rangle$ has the same orbit structure on Ω as G.*

(iii) *Let $x \in \Omega$, T a transversal of G_x in G, and $\overline{H} < G_x$ is a proper subgroup then the probability that a random Schreier subproduct, r_x, formed using S and T, is not in $\overline{H}$, is at least $1/4$.*

The proof of (i) and (ii) can be found in [**BCF$^+$91**] or [**CF92**]. The next theorem is a generalization of Theorem 2.3 in [**BCF$^+$91**]. For convenience, we base the proof on the version appearing as Theorem 2.9 in [**CF92**].

THEOREM 3.3. *Let $G = \langle S\rangle$ be a finite group. Let $H \leq G$, and let $\widehat{L}$ be a known upper bound on the length of all subgroup chains from H to G. Then for any fixed parameter p such that $0 < p < 1$, with probability at least p one can find a set $\widehat{S}$ with $|\widehat{S}| = O(\widehat{L}\log(1/(1-p)))$ using $O(|S|\,(\log\widehat{L})\,\log(1/(1-p)))$ group operations such that $\langle\widehat{S}, H\rangle = G$.*

PROOF. The proof is given as a modification of the proof of Theorem 2.9 in [**CF92**] (this volume). One must initialize the S' of the original proof to T, a generating set for H, instead of the empty set, and interpret the L of the original proof as $|T| + \widehat{L}$. Further, if one replaces $\langle g_1, \dots, g_{i-1}\rangle$ by $\langle H, g_1, \dots, g_{i-1}\rangle$ everywhere, then the logic of the proof generalizes verbatim to the current theorem. The new generating set $\widehat{S}$ is then chosen as $S' \setminus T$, and $\langle\widehat{S}, H\rangle = G$. □

THEOREM 3.4. (*Cameron et al.* [**CST89**]*) Any subgroup chain of a permutation group of degree n has length at most $3n/2$.*

Another variation of the above theorem was proved by Babai [**Bab86**] with a bound of $2n-3$.

LEMMA 3.5. (*Scott* [**Sco79**]) *Let* $G = H_1 \times \ldots \times H_r$ *be the direct product of non-abelian simple groups* H_i, $1 \le i \le r$. *Let* M *be a subgroup of* G *which projects onto each* H_i *for* $1 \le i \le r$. *Then there exists a partition* $\{B_1, \ldots, B_s\}$ *of* $\{1, \ldots, r\}$ *such that* $M = D_1 \times \ldots \times D_s$ *where each* D_i *is a diagonal subgroup of* $\times_{j \in B_i} H_j$.

COROLLARY 3.6. *Let* W *be the disjoint union of* r *sets* W_i, *each of size at least 2. Suppose that* H *is a subgroup of* $G = \mathrm{Alt}(W_1) \times \cdots \times \mathrm{Alt}(W_r)$ *and* H *projects onto* $\mathrm{Alt}(W_i)$ *for each* i. *Then the length of any chain of proper subgroups from* H *to* G *is at most* $2r$.

The proof follows easily from Lemma 3.5. If $|W_i| \neq 4$ for all i, then the alternating groups are simple, and a bound of at most r on the chain can be found. Otherwise, $2r$ subgroups may be needed.

3.2. Cameron Groups. We begin by defining a class of primitive groups referred to as *Cameron* groups.

DEFINITION 3.7. Let (k, r, s) be a triple of positive integers such that $k \ge 5$ and $s \le k/2$. Let C be the disjoint union of r sets $B_1, \ldots, B_r$ each of size k and let $A = \{X \subset C \mid \forall i, |X \cap B_i| = s\}$. If G is a transitive subgroup of $\mathrm{Sym}(C)$ for which each B_i, $1 \le i \le r$, is a block of imprimitivity and if G contains $\mathrm{Alt}(B_1) \times \cdots \times \mathrm{Alt}(B_r)$, then G^A is a primitive group (as can be seen by noting that the point stabilizer subgroup must be maximal in G). G^A is said to be a *Cameron type group* with parameters (k, r, s). G^C is called the *natural* or *imprimitive Cameron action.*

In the discussion to follow, we often directly identify elements $X \in A$ with subsets of C satisfying the property $|X \cap B_i| = s$ for $1 \le i \le r$. Note that the socle, $\mathrm{Soc}(G)$, has a faithful representation on C as $\mathrm{Alt}(B_1) \times \cdots \times \mathrm{Alt}(B_r)$, and that $\mathrm{Soc}(G)$ is a single minimal, normal subgroup. It is easy to verify that the action of G on the set $\{X \subset C \mid |X \cap B_i| = s\}$ is primitive, by noting that the point stabilizer subgroup must be maximal in G. Maximality follows from the characterization of the point stabilizer subgroup in Lemma 3.14.

The motivation for defining Cameron groups is the following theorem due to Cameron [**Cam81**]. It is based on the classification of finite simple groups, and on work by Kantor on permutation representations of classical groups [**Kan79**].

THEOREM 3.8. [**Cam81**] *Let* G *be a primitive group acting on* n *points. If* $|G| \ge n^{5 \log n}$, *then* $n = \binom{k}{s}^r$ *and* G *is a Cameron group of type* (k, r, s).

The following lemma gives some properties of a Cameron group. The calculations are based on those in [**BLS87, BLS88**], but have been re-calculated to achieve tighter bounds. The logarithmic base is 2 unless otherwise specified.

LEMMA 3.9. *Let G be a Cameron type group with parameters (k, r, s) acting on A. Let $|A| = n$. Then,*

(i) $n = \binom{k}{s}^r$;

(ii) $rs \log(k/s) \leq \log n$;

(iii) $k \leq \sqrt{2n}$ *when* $r > 1$ *or* $s > 1$, *otherwise* $k = n$;

(iv) $\log |G| < \frac{(\log n) \log\log n}{s \log(k/s)} + \frac{k}{s} \log n + \frac{k}{s} \log n \frac{\log s}{\log(k/s)}$;

PROOF. Part (i) is clear. Part (ii) follows from $(\frac{k}{s})^{rs} \leq \binom{k}{s}^r$. In part (iii), for fixed n, k is maximized when $r = 1$ and $s = 2$, and the bound follows from (i). For part (iv), the inequality follows from $|G| \leq r!(k!)^r \leq r!\binom{k}{s}^{rk/s}(s!)^{rk/s} = r!n^{k/s}(s!)^{rk/s}$. After applying part (ii), $\log |G| \leq r \log r + k/s \log n + rk/s \log(s!) < \frac{(\log n) \log\log n}{s \log(k/s)} + \frac{k}{s} \log n + \frac{k}{s} \log n \frac{\log s}{\log(k/s)}$. □

DEFINITION 3.10. Let $G = G^A$ be a Cameron type group with parameters (k, r, s). For $X \in A$ (and $B_1, \ldots, B_r$ as in Definition 3.7), define the G_X-orbit $\Sigma_1(X)$ to be the set

$$\{Y \in A \mid \exists j,\, 1 \leq j \leq r,\, \forall i \neq j,\, |Y \cap X \cap B_j| = s - 1,\ Y \cap B_i = X \cap B_i\}.$$

It is clear that $\Sigma_1(X) \subset A$ is an orbit of the point stabilizer subgroup G_X. The next lemma provides a finer estimate of calculations based on Theorem 4.1 of [**BLS87**].

LEMMA 3.11. *Let $G = G^A$ be a Cameron type group. For arbitrary $X \in A$, the action of G_X on $\Sigma_1(X)$ is faithful. If G is not a giant, then $|\Sigma_1(X)| \leq 3\sqrt{n} - 1$, and therefore $\Sigma_1(X) \cup \{X\}$ is a base for G of size at most $3\sqrt{n}$. If $\log |G| > 5(\log n) \log\log n$, then $\Sigma_1(X)$ is the smallest orbit of G_X in $A \setminus \{X\}$.*

PROOF. The action of G_X on C has orbits $X \cap B_j$ and $(C \setminus X) \cap B_j$ for $1 \leq j \leq r$. To see that the action is faithful, we show that the kernel is trivial. Let $g \in G_X^C$ be non-trivial. Let $g' \in G_X^A$ correspond to g under the permutation equivalence. There is a $y \in B_j \subseteq C$ for some j, such that $y^g \neq y$. If $y \in X \cap B_j$, choose $Y \in \Sigma_1(X)$ such that $Y \cap B_i = X \cap B_i$ for $i \neq j$, and $Y \cap B_j = ((B_j \cap X) \setminus \{y\}) \cup \{z\}$ for arbitrary $z \in B_j \setminus X$. If $y \in B_j \setminus X$, choose $Y \in \Sigma_1(X)$ such that $Y \cap B_i = X \cap B_i$ for $i \neq j$, but $Y \cap B_j = ((B_j \cap X) \setminus \{z\}) \cup \{y\}$ for arbitrary $z \in B_j \cap X$. Since $y^g \neq y$, it is clear that $Y^{g'} \neq Y$, and so G_X is faithful on $\Sigma_1(X)$.

The length of $\Sigma_1(X)$ is $rs(k - s)$. If G is not a giant, then $r > 1$ or $s > 1$, and Lemma 3.9(ii, iii) yields $rs(k - s) \leq (k - s)(\log n)/\log(k - s) < k\,(\log n)/\log k \leq 2^{3/2}\sqrt{n} < 3\sqrt{n}$. Direct calculation shows that $\Sigma_1(X)$ is the smallest orbit of G_X in $A \setminus \{X\}$ except possibly when $s \geq \lfloor (k - 1)/2 \rfloor$. Consider an orbit $\Sigma_2(X)$ with parameters t_1 and t_2 such that for all $Y \in \Sigma_2(X)$, $|Y \cap B_i \setminus X| = t_1 > 0$ and $|Y \cap B_j \setminus X| = t_2 > 0$ for some $i \neq j$. (It is clear that this property will be preserved for all points in the same orbit.) Then $|\Sigma_2(X)| \geq q\binom{s}{t_1}\binom{k-s}{t_1}\binom{s}{t_2}\binom{k-s}{t_2} > |\Sigma_1(X)|$ when $s = t < \lfloor (k - 1)/2 \rfloor$. Here, q is the number of distinct intersection patterns, $(|Y \cap B_1 \setminus X|, \ldots, |Y \cap B_r \setminus X|)$, for $Y \in \Sigma_2(X)$. It is easy to show that $q \geq r$, using the transitivity of the action of G on $\{B_1, \ldots, B_r\}$. Next, consider an orbit $\Sigma_3(X)$ with parameter t such that for all $Y \in \Sigma_3(X)$, $|Y \cap B_i \setminus X| = t > 1$ and $|Y \cap B_j \setminus X| = 0$ for some i and for all $j \neq i$. Then $|\Sigma_3(X)| = r\binom{s}{t}\binom{k-s}{t} > |\Sigma_1(X)|$ when $s = t < \lfloor (k-1)/2 \rfloor$.

If $\Sigma_1(X)$ is not the smallest orbit, then $s \geq \lfloor (k-1)/2 \rfloor$, and one has the estimates $s \geq 2$, $2 \leq k/s \leq 5/2$, $n \geq 10$, and $\log s \leq \log\log n$. (The last inequality follows from Lemma 3.9(ii).) Applying Lemma 3.9(iv) shows that $\log|G| \leq 5(\log n)\log\log n$. So, $\Sigma_1(X)$ is the smallest orbit of G_X when $\log|G| > 5(\log n)\log\log n$. □

COROLLARY 3.12. *If G is primitive of degree $n \geq 2$, and $\log|G| > 5\log^2 n$, then G is of Cameron type and the second smallest orbit of its point stabilizer subgroup, G_x, is $\Sigma_1(x)$. Further, the action of G_x on $\Sigma_1(x)$ is faithful. If G is not a giant, then $|\Sigma_1(x)| < 3\sqrt{n}$.*

PROOF. For $n \geq 2$, Theorem 3.8 shows that G is of Cameron type. Since $5\log^2 n > 5(\log n)\log\log n$ for $n \geq 2$, the remaining conclusions follow from Lemma 3.11. □

This motivates the next definition. The restriction of *giants* to groups of degree n at least 35 avoids certain pathological case that would have arisen in section 5 if $n = 4$ or $n = 6$.

DEFINITION 3.13. A *small base group* is a group of degree n such that $\log|G| \leq 5\log^2 n$. A *giant* is a group of the form S_n or A_n that is not of type small base. This implies that $n > 10$ for giants. An orbit of G is identified as large, small base, or Cameron type according to whether the action of G on that orbit is a group of the corresponding type.

LEMMA 3.14. *Let $G = G^A$ be a primitive Cameron group with parameters (k, r, s) such that $|A| = n$ and let G^C be the imprimitive Cameron action. Let $X \in A$ and let $Y_1, \ldots, Y_m$ be a sequence of points in $\Sigma_1(X)$. Let $H = G_{X,Y_1,\ldots,Y_m}$. Then H has a normal subgroup K such that K^C has orbits $U_i = X_i \cap Y_1 \cap \cdots \cap Y_m$ and $V_i = (B_i \setminus X_i) \cap (B_i \setminus Y_1) \cap \cdots \cap (B_i \setminus Y_m)$, $1 \leq i \leq r$ and $K^C = \mathrm{Alt}(U_1) \times \mathrm{Alt}(V_1) \times \cdots \times \mathrm{Alt}(U_r) \times \mathrm{Alt}(V_r)$. Furthermore, H^C acts faithfully on $(\cup_{i=1}^r U_i) \cup (\cup_{i=1}^r V_i)$ (and hence permutes the set $\{U_1, V_1, \ldots, U_r, V_r\}$).*

PROOF. Note that G_X satisfies the conclusion. Note that $X \setminus Y_i$ and $Y_i \setminus X$ are trivial orbits of $G_{X,Y_1,\ldots,Y_i}$ for $1 \leq i \leq m$. Each non-trivial orbit of $G_{X,Y_1,\ldots,Y_{i-1}}$ is formed from the set difference of an orbit of $G_{X,Y_1,\ldots,Y_i}$ and the union of the two trivial orbits. So, the conclusion will continue to be satisfied for all i. □

3.3. Main Result. There are two situations where we require reduction of generators. The first is in `SGS_Primitive_Orbits` when we are given the initial generating set S for G and want to reduce it to one of size $O^{\sim}(t)$. The second occurs in `SGS_Non-Giant_&_Orbit_Stabilizer` when we construct the Schreier generators for a point stabilizer and want to reduce it to one of size $O^{\sim}(t)$ for the next round. The hypotheses of the following result are designed to capture both instances.

THEOREM 3.15. *Let G acting on Ω have t primitive orbits. Let $H = \langle S \rangle$ be a subgroup of G that acts faithfully on $A \subseteq \Omega$, where A is a union of some of the G-orbits, $O_1, \ldots, O_t$. Each O_i is either a small base orbit or a Cameron orbit or both. Without loss of generality, assume that the Cameron orbits appear as $O_1, \ldots, O_r$ for $r \leq t$. Assume that if G^{O_i} is a Cameron orbit with parameters (k_i, r_i, s_i) and C_i supports the natural action of G^{O_i}, then the following holds for H^{C_i}.*

(i) *There exist mutually disjoint subsets $V_1^i, \ldots, V_{\ell_i}^i$ of C_i such that H^{C_i} acts faithfully on $U_i = \cup_{j=1}^{\ell_i} V_j^i$.*
(ii) $\mathrm{Alt}(V_1^i) \times \cdots \times \mathrm{Alt}(V_{\ell_i}^i)$ *is normal in H^{U^i} (and so H^{U^i} permutes the sets $\{V_1^i, \cdots, V_{\ell_i}^i\}$).*
(iii) $\ell_i \leq 2r_i \leq 2\log n$.

Then in $O^{\sim}(|S|n)$ Monte Carlo time, one can construct a generating set S' for H such that $|S'| = O(t \log^2 n)$.

The proof of Theorem 3.15 follows directly from Lemma 3.16 and Theorem 3.3.

LEMMA 3.16. *Assume the hypotheses of Theorem 3.15 Let M be a subgroup of H generated by $O^{\sim}(1)$ random subproducts of S. Then $L(M, H) = O(t \log^2 n)$.*

PROOF. Assume first that each of the orbits O_i is a Cameron orbit. Let $U = \cup_{i=1}^t (\cup_{j=1}^{\ell_i} V_j^i)$, so that G acts faithfully on U and permutes the sets $\{V_j^i \mid 1 \leq j \leq \ell_i, 1 \leq i \leq t\}$.

For each V_j^i, let $d_j^i = \min(|V_j^i| - 2, 6)$ and define the set $\binom{V_j^i}{d_j^i}$ to be the set of all subsets of V_j^i of size d_j^i. Let $\mathcal{U} = \cup_{i=1}^t (\cup_{j=1}^{\ell_i} \binom{V_j^i}{d_j^i})$. Then H acts faithfully on $\mathcal{U}$ and $|\mathcal{U}| \leq \sum_{i=1}^t (\sum_{j=1}^{\ell_i} |V_j^i|^6) \leq \sum_{i=1}^t |U_i|^6 \leq \sum_{i=1}^t |C_i|^6 \leq \sum_{i=1}^t |O_i|^6 \leq n^6$. By Lemma 3.2, $O^{\sim}(1)$ random subproducts on S will generate a subgroup M which has high probability of having the same orbits on $\mathcal{U}$ as H. Since $H_{\{V_j^i\}}^{V_j^i}$ contains $\mathrm{Alt}(V_j^i)$ it follows that $\binom{V_j^i}{d_j^i}$ is contained in an orbit of $M^{\mathcal{U}}$. But any element of M^U that which takes one d_j^i subset of V_j^i to another must stabilize V_j^i. So the set stabilizer $M_{\{V_j^i\}}$ is d_j^i-transitive on V_j^i. By 3.4, $M_{V_j^i}^{V_j^i}$ contains $\mathrm{Alt}(V_j^i)$. We will use this property to show that $L(M, H) = O^{\sim}(t)$. Note that the orbits of M on U are the same as those of H on U.

Define $\widehat{H} = H^{U_1} \times \cdots \times H^{U_t} \leq \mathrm{Sym}(U)$. We will show that $L(M, \widehat{H}) = O^{\sim}(t)$, which suffices since $M \leq H \leq \widehat{H}$. Let

$$\widehat{K} = \bigtimes_{\substack{1 \leq i \leq t \\ 1 \leq j \leq \ell_i}} \mathrm{Alt}(V_j^i),$$

so that $\widehat{K} \leq \widehat{H}$. Consider the (possibly unfaithful) action of $\widehat{H}$ on $\{V_j^i \mid 1 \leq j \leq \ell_i, 1 \leq i \leq t\}$. By (iii), there are at most $2t \log n$ elements in this set. Moreover, the kernel of the action contains $\widehat{K}$ with index at most $2^{2t \log n}$. It then follows from this and the bound on the length of subgroup chains in symmetric groups (Theorem 3.4), that $L(\widehat{K}, \widehat{L}) = L(\widehat{H}/\widehat{K}) \leq 2t \log n + 3/2(2t \log n) \leq 5t \log n = O^{\sim}(t)$.

Since $\widehat{K}$ is normal in $\widehat{H}$, it remains to estimate $L(M, \widehat{K})$. Let $M = M_0 < M_1 < \cdots < M_p = \widehat{H}$ be a proper chain of subgroups. By retaining only the subgroups in the chain that satisfy $M_{i+1} \cap \widehat{K} > M_i \cap \widehat{K}$, we construct a new chain

$$M \cap \widehat{K} = K_0 < K_1 < \cdots < K_{p'} \leq \widehat{K}$$

for $p' \leq p$, such that for each r with $1 \leq r \leq p$ there is a unique s such that $K_r = M_s \cap \widehat{K}$. Note that $L(M \cap \widehat{K}, \widehat{H}) \leq p' + L(\widehat{K}, \widehat{H})$. So, we must estimate p', which will be an upper bound on $L(M, \widehat{K})$.

We show that $(K_r)^{V_j^i}$ is either the alternating or trivial group. Let s correspond to r, so that $M_s \cap \widehat{K} = K_r$. Note that $M_s \cap \widehat{K} \triangleleft M_s$ and $(M_s)_{\{V_j^i\}}^{V_j^i} \geq \mathrm{Alt}(V_j^i)$. So, $(M_s \cap \widehat{K})^{V_j^i} \triangleleft (M_s)^{V_j^i} \cap \mathrm{Alt}(V_j^i)$. Thus, if $d_j^i \neq 4$, then $(K_r)^{V_j^i} = (M_s \cap \widehat{K})^{V_j^i}$ is either $\mathrm{Alt}(V_j^i)$ or the trivial group. Hence, this is true for all K_r, $1 \leq r \leq p'$.

For each K_r, the action on each component with $d_j^i \neq 4$ must be alternating or trivial. Let u_r be the number of components on which there is an alternating action. By Lemma 3.5, the action on the u_r alternating components decomposes into the direct product of $v_r \leq u_r$ actions, where each of the v_r actions is a diagonal of alternating actions on a subset of the $\{V_j^i\}$. Further, if $K_{r+1} < K_r$, then $u_{r+1} > u_r$ or $v_{r+1} > v_r$. Since $u_{p'} \leq 2t \log n$ and $v_{p'} \leq 2t \log n$, $p' \leq 4t \log n$. On the other hand, if each of the orbits have length 4, then a similar argument, in conjunction with Corollary 3.6 shows that $v_{p'} \leq 4t \log n$ and so $p' \leq 6t \log n$.

We have shown that the result holds in the case where all the orbits O_i are Cameron orbits. The general case now follows easily by applying this result to the action of G on the union of the Cameron orbits and observing that the kernel of the action is faithfully represented on a union of small base group orbits of G on Ω. The kernel has order at most $2^{2t \log^2 n}$ by Theorem 3.8, yielding a bound of $2t \log^2 n$ for the subgroup chain length. □

The final result in this section describes the situation in which Theorem 3.15 is invoked.

LEMMA 3.17. *Let G acting on Ω have t primitive orbits and let N be a normal subgroup of G. Let O be a G-orbit and $x_1, \ldots, x_\ell$ a sequence of base points in O for N^O chosen so that x_i, $2 \le i \le \ell$ is in the smallest orbit of G_{x_1} acting on O. Let A be the union of O and all G-orbits O_j such that $N_O^{O_j}$ is non-trivial. Then $N_{x_1,\ldots,x_{i-1}}$ and A satisfy the hypotheses of Theorem 3.15, where $N_{x_1,\ldots,x_{i-1}}$ is identified with H.*

PROOF. It is clear that N acts faithfully on A. Let $O_j \subseteq A$ be a Cameron type orbit with $O \neq O_j$. Since $N_O^{O_j}$ is a non-trivial normal subgroup of G^{O_j}, $N_O^{O_j}$ contains the socle of G^{O_j}. This is also true for $N_{x_1,\ldots,x_\ell}^{O_j}$, and so the hypotheses of Theorem 3.15 are satisfied with regard to $N_{x_1,\ldots,x_\ell}$ and O_j whenever O_j is a non-giant Cameron orbit. It remains to verify the hypotheses when O is a non-giant Cameron orbit. But this follows directly from Lemma 3.14 and the assumption on how the sequence $x_1, \ldots, x_\ell$ is chosen. □

4. A Monte Carlo Test for Giant Action

In the main procedure `SGS_Primitive_Orbits`, it is necessary to classify the t primitive orbits of G into types giant and non-giant. In this section we present a $O^\sim(tn^2)$ Monte Carlo time algorithm for accomplishing this task. The procedure `Test_Giants` is based on repeated application of Theorem 3.15. This allows for a simple exposition of the algorithm.

Procedure `Test_Giants`
  Input: (S, Ω) where $S \subseteq \mathrm{Sym}(\Omega)$, $|S| = O^\sim(t)$,
    $G = \langle S \rangle$ and G has t primitive orbits.
  Output: Each orbit is identified as type giant or non-giant

For each orbit O of G do
  If $|O| < 35$ then mark O as non-giant
  Else set $S' \leftarrow S$
    Choose points $x_1, \ldots, x_6$ in O
    For $i \leftarrow 1$ to 6 do
      Let T be $O^\sim(1)$ random subproducts of $\langle S' \rangle$ such that
        $x_i^{\langle T \rangle} = x_i^{G_{x_1,\ldots,x_{i-1}}}$
      If $|x_i^{\langle T \rangle}| \neq |O| - i + 1$ then
        Mark O as a non-giant and break from the loop
      Let C be a transversal for $G_{x_1,\ldots,x_i}$ in $G_{x_1,\ldots,x_{i-1}}$ built from T
      Let S'' be a set of Schreier generators for $G_{x_1,\ldots,x_i}$ built
        from S'' and C
      Replace S' by a generating set for $G_{x_1,\ldots,x_i}$ of size $O^\sim(t)$ built
        from S''

THEOREM 4.1. *Procedure* **Test_Giants** *is correct and runs in Monte Carlo time* $O(tn^2)$.

PROOF. The algorithm simply tests whether G^O is 6-transitive. The correctness of this approach is a consequence of the classification of finite simple groups [**Gor82**]. In addition, we need to invoke Lemma 3.17 in order to ensure that the the hypotheses of Theorem 3.15 apply. The time for each orbit is easily seen to be $O^{\sim}(t|O|n)$ from Theorem 3.15 and this then leads to the stated time bound.
□

5. Giant Orbits

In this section, we describe the procedure `SGS_Giant_Orbits`. When this procedure is called, `SGS_Primitive_Orbits` has computed a set of $O^{\sim}(t)$ generators for the pointwise stabilizer of all the non-giant orbits. Since this subgroup is normal, the action on the remaining non-trivial orbits must be of giant type. This justifies the assumptions made for the input to `SGS_Giant_Orbits`. The procedure constructs a strong generating set for the input group in two stages. First, `SGS_Alt_Orbits` (described later) is invoked to compute a strong generating set for the alternating action on each orbit and then a standard linear algebra argument is invoked in order to complete the strong generating set to one for the entire group. Corollary 5.1 reveals the structure of G as a direct product of alternating groups which act diagonally on subsets of the orbits.

Procedure `SGS_Giant_Orbits`
  Input: (S, A) where $\langle S\rangle = G \leq Sym(\Omega)$, G acts faithfully on $A \subseteq \Omega$,
    A is a union of giant orbits $O_1, \ldots, O_t$ of G and $|S| = O^{\sim}(t)$.
  Output: T such that T is a SGS for G with respect to an ordering
    determined by the algorithm.

  $T \leftarrow$ `SGS_Alt_Orbits`(S, A)
    [T is a SGS for $G \cap (\mathrm{Alt}(O_1) \times \cdots \times \mathrm{Alt}(O_t))$]
  Let $S' = \{s \in S \mid s|_{O_i}$ is odd for some $O_i\}$
  [Simulate Gaussian elimination on G/G']
  For $i \leftarrow 1$ to t do
    Let s' be the first element of S' such that $s'|_{O_i}$ is odd (if it exists)
    If such an s' exists then
      Set $S' \leftarrow S' \setminus \{s'\}$
      Set $s' \leftarrow$ the residue of sifting s' through T on O_i
      [$s'|_{O_i}$ is now a 2-cycle]
      Set $T \leftarrow T \cup \{s'\}$
      Set $S' \leftarrow S's'^{-1}$
    Replace each $s \in S'$ by the residue of sifting s through T on O_i
      [S'^{O_i} is now trivial]

Return T

The next procedure makes use of a fast normal closure routine, developed in [**BCF⁺91**], and described further in [**CF92**] (this volume). It also refers to a procedure `Three_Cycle` due to Babai, Luks and Seress and is described in [**BLS88**]. If a group specified by a generating set of size $O^{\sim}(n)$ contains the alternating group, then this procedure will construct a three-cycle in $O^{\sim}(n^2)$ Las Vegas time.

Procedure `SGS_Alt_Orbits`
  Input: (S, A) as in `SGS_Giant_Orbits`
  Output: a SGS, T, for $G \cap (\mathrm{Alt}(O_1) \times \cdots \times \mathrm{Alt}(O_t))$
    with respect to an ordering determined by the algorithm.

  Initialize $T \leftarrow \emptyset$
  Let all orbits be unmarked.
  Mark the first orbit, O_1, of G
  Let $x \in O_1$ and S' a set of Schreier generators for G_{x_1} formed from S
  For $i \leftarrow 2$ to t do
    If S' has a fixed point on O_i then mark O_i
      [In our setting, this is true if and only if $G_{O_1}^{O_i}$ is trivial.]
  Let T' be a SGS for $G^{O_1} \cap \mathrm{Alt}(O_1)$ whose elements act as 3-cycles on O_1
    [By invoking `3-Cycle`]
  Replace each element in T' by its square [This ensures that $T' \subseteq \mathrm{Alt}(\Omega)$]
  For each unmarked orbit O of G do
    Let S'' be a generating set for $S^{O_1 \cup O}$ of size $O^{\sim}(1)$
      [Found by applying Theorem 3.15]
    Set $\sigma_{\overline{O}} \leftarrow$ `Element_Fixing_an_Orbit`(S'', O_1, O)
    Let $\mathcal{O}$ be the set unmarked orbits on which $\sigma_{\overline{0}}$ acts non-trivially
    Let $\overline{O} = \cup_{O \in \mathcal{O}} O$ and $n_{\overline{O}} = |\mathcal{O}|$
    Let $\sigma'_{\overline{O}}$ be the residue of sifting $\sigma_{\overline{O}}$ through T
      [$\sigma'_{\overline{O}}|_{\overline{O}} = \sigma_{\overline{O}}|_{\overline{O}}$ and is trivial elsewhere.]
    Let $S'_{\overline{O}}$ be a set of $O(n)$ generators for the normal closure of $\langle \sigma'_{\overline{O}} \rangle$ in G
      [The "fast normal closure" routine is used"]
    Let $S''_{\overline{O}}$ be a reduced set of $O^{\sim}(n_{\overline{0}})$ generators for $\langle S'_{\overline{O}} \rangle$
    Set $T \leftarrow$ `Prepend (SGS_Alt_Orbits(`$S''_{\overline{O}}, \overline{O}$`)`, T)
    Mark all orbits in $\overline{O}$
  Set $T \leftarrow$ `Prepend`(T', T)
  Return T

Procedure `Element_Fixing_an_Orbit`
  Input: (S, O_1, O_2) where O_1 and O_2 are unlinked giant

orbits for $G = \langle S \rangle$ and $|S| = O^{\sim}(1)$.
Output: An element of G_{O_1} which acts non-trivially on O_2.

$\sigma \leftarrow$ `Three_Cycle` (S, O_1)
Set $\sigma \leftarrow \sigma^2$ [to ensure that $\sigma \in \mathrm{Alt}(O_1 \cup O_2)$]
Let $\sigma|_{O_1} = (a\ b\ c)$, $\sigma|_{O_2} = \delta$
If δ is non-trivial then
  Let U be a set of $O^{\sim}(1)$ generators for $G_{a,b,c}$
    [Found by repeated application of Theorem 3.15]
  Let g be an element of U such that $[g, \sigma] = [g, \delta] \neq 1$
    [Since $\mathrm{Alt}(O_2) \leq \langle S \rangle^{O_2}$, δ is not centralized by S]
 Return($[g, \sigma]$)
Else [$\delta = (a\ b\ c)$]
 Let T be a SGS for G^{O_1} constructed from $(a\ b\ c)$ with $T \subseteq G_{O_2}$
 For each $s \in S$ do
  Let $s' \in G_{O_1}$ be the residue of s sifted through T
  If $s' \neq 1$ return(s')

5.1. Giant Orbits: Proof of Correctness. The proof of correctness for `SGS_Alt_Orbits` follows directly from the following result.

COROLLARY 5.1. *Assume that $G \leq \mathrm{Sym}(\Omega)$ and that $\mathrm{Soc}(G)$ acts as the full alternating group on each orbit O_i, $1 \leq i \leq t$ of G. Then there exists a partition of $\{O_1, \ldots, O_t\}$ into subsets $\{B_1, \ldots, B_s\}$ such that $\mathrm{Soc}(G)$ is the direct product of alternating groups, with each direct factor acting as a diagonal subgroup on the orbits in precisely one of the blocks B_i and fixes pointwise the orbits in the other blocks. Furthermore, if $\sigma \in \mathrm{Soc}(G)$, then the normal closure of $\langle \sigma \rangle$ in G is the product of the diagonal subgroups corresponding to the blocks B_i on whose orbits σ acts non-trivially.*

THEOREM 5.2. *Let $G = \langle S \rangle$ act on Ω with $|\Omega| = n$. Assume that G has t giant orbits and that $|S| = O^{\sim}(t)$. Then the procedure `SGS_Giant_Orbits` constructs a SGS for G in $O^{\sim}(tn^2)$ Monte Carlo time.*

PROOF. The proof of correctness and the timing for `SGS_Giant_Orbits` is clear once the corresponding results have been proved for `SGS_Alt_Orbits`.

The correctness of `SGS_Alt_Orbits` relies on Corollary 5.1. As long as there are unmarked orbits, the procedure will construct a non-identity element $\sigma'_{\overline{O}}$ which fixes pointwise each marked orbit and acts non-trivially on the union $\overline{O}$ of a subset of unmarked orbits. In the notation of Corollary 5.1, $\overline{O}$ is the union of a certain number of the $B_i's$. Thus the normal closure of $\sigma'_{\overline{O}}$ in G is a direct factor of $\mathrm{Soc}(G)$ consisting of the pointwise stabilizer in $\mathrm{Soc}(G)$ of $\Omega \setminus \overline{O}$. A recursive call is made and once a strong generating set has been computed it is prepended to the set T which holds elements of the strong generating set currently being accumulated.

In analyzing the timing, it is straightforward to show that the top-level call to `SGS_Alt_Orbits` takes time $O(tn^2)$ aside from the recursive calls to smaller problems. By induction, each recursive call to a subproblem takes time $O(n_{\overline{O}}n^2)$, where $n_{\overline{O}}$ is the number of orbits O_i in $\overline{O}$ and $\overline{O}$ is a union of blocks B_j as given in Corollary 5.1. Since a recursive call at the top-level never involves an orbit O_i more than once, the total time spent on the recursive calls is $O^{\sim}(tn^2)$. □

References

[Bab86] L. Babai. On the length of chains of subgroups in the symmetric group. *Comm. in Algebra*, 14:1729–1736, 1986.

[BCF+91] L. Babai, G. Cooperman, L. Finkelstein, E.M. Luks, and Á. Seress. Fast Monte Carlo algorithms for permutation groups. In *Proc. 23rd ACM STOC*, pages 90–100, 1991.

[BCF++91] L. Babai, G. Cooperman, L. Finkelstein, and Á. Seress. Nearly Linear Time Algorithms for Permutation Groups with a Small Base. In *Proc. ISSAC '91* (International Symposium on Symbolic and Algebraic Computation '91), ACM Press, pages 200–209, 1991.

[BFP89] C.A. Brown, L. Finkelstein, and P.W. Purdom. A new base change algorithm for permutation group. *SIAM J. Computing*, 18:1037–1047, 1989.

[BLS87] L. Babai, E. Luks, and Á. Seress. Permutation groups in NC. In *Proc. 19th ACM STOC*, pages 409–420, 1987.

[BLS88] L. Babai, E. Luks, and Á. Seress. Fast management of permutation groups. In *Proc. 19th IEEE FOCS*, pages 272–282, 1988.

[Cam81] P.J. Cameron. Finite permutation groups and finite simple groups. *Bull. London Math. Soc.*, 13:1–22, 1981.

[CF92] G. Cooperman and L. Finkelstein. Combinatorial tools for computational group theory. In *DIMACS Proceedings* (this volume), 1992.

[CFS90] G. Cooperman, L. Finkelstein, and N. Sarawagi. A random base change algorithm for permutation groups. In *Proc. of 1990 International Symposium on Symbolic and Algebraic Computation*, pages 161–168, Tokyo, Japan, August 1990. ACM Press and Addison-Wesley.

[CST89] P.J. Cameron, R. Solomon, and A. Turull. Chains of subgroups in symmetric groups. *J. of Algebra*, 127:340–352, 1989.

[Gor82] D. Gorenstein. The Classification of Finite Simple Groups, Vol. 1. Plenum Press, New York, 1982.

[Kan79] W.M. Kantor. Permutation representations of the finite classical groups of small degree or rank. *J. Algebra*, 60:158–168, 1979.

[Luk90] E. Luks. Lectures on polynomial-time computation in groups. Technical Report NU-CCS-90-16, College of Computer Science, Northeastern University, 1990.

[Sar92] N. Sarawagi. Computational Group Theory and Applications to Search. Ph.D. Thesis, Northeastern University, 1992.

[Sco79] L. Scott. Representations in characteristic p. The Santa Cruz Conference on Finite Groups, 1980, Amer. Math. Soc., 319–322.

[Sim] C.C. Sims. Computation with permutation groups. In *Proc. Second Symposium on Symbolic and Algebraic Manipulation*, ACM Press, New York, 1971, pages 23–28.

College of Computer Science, Northeastern University, Boston, Massachusetts 02115

E-mail address: namita@ccs.northeastern.edu, gene@ccs.northeastern.edu, and laf@ccs.northeastern.edu

DIMACS Series in Discrete Mathematics
and Theoretical Computer Science
Volume **11**, 1993

PERM: A Program Computing Strong Generating Sets

ÁKOS SERESS AND IVÁN WEISZ

December 16, 1992

ABSTRACT. A *base* of a permutation group G is a subset B of the permutation domain such that only the identity of G fixes B pointwise. We call a family of groups *small base groups* if they admit bases of size $O(\log^c n)$ for some fixed constant c, where n is the size of the permutation domain. For example, all permutation representations of all finite simple groups other than the alternating groups admit $O(\log n)$ size bases. Groups with very small bases dominate the work on permutation groups in much of computational group theory.

In [**3**], Babai, Cooperman, Finkelstein, and Seress presented a Monte Carlo algorithm for the fundamental task of constructing a strong generating set for small base groups which runs in *almost linear*, $O(n \log^{c'} n)$ time. The purpose of this note is to report an implementation of this algorithm.

1. Introduction

In his seminal work [**8**], Sims introduced the notions of a *base* and *strong generating set* as the fundamental data structures for computing with permutation groups. We consider permutation groups given by a list of generators. A *base* for a permutation group G of degree n is a subset $B = \{\beta_1, \beta_2, .., \beta_M\}$ of the permutation domain, Ω, with the property that only the identity of G fixes each point of B. The point stabilizer sequence for G relative to B, considered as an ordered subset, is the chain of subgroups

$$G = G^{(1)} \geq G^{(2)} \geq \ldots \geq G^{(M+1)} = 1,$$

1991 *Mathematics Subject Classification.* Primary 20B40, 68Q40; Secondary 68P05.

The first author was supported in part by NSF Grant CCR-9201303 and NSA Grant MDA904-92-H-3046.

This paper is in final form and no version of it will be submitted for publication elsewhere.

where $G^{(i)} = G_{\{\beta_1,..,\beta_{i-1}\}}$ is the subgroup of G which fixes pointwise the set $\{\beta_1, \dots, \beta_{i-1}\}$, $1 \le i \le M+1$. A *strong generating set* (SGS) for G relative to B is a generating set S for G with the property that

$$\langle S \cap G^{(i)} \rangle = G^{(i)}, \text{ for } 1 \le i \le M+1.$$

An (infinite) family $\mathcal{G}$ of groups is called a family of *small base groups* if all its members G of degree n admit bases of size $O(\log^c n)$, for some fixed constant c. Equivalently, $\log|G| = O(\log^{c'} n)$ for a fixed constant c'.

In [**3**], an $O(n \log^3 |G| \log\log^2 |G|)$ Monte Carlo algorithm is presented for the construction of an SGS. (A Monte Carlo algorithm is a random algorithm which may return an incorrect answer but the probability of error can be made less than any bound prescribed by the user.) In particular, for families of small base groups, an SGS can be computed in *almost linear*, $O(n \log^{c''} n)$, time. The purpose of this note is to report an implementation called PERM of the BCFS algorithm.

Our goal was twofold. First, we wanted to test whether the new theoretical ideas of [**3**] could be exploited in a practical algorithm. Our second goal was to provide an environment to examine the sharpness of the theoretical estimates. We hope that heuristics obtained from experiments using this implementation will suggest directions in which the theoretical results can be improved. In turn, better theoretical estimates would allow us to change parameters in the algorithm, lowering the running time while maintaining the same small probability of error. In particular, a strengthening of Babai's local expansion lemma (cf. Section 2) would be useful.

The algorithm BCFS is the centerpiece of the rapidly growing family of almost linear time algorithms for small base groups. Recently, Beals [**5**] gave an algorithm for computing blocks of imprimitivity; Beals and Seress [**6**] for the composition factors and center; and Luks and Seress for the centralizer of a normal subgroup, the Fitting subgroup and the maximal solvable normal subgroup. The implementation of some of these algorithms is in progress.

In Section 2, we introduce some notation and definitions. Section 3 contains the outline of the algorithm while in Section 4 we present experimental data which indicates why the actual running time and error probability of the algorithm is much lower than the theoretical estimates.

2. Schreier trees, random subproducts, and local expansion

Let $B = \{\beta_1, ..., \beta_M\}$ be a base of the group G and let $G = G^{(1)} \ge G^{(2)} \ge ... \ge G^{(M+1)} = 1$ be the corresponding point stabilizer subgroup chain. Moreover, let R_i denote a *transversal* (complete set of right coset representatives) for $G^{(i+1)}$ in $G^{(i)}$, $1 \le i \le M$. Each $g \in G$ can be written uniquely in the form $g = \rho_M \rho_{M-1} \cdots \rho_2 \rho_1$, $\rho_i \in R_i$. The process of factorizing g in this form is called *sifting* or *stripping*.

The coset representatives will be stored in an especially efficient version of Sims's Schreier vector data structure [**8**]. Given a permutation group $G = \langle S \rangle \leq S_n$, a *Schreier vector data structure* for G is a sequence of pairs (S_i, T_i) called *Schreier trees*, one for each base point β_i, $1 \leq i \leq M$. T_i is a directed labelled tree, rooted at β_i, with edge labels from the set $S_i = \{g_1, \ldots, g_k\} \subseteq G^{(i)}$. The nodes of T_i are the points of the orbit $\beta_i^{\langle S_i \rangle}$. For a directed edge from u to v with label h, $u^h = v$. If v is a node of T_i, then the product of the edge labels along the path from β_i to v in T_i is a word in the elements of S_i whose corresponding permutation moves β_i to v. Thus each Schreier tree (S_i, T_i) defines a set of coset representatives for $G^{(i+1)}$ in $G^{(i)}$.

For the efficiency of the algorithm, it is crucial that the Schreier trees have small depth so the coset representatives can be written as a short product of elements of S_i. That's why PERM provides seven options for the user to construct Schreier trees. The default option, described below, guarantees that the depth of trees is $\leq 2 \log |G|$; however, empirical evidence shows that the actual depth is significantly smaller (cf. Section 4).

The default option is the following. Given generators for $H = \langle S \rangle$ and a point β in the permutation domain, construct the edge labels $g_1, g_2, \ldots$ for a Schreier tree coding the coset representatives for H_β by the following rule. Suppose that $h_1, \ldots, h_k$ are already defined and let P_k denote the set of $\leq 2k$-long products built from $h_1, \ldots, h_k$ and their inverses. If $\beta^S = \beta^{P_k}$ then stop; otherwise, let $h_{k+1} \in P_k S$ such that $\beta^{h_{k+1}} \notin \beta^{P_k}$. Note that β^{P_k} can be constructed as the vertex set of a breadth-first-search tree of depth $2k$ with root β. It is proven in [**3**, Lemma 2.2] that this procedure stops for some $k \leq \log |H|$. During the algorithm, this procedure is applied with for $H = \langle S_i \rangle$, where S_i is a set of generators for the currently known subgroup of $G^{(i)}$.

An important subproblem in SGS constructions is the following. Given $G = \langle S \rangle$, coset representatives R_β for G_β in G, and an SGS for $H \leq G_\beta$, decide whether $H = G_\beta$. The standard method is to check whether the *Schreier generators* $\{\rho s \bar{\rho}^{-1} : \rho, \bar{\rho} \in R_\beta, s \in S, \rho s \bar{\rho}^{-1} \in G_\beta\}$ are in H. In [**3**] and PERM, this procedure is sped up in the following way.

Let $S = \{s_1, \ldots, s_m\}$. A *random subproduct* of the sequence $(s_1, \ldots, s_m)$ is a product of the form $s_1^{\varepsilon_1} s_2^{\varepsilon_2} \cdots s_m^{\varepsilon_m}$ where the ε_i are chosen independently from the uniform distribution over $\{0, 1\}$. Random subproducts were introduced in [**4**] while most applications can be found in [**2**]. Here, we form a pair of subproducts w_1, w_2: w_1 is a random subproduct of a random ordering of $\{s_1, \ldots, s_m\}$ while w_2 is a random subproduct of a random ordering of a random subset of size k, with k uniformly distributed in the interval $[0, m/2]$.

Let t be the sum of tree depths in the Schreier vector data structure for H and for R_β. The algorithm forms c_1 pairs of subproducts w_1, w_2 and combines them with $c_2 t$ randomly chosen elements of R_β to form Schreier generators for G_β. (c_1, c_2 are constants chosen by the user, depending on the desired error probability.) As it is shown in [**3**, Section 3], if $H \neq G_\beta$ then with high probability one of these Schreier generators will be in $G_\beta \setminus H$.

The correctness of the above procedure, as well as of the strong generating test described in the next section, depends on Babai's local expansion lemma. We state this lemma explicitly because this is the major bottleneck for improving the algorithm. A strengthening of this lemma would allow us to change the parameters, thereby lowering the running time and/or improving the error probability.

Lemma (Babai [**1**]) *Let S denote a set of generators of the group G and set $T = S \cup S^{-1} \cup \{1\}$. Let D be a finite subset of T^t, the set of t-term products of members of T (in any order), such that $|D| \le |G|/2$. Then for at least one generator $g \in S$, $|Dg \setminus D| \ge |D|/4t$.*

To apply the lemma, note that if $H \neq G_\beta$ then $|H| \le |G_\beta|$ and, consequently, the set $D = HR_\beta$ satisfies $|D| \le |G|/2$. The sum of depths of Schreier trees is an upper estimate for t and it is easy to see that $dg \in D$ for some $d = hr$, $h \in H$, $r \in R_\beta$, $g \in S$ if and only if the Schreier generator formed from r and g is in H. Using random subproducts instead of elements of S further improves the running time.

3. Description of the algorithm

In this section, we outline the version of the BCFS algorithm which is implemented in PERM. For more details, we refer the reader to [**3**]. The algorithm is a variant of the basic random Schreier-Sims method, with some parts also influenced by Knuth [**7**]. It consists of two parts: strong generator construction (SGC) and strong generating test (SGT).

Let $O_1, O_2, \ldots$ be the orbits of G. SGC works orbitwise, i.e., first it computes a non-redundant base and an SGS for the G-action on O_1. Then the strong generating test is called. There are three possibilities:

1) The SGT returns a group element g acting nontrivially on O_1 which shows that the output of the SGC is incorrect. In this case, the SGC continues on O_1, adding g to the list of generators.
2) The SGT confirms that the SGS on O_1 is computed correctly and returns a group element $g \neq 1$ such that g acts trivially on O_1. Then SGC is invoked on an orbit O_{i_2} where g acts nontrivially and computes a base and SGS for the O_{i_2}-action of G_{O_1}.
3) The SGT confirms the SGC and reports that G_{O_1} fixes all other orbits as well. In this case, the algorithm terminates.

Similarly, the strong generating test is called after the SGC returns a non-redundant base and SGS for the G-action on $O_1 \cup O_{i_2} \cup \ldots \cup O_{i_k}$, $k \ge 1$. One of the above three cases will occur and the algorithm terminates when we are in Case 3.

First, we describe the strong generator construction. Suppose that base points

$\beta_1, ..., \beta_b$ and a set T of generators, part of a strong generating set, are already computed. Denote by O^* the orbit the SGC currently works in and by b^* the number of base points at the end of the sequence $\beta_1, ..., \beta_b$ which are in O^*. We say that T is *up-to-date below the level* i if $T \cap G^{(i+1)}$ is an SGS for the group $\langle T \cap G^{(i+1)} \rangle$ with $T \cap G^{(i+1)}$ being the union of edge labels S_j in Schreier trees (S_j, T_j) below the level i.

If T is up-to-date below the level i then do the following. (Note that β_i is not necessarily in O^*.) If it does not exist already, compute the Schreier tree (S_i, T_i) for the coset representatives of $G^{(i+1)}$ in $G^{(i)}$, as described in the previous section. Compute the random subproducts w_1, w_2 from $\bigcup_{j \geq i} S_j$ and the random subset of Schreier generators for $G^{(i+1)}$, also described in the previous section. Write the Schreier generators *only as a word* in w_1 or w_2 and the labels in S_i.

Sift these Schreier generators through the SGS and write the residues as *words* in the edge labels $\bigcup_{j \geq i} S_j$ (cf. [**3**, section 2]). If a Schreier generator h does not sift through, i.e., on some level $j^* > i$ the necessary coset representative is not in $\langle T \cap G^{(i+1)} \rangle$ then compute the residue h^* of h as a permutation acting on all n points, add h^* to T, and consider T up-to-date below the level j^*. If h sifts through and the residue h^* is written as a word in the edge labels then distinguish three cases:

(i) if $|O^*| \leq 50$ then evaluate the word for h^* on all points of O^*;

(ii) if $|O^*| > 50$ and $b^* \leq 5$ then evaluate the word for h^* on 5 randomly chosen points of O^*;

(iii) if $|O^*| > 50$ and $b^* > 5$ then evaluate the word for h^* on b^* randomly chosen points of O^*.

(Note that the values 50 and 5 are quite arbitrary and the user can change them. The special treatment of small $|O^*|, b^*$ are heuristics which improve the running time and reliability of PERM.)

If h^* fixes all of the evaluated points then discard h; otherwise, compute h^* as a permutation acting on all n points, add h^* to T, add a point β_{b+1} of O^* not fixed by h^* to the sequence of base points, and consider T up-to-date below the level $b+1$. If all Schreier generators are discarded then consider T up-to-date below the level $i-1$. The SGC terminates when T is up-to-date below the level 0.

Finally, we describe the strong generating test. The input of the SGT is a Schreier vector data structure $\{(S_i, T_i) | 1 \leq i \leq b\}$, coding coset representative sets R_i. Let t denote the sum of depths of all trees in the data structure. Compute c_3 pairs of random subproducts w_1, w_2 from the set $\bigcup_{i \geq 1} S_i$ as described in Section 2 and $c_4 t$ uniformly distributed random elements g from the set $R_b R_{b-1} \cdots R_2 R_1$. Sift the original generators of G and all products of the form gw_j, $j = 1, 2$ through the input Schreier vector data structure, evaluating the residues on all n points. If there is a nontrivial residue then the SGT invokes the SGC in one of cases 1), 2); otherwise, PERM terminates in case 3).

Group	n	$\lvert B\rvert$	PERM SGC	PERM SGT	Number of multiplications
Suzuki	1782	4	2.30	12.64	9816
Conway 2	2300	6	4.51	22.26	13855
GU(7,4)	16384	7	11.05	288.40	14467
Sp(12,2)	4096	12	17.38	121.72	37190
GL(4,11)	14641	4	19.08	231.58	14130
GL(6,5)	15625	6	45.15	585.17	29833

TABLE 1. Running times in seconds

Group	n	Theor. bound $2\log\lvert G\rvert$	PERM
Suzuki	1782	78	6
Conway 2	2300	92	6
GU(7,4)	16384	100	6
Sp(12,2)	4096	156	6
GL(4,11)	14641	112	8
GL(6,5)	15625	168	9

TABLE 2. Depth of trees

4. Some data

PERM is written in C. Table 1 contains some sample running times, obtained on a SPARCstation 2 workstation with 64 Mb RAM. The reported times are averages of three runs, with the group generated by two random elements. For possible comparison with other methods, we also report the number of performed permutation multiplications. In PERM, the worst case estimate for the running time (with the default setting for constructing Schreier trees) is worse by a factor of $\log\lvert G\rvert$ than the theoretically best version of [**3**]; however, it seems that the practical performance is much better than even the estimate in [**3**]. With the default parameter setting, the theoretically estimated error probability is about 15%. (Note that doubling the running time squares the error probability.) Also, the practical performance seems to be much better: already the SGC part seldom returns an incorrect answer and in these cases the SGT part "captures" the mistake very fast. The SGT was also tested by setting the parameters of the SGC very low (so increasing the error probability); even in these cases, the SGT found the error within a couple of seconds.

Table 2 may indicate why the actual running times are much better than the theoretical estimates. It contains a comparison of depths of Schreier trees in PERM runs and in theoretical estimates.

Finally, Table 3 contains the estimated and actual percentage of "good" coset representatives, i.e. coset representatives such that the Schreier generators constructed from them augment the corresponding point stabilizer subgroup. If the

Level	Theor. estimate	PERM
1	12.50	100
2	0	0
1	10.00	100.00
2	20.00	96.05
3	0	0
2	12.50	100.00
3	20.00	88.28
4	0	0
3	16.67	99.90
4	0	0
3	12.50	99.32
4	20.00	46.67
5	0	0
4	0	0
3	0	0
2	0	0
1	0	0

TABLE 3. Percentage of "good" coset representatives in a typical run on the Suzuki group

level was up-to-date then of course both of these numbers are 0; on non-trivial levels, the majority of coset representatives is good, explaining why already the SGC part gives the correct result. This feature was not tested extensively; in limited tests, the actual percentage was never lower than 40%.

REFERENCES

1. L. Babai, *Bounded Round Interactive Proofs in Finite Groups*, SIAM J. Discr. Math., to appear.
2. L. Babai, G. Cooperman, L. Finkelstein, E.M. Luks, and Á. Seress, *Fast Monte Carlo Algorithms for Permutation Groups*, Proc. 23rd ACM STOC (1991), pp. 90–100.
3. L. Babai, G. Cooperman, L. Finkelstein, Á. Seress, *Nearly Linear Time Algorithms for Permutation Groups With a Small Base*, Proc ISSAC'91 (Internat. Symp. on Symbolic and Algebraic Computation), Bonn 1991, pp. 200–209.
4. L. Babai, E. Luks, and Á. Seress, *Fast Management of Permutation Groups*, Proc. 29th IEEE FOCS (1988), pp. 272–282.
5. R. Beals, *Structure Forest in Nearly Linear Time for Small Base Groups*, this volume.
6. R. Beals and Á. Seress, *Structure Forest and Composition Factors for Small Base Groups in Nearly Linear Time*, Proc. 24th ACM STOC (1992), pp. 116–125.
7. D.E. Knuth, *Notes on Efficient Representation of Perm Groups*, Combinatorica **11** (1991), pp. 57–68 (preliminary version circulated since 1981).
8. C.C. Sims, *Computation with Permutation Groups*, in Proc. Second Symposium on Symbolic and Algebraic Manipulation, edited by S.R. Petrick, ACM, New York, 1971.

Department of Mathematics, Ohio State University, Columbus, Ohio 43210
E-mail address: akos@function.mps.ohio-state.edu

Department of Mathematics, Ohio State University, Columbus, Ohio 43210
E-mail address: weisz@function.mps.ohio-state.edu

DIMACS Series in Discrete Mathematics
and Theoretical Computer Science
Volume 11, 1993

Complexity Issues in Infinite Group Theory

CHARLES C. SIMS

ABSTRACT. In computational group theory the techniques of asymptotic complexity analysis have so far been applied only to problems involving finite groups. This paper suggests ways in which these techniques might be used to help better understand problems connected with infinite groups.

This workshop has brought together researchers from two quite distinct traditions. For some, the primary reason for their interest in group-theoretic algorithms is their desire to produce software which will permit them or their colleagues to carry out some specific computations with groups. Others, coming to the subject with the perspective of theoretical computer science, have as their main goal the formulation of group-theoretic algorithms which have good asymptotic complexity. These two traditions have different cultures and different ways of evaluating algorithmic research. Ideally, the algorithms with the best asymptotic complexity should also be the ones which are most practical for implementation on current hardware. Unfortunately, this ideal does not always seem to be realized. This fact has led in the past to a certain amount of tension between these two traditions. One of the stated goals of this workshop is to stimulate a dialogue among representatives of these two segments of the computational group theory community. This talk is intended to be a modest effort in that direction.

Specifically, I have two purposes in mind. They are:

(1) To suggest ways in which complexity theory could be made more relevant to the development of group-theoretic software without abandoning its point of view or its esthetic sense.

(2) To mention some problems which arise in attempts to work with infinite groups, or groups not obviously finite, which might be of interest to researchers in complexity theory.

1991 *Mathematics Subject Classification.* Primary 20-04, 20F05, 68Q25; Secondary 20F10, 68Q42.

The author's research is supported in part by the National Science Foundation.

This paper is in final form and no version of it will be submitted for publication elsewhere.

Before addressing these points, however, I want to describe my own involvement with computational group theory. Much of my early work on computing with permutation groups was motivated by a desire to prove the existence of several of the larger sporadic finite simple groups. More recently, my research has been focused on computational problems related to some specific groups defined by presentations. However, I must admit that on occasions I have simply been seeking that exhilaration I experience when I can get a computer to perform a group-theoretic calculation faster or with less memory than had been previously possible.

Although I still consider myself a group theorist, the articles and books I read deal at least as much with topics in computer science as they do with group theory. In fact, I consider that I still have a lot to learn from undergraduate computer science texts such as [**7**] and its earlier edition. I probably spend more time trying to develop space-efficient data structures than on any other activity.

Despite what I hope is an openness to the role of computer science in computational group theory, I am sometimes put off by statements of the complexity theory point of view. An observation by Hendrik Lenstra may be useful at this point. Lenstra is a strong advocate of the value of asymptotic complexity theory. However, in [**6**] he pointed out that many cosmologist believe the universe to be finite. If true, this means that the number of elementary particles in the universe is finite. Since it is unlikely that the input to a computer program can involve more bits in its description than there are elementary particles, there are only finitely many computational problems which can be submitted to a physical computer. Thus it is at least conceivable that a complexity theorist may assert the superiority of an algorithm even though it is not faster than competing algorithms on any physically realizable problem instance.

Since I want to encourage dialogue and not increase tensions further, let me move on quickly to the first of the points mentioned above: How can complexity theory help the software developers among us. One important way is by emphasizing average-case complexity where possible. A piece of software can be quite valuable even if there are a few problem instances for which termination can not be expected to occur in a reasonable amount of time. As long as the program can handle most problems satisfactorily, it will be used. Perhaps the best known example of a practical algorithm with poor worst-case complexity is the simplex method for linear programming. Despite its exponential running time on a few pathological inputs, it has been used for decades to solve important problems in industry. I am well aware of the difficulty of carrying out average-case analyses on many of the algorithms of interest to group theorists. Nevertheless, when such analyses are possible, they provide very valuable information to the implementer.

In some cases work needs to be done even to make an appropriate definition of the average case. In [**2**], John Dixon has shown that two or more randomly selected permutations generate the alternating or the symmetric group with probability approaching 1 as the degree goes to infinity. Personal observation suggests that "typical" sets of generating permutations which arises in practice are not random.

Defining a random group presentation requires careful thought. Simplifying the problem somewhat, let us consider presentations consisting of n relators of length m on a fixed set of generators. If m and n are allowed to go to infinity at roughly the same rate, then these presentations are almost certain to define small cancellation groups and the associated groups will have solvable word problems. However, if n grows much more rapidly than m, then this need not be the case.

The use of the "big Oh" notation in complexity results makes possible succinct statements of theorems. However, it would be very helpful to implementers reading the papers if comments could be included which give some information about the size of the constants implied by the O's. This would make it easier to gain insight concerning the practicality of the algorithms being discussed. Precise values are not necessary, just some indication as to whether the constants are closer to 3 or to the number of elementary particles in the universe. Sometimes constants must be made large just to take care of a few small problem instances. For implementation purposes it is probably most useful to know what values the constants would have if these small exceptional cases did not exist. The recently introduced notation $O^{\tilde{}}$, which hides powers of logarithms, not just constants, makes it even more difficult for an implementer to assess the practicality of an algorithm.

While I do not want to belabor the point, I must admit to experiencing some uneasiness when I encounter O's in the hypothesis of a theorem, such as one of the form:

THEOREM. *Given a set of ... of size $O(\dots)$, we can compute ... in time $O(\dots)$.*

Two questions come up immediately. How are the two constants masked by the O's related? In particular, does a larger input set make the computation harder or easier? When two or more O's occur in the hypothesis, I get even more uneasy. If I see something like

THEOREM. *Given a set of ... of size $O(\dots)$ and a set of ... of size $O(\dots)$, we can compute ... in time $O(\dots)$.*

I begin looking frantically for a comment describing how the constant in the conclusion depends on the two constants in the hypothesis.

Complexity results are always obtained relative to some model of computation. In my opinion, the current models do not put a high enough premium on the total amount of memory in use at a given time. In symbolic computation in general and in computational group theory in particular, space is frequently scarce. A simple example can be found in multiple-precision arithmetic packages. To multiply two n-bit numbers using the algorithm we learn in grade school takes time $O(n^2)$ and uses $2n$ bits of memory. There is a relative simple algorithm attributed to Karatsuba, see **[4]**, which can be used to multiply two n-bit numbers in time $O(n^{\log_2 3})$. This algorithm is practical, since on most serial architectures it becomes faster than the grade school algorithm at somewhere around 1000 bits. Despite its superiority, the Karatsuba algorithm is not implemented in most general-purpose systems for symbolic computation. The reason appears to be in part that the Karatsuba algorithm requires at least $6n$ bits, three times as much as the grade school method. This factor of three is considered too great a price to pay for the improvement in running time.

It is possible to connect gigabytes of disk space to current computers. Why is space such a problem? The answer is access time. Experiments I have performed on a number of serial architectures suggest that a tight loop in which each iteration causes a page fault runs 30 to 50 times slower than the same loop does when the data needed are all in real memory. Thus a program that might terminate in a day if all its data could be squeezed into real memory can easily take a month when its data exceeds the capacity of real memory by even a relatively small amount. Waiting an extra day or two might be acceptable, but not waiting a month. Trying to improve on system paging software is not a good idea, since writing portable code which accesses large external files is very difficult.

The next point finally gets around to something related to the title of this talk. Most natural questions about a finitely presented group G are algorithmically unsolvable. Many of these questions call for a yes or no answer. For example: Is G trivial? Is G finite? Is G nilpotent? For these and many similar questions, if the answer is yes, then that fact can be verified. However, if the answer is no, then there is no general algorithmic way to confirm that answer. This means that we can devise a procedure which takes as input a finite presentation and, if the answer to our question is yes, the procedure terminates with a proof of the answer. If the answer is no, the procedure does not terminate.

It may be possible to devise quite different procedures for studying the same question. What should it mean to analyze and compare such procedures, which are not guaranteed to terminate for all inputs? Analysis of algorithms is one of the fundamental themes of computer science. Can such analyses be extended to cover procedures which may not terminate?

Let me illustrate what I have in mind with an example. Two important tools in studying groups given by finite presentations are coset enumeration and the Knuth-Bendix procedure for strings. Although they do not do exactly the same thing, they may be loosely described as procedures for determining membership in a finitely generated subgroup H of a finitely presented group G. Coset enumeration terminates if and only if the index of H in G is finite. The Knuth-Bendix procedure for strings terminates in this case as well, but it also terminates in some cases in which H has infinite index in G. Thus the Knuth-Bendix procedure for strings is more powerful in the sense that it terminates more often than coset enumeration. However, on problems for which current implementations of both procedures terminate, coset enumeration frequently arrives at the final answer more quickly. See **[9]** for two examples in which the Knuth-Bendix procedure is superior to coset enumeration.

In practice, programs implementing coset enumeration and the Knuth-Bendix procedure for strings are often aborted because they run out of memory. Both procedures have the property that with a given amount of memory they must terminate, either normally or abnormally. One way to compare them is to study how their running times depend on the amount of memory available. When these procedures terminate abnormally, the data which has been accumulated so far can sometimes be used to gain insight into the structure of the group under investigation. Thus another way to compare these two procedures is to analyze the amount of information contained in the data structures they employ. From this point of view I think it is fair to say that the data structures for the Knuth-Bendix procedure for strings have a higher density of information per megabyte.

The operation of the Knuth-Bendix procedure for strings poses a more traditional type of complexity question. The Knuth-Bendix procedure builds up large rewriting systems, which consist of rules, pairs (P, Q) of words over a fixed finite alphabet X. During the operation of the procedure, words are rewritten with respect to the current rewriting system. This means that a word of the form APB is replaced by AQB and the process repeated until no left side of any rule is a subword of the word.

To rewrite efficiently with respect to a rewriting system $\mathcal{R}$ we must have a fast method for answering the following question: Given a word W, is there a subword of W which is a left side of a rule in $\mathcal{R}$? Of course, if the answer is yes, then we need to know one such subword. For a fixed alphabet size, there is a linear-time solution to this problem. It involves first building an index structure for $\mathcal{R}$ in time linear in the size of $\mathcal{R}$. Once the index is available, the determination of whether W contains a left side as a subword takes time linear in the length of W. It frequently happens that, if (P, Q) is a rule in $\mathcal{R}$, then P is not a subword of the left side of any other rule. With this assumption, one linear-time index structure is essentially a finite automaton. The states of the automaton correspond to the proper prefixes of the left sides and the transitions from a given state correspond to elements of the generating set. If A is a proper prefix of a left side and x is a generator, then the associated transition is defined as follows: If Ax does not contain a left side, then the transition is to the state defined by the longest suffix of Ax which is a prefix of some left side. If Ax contains the left side of a rule (P, Q), then (P, Q) is unique and P is a suffix of Ax. In this case, the transition points not to another state but to (P, Q).

The following rewriting system on two monoid generators a and b defines multiplication in an elementary abelian group of order 25:

$$(ba, ab), \quad (a^5, \varepsilon), \quad (b^5, \varepsilon).$$

Here ε denotes the empty word. There are nine proper prefixes of the left sides and the index automaton is described by the following table:

Prefix		Number	a	b
ε	$\longleftrightarrow$	1	2	3
a	$\longleftrightarrow$	2	4	3
b	$\longleftrightarrow$	3	-1	5
a^2	$\longleftrightarrow$	4	6	3
b^2	$\longleftrightarrow$	5	-1	7
a^3	$\longleftrightarrow$	6	8	3
b^3	$\longleftrightarrow$	7	-1	9
a^4	$\longleftrightarrow$	8	-2	3
b^4	$\longleftrightarrow$	9	-1	-3

The transitions are given in the last two columns. Positive entries in those columns refer to states while negative entries are pointers to rules.

For a fixed alphabet size, the space required for the index automaton is linear in the space required for the rules, provided space is measured in words and words are large enough to store an entry in the transition table. (If bits are counted, then there is an extra logarithmic factor.)

As the number of generators increases, the index automaton takes more and more space compared to the size of the rewriting system. Let us consider a simple example. Suppose that there are n left sides which are words of length m over an alphabet X with $2^{r-1} < |X| \leq 2^r$, so that r bits are needed to specify an element of X. The left sides require nmr bits. The number of proper prefixes of left sides could be almost as large as mn. The transition table for the index automaton will have $|X|$ entries per prefix and these entries could require $\log_2(mn)$ bits each. Thus the transition table might need as many as $mn2^r \log_2(mn)$ bits. This bound can be improved by considering words to be strings of mr bits. Now the index automaton will have only two entries per prefix, but the number of prefixes could be nearly nmr. In the worst case, the number of bits required for the transition table is about $2mnr \log_2(nmr)$, which is a factor of $2 \log_2(nmr)$ times as large as the number of bits needed for the left sides. A factor this large is not likely to occur in practice, but even with alphabets of size 16 one can expect that the index automaton will require too much space to be of use with large rewriting systems.

When space is very tight, we can sort the rules lexicographically by their left sides. Given a word W, we can then use binary searches to look up the m-character subwords of W in the list of left sides. This approach uses very little space, but now the time to decide whether W contains a left side as a subword is proportional to the product of $N(|W| - m + 1) \log_2 n$, where N is the average number of character comparisons needed to compare two words of length m during a binary search. The value of N depends on the set of left sides, but N can be expected to grow with increasing m and n.

Between these extremes are a number of other possibilities. A Patricia trie, see [7], takes space proportional to the number of left sides and makes the time for deciding whether there are any left sides in W typically proportional to $(|W| - m + 1)(\log_2 n + m)$, which is slightly better than using a lexicographic sort, but the time still depends on n.

The Rabin-Karp probabilistic string matching algorithm can be modified to produce a probabilistic algorithm for locating left sides in W. Under reasonable assumptions, the expected time to decide whether W contains a left side is proportional to $|W|$ and independent of n.

I am not aware of any results describing the tradeoffs for deterministic algorithms between size of index structures and time to find left sides. In particular, it would be interesting to know the minimum amount of memory needed for an index structure which will make the time to deterministically check W for left sides depend only on m and $|W|$, and not on n.

There is another problem related to devising good index structures for use with the Knuth-Bendix procedure for strings. Almost all the time in this procedure is spent in rewriting. However, the set of rewriting rules is constantly changing, at least in the early stages. The index structures which permit fast rewriting are usually difficult to update quickly. The tradeoffs between speed of rewriting and ease of modification of the index are not well understood.

Now I would like to mention some computational problems in infinite groups for which termination is guaranteed. To do so, I need to define one term. A group G is said to be *polycyclic* if there is a series of subgroups

$$G = G_1 \supseteq G_2 \supseteq \ldots \supseteq G_{r+1} = 1,$$

such that for $1 \leq i \leq r$ the subgroup G_{i+1} is normal in G_i and the quotient G_i/G_{i+1} is cyclic. Such a series is called a *polycyclic series* for G. Polycyclic groups are finitely presented and many natural problems involving elements and subgroups of polycyclic groups have algorithmic solutions. See [1] and [8]. The following result provides a useful characterization of polycyclic groups.

THEOREM. *A group G is polycyclic if and only if G is solvable and all subgroups of G are finitely generated.*

It is easy to show that the class of polycyclic groups lies strictly between the class of finitely generated nilpotent groups and the class of finitely generated solvable groups.

Let $G = G_1, \ldots, G_{r+1} = 1$ be a polycyclic series for G. For $1 \leq i \leq r$, let a_i be an element of G_i which maps onto a generator of G_i/G_{i+1}. If G_i/G_{i+1} is finite, let m_i be its order. Then an element g of G can be uniquely written in the canonical form $a_1^{\alpha_1} a_2^{\alpha_2} \ldots a_r^{\alpha_r}$, where the α_i are integers and $0 \leq \alpha_i < m_i$ if m_i is defined. Thus G may be identified with a subset of $\mathbf{Z}^r$. Suppose that $1 \leq i < j \leq r$. Then ${a_i}^{-1} a_j a_i$ is in G_{i+1} and can be described by a word V_{ij} involving only $a_{i+1}, \ldots, a_r$ and their inverses. Similarly, if m_i is defined, then ${a_i}^{m_i}$ can be described by a word W_i involving only $a_{i+1}, \ldots, a_r$ and their inverses. The relations

$$a_j a_i = a_i V_{ij}, \qquad 1 \le i < j \le r,$$
$$a_i^{m_i} = W_i, \qquad m_i \text{ defined},$$

define G as a group and permit computation of canonical forms.

Let U be the subgroup of $\mathrm{GL}(n, \mathbf{Z})$ consisting of all upper triangular matrices with 1's on the main diagonal. Then U is finitely generated and nilpotent. Therefore U is polycyclic. We can get a polycyclic series for U which is also a central series as follows: Let s and j be integers with $1 \le s < j \le n$ and let U_{sj} be the subgroup consisting of those matrices $A = [a_{pq}]$ in U such that $a_{pq} = 0$ if $0 < q - p < s$ or if $q - p = s$ and $q < j$. For example, if $n = 4$, then U_{24} consists of all matrices of the form

$$\begin{bmatrix} 1 & 0 & 0 & * \\ 0 & 1 & 0 & * \\ 0 & 0 & 1 & 0 \\ 0 & 0 & 0 & 1 \end{bmatrix},$$

where $*$ denotes any integer. The series $U = U_{12} \supset U_{13} \supset \cdots \supset U_{1n} \supset U_{23} \supset \cdots \supset U_{2n} \supset \cdots \supset U_{n-1n} \supset 1$ is a central, polycyclic series.

The entries in the product of two matrices are defined by quadratic polynomials in the entries of the two factors. In [**3**] Philip Hall showed that multiplication in any torsion free, finitely generated nilpotent group can be defined by polynomials. Suppose that G is a torsion free, finitely generated nilpotent group. It is possible to find a central, polycyclic series $G = G_1 \supset \cdots \supset G_{r+1} = 1$ for G such that all the quotients are infinite. Suppose that $a_1, \ldots, a_r$ is a polycyclic generating sequence for G relative to such a series and that $(\alpha_1, \ldots, \alpha_r)$, $(\beta_1, \ldots, \beta_r)$, $(\sigma_1, \ldots, \sigma_r)$, and $(\tau_1, \ldots, \tau_r)$ are, respectively, the vectors of exponents for the canonical forms of elements g, h, gh, and g^{-1} in G. Then the σ_i and the τ_i are polynomial functions of the α's and β's.

For example, if F is the free group on generators a and b and $G = F/\gamma_6(F)$, the corresponding free nilpotent group of class 5, then $r = 14$ and $a_1, \ldots, a_{14}$ may be defined as follows:

$$\begin{aligned}
a_1 &= a, & a_8 &= [a_5, a_2], \\
a_2 &= b, & a_9 &= [a_4, a_3], \\
a_3 &= [a_2, a_1], & a_{10} &= [a_5, a_3], \\
a_4 &= [a_3, a_1], & a_{11} &= [a_6, a_1], \\
a_5 &= [a_3, a_2], & a_{12} &= [a_6, a_2], \\
a_6 &= [a_4, a_1], & a_{13} &= [a_7, a_2], \\
a_7 &= [a_4, a_2], & a_{14} &= [a_8, a_2],
\end{aligned}$$

where $[x, y]$ denotes the commutator $x^{-1}y^{-1}xy$. In terms of this generating set, the Hall polynomials defining multiplication in G are

$$\sigma_1 = \alpha_1 + \beta_1,$$
$$\sigma_2 = \alpha_2 + \beta_2,$$
$$\sigma_3 = \alpha_3 + \beta_3 + \alpha_2\beta_1,$$
$$\sigma_4 = \alpha_4 + \beta_4 + \alpha_3\beta_1 + \alpha_2\binom{\beta_1}{2},$$
$$\sigma_5 = \alpha_5 + \beta_5 + \alpha_3\beta_2 + \binom{\alpha_2}{2}\beta_1 + \alpha_2\beta_1\beta_2,$$
$$\sigma_6 = \alpha_6 + \beta_6 + \alpha_4\beta_1 + \alpha_3\binom{\beta_1}{2} + \alpha_2\binom{\beta_1}{3},$$
$$\sigma_7 = \alpha_7 + \beta_7 + \alpha_5\beta_1 + \alpha_4\beta_2 + \alpha_3\beta_1\beta_2 + \binom{\alpha_2}{2}\binom{\beta_1}{2} + \alpha_2\binom{\beta_1}{2}\beta_2,$$
$$\sigma_8 = \alpha_8 + \beta_8 + \alpha_5\beta_2 + \alpha_3\binom{\beta_2}{2} + \binom{\alpha_2}{3}\beta_1 + \binom{\alpha_2}{2}\beta_1\beta_2 + \alpha_2\beta_1\binom{\beta_2}{2},$$
$$\sigma_9 = \alpha_9 + \beta_9 + \alpha_4\beta_3 + \alpha_5\beta_1 + \alpha_7\beta_1 + \binom{\alpha_3}{2}\beta_1 + \alpha_5\binom{\beta_1}{2} + 2\binom{\alpha_2}{2}\binom{\beta_1}{2} +$$
$$\alpha_2\alpha_3\binom{\beta_1}{2} + \alpha_2\binom{\beta_1}{3} + \alpha_2\binom{\beta_1}{2}\beta_3 + \alpha_3\beta_1\beta_3 + 3\binom{\alpha_2}{2}\binom{\beta_1}{3},$$
$$\sigma_{10} = \alpha_{10} + \beta_{10} + \alpha_5\beta_1 + \alpha_5\beta_3 + \alpha_8\beta_1 + \binom{\alpha_2}{2}\beta_1 + \binom{\alpha_3}{2}\beta_2 + \alpha_3\beta_2\beta_3 +$$
$$2\binom{\alpha_2}{3}\beta_1 + \binom{\alpha_2}{2}\alpha_3\beta_1 + 3\binom{\alpha_2}{2}\binom{\beta_1}{2} + \binom{\alpha_2}{2}\beta_1\beta_2 + \binom{\alpha_2}{2}\beta_1\beta_3 +$$
$$\alpha_2\alpha_3\beta_1\beta_2 + \alpha_2\binom{\beta_1}{2}\beta_2 + \alpha_2\beta_1\beta_2\beta_3 + 4\binom{\alpha_2}{3}\binom{\beta_1}{2} + 2\binom{\alpha_2}{2}\binom{\beta_1}{2}\beta_2,$$
$$\sigma_{11} = \alpha_{11} + \beta_{11} + \alpha_6\beta_1 + \alpha_4\binom{\beta_1}{2} + \alpha_3\binom{\beta_1}{3} + \alpha_2\binom{\beta_1}{4},$$
$$\sigma_{12} = \alpha_{12} + \beta_{12} + \alpha_6\beta_2 + \alpha_7\beta_1 + \alpha_4\beta_1\beta_2 + \alpha_5\binom{\beta_1}{2} + \binom{\alpha_2}{2}\binom{\beta_1}{3} +$$
$$\alpha_2\binom{\beta_1}{3}\beta_2 + \alpha_3\binom{\beta_1}{2}\beta_2,$$
$$\sigma_{13} = \alpha_{13} + \beta_{13} + \alpha_7\beta_2 + \alpha_8\beta_1 + \alpha_4\binom{\beta_2}{2} + \alpha_5\beta_1\beta_2 + \binom{\alpha_2}{3}\binom{\beta_1}{2} +$$
$$\binom{\alpha_2}{2}\binom{\beta_1}{2}\beta_2 + \alpha_2\binom{\beta_1}{2}\binom{\beta_2}{2} + \alpha_3\beta_1\binom{\beta_2}{2},$$
$$\sigma_{14} = \alpha_{14} + \beta_{14} + \alpha_8\beta_2 + \alpha_5\binom{\beta_2}{2} + \alpha_3\binom{\beta_2}{3} + \binom{\alpha_2}{4}\beta_1 + \binom{\alpha_2}{3}\beta_1\beta_2 +$$
$$\binom{\alpha_2}{2}\beta_1\binom{\beta_2}{2} + \alpha_2\beta_1\binom{\beta_2}{3}.$$

Here $\binom{n}{m}$ is a binomial coefficient.

In nonnilpotent polycyclic groups we can sometimes develop formulas for multiplication, but they are likely to involve exponentials. The relations

$$a_3a_2 = a_2a_3, \quad a_2a_1 = a_1{a_2}^{-1}{a_3}^{-1}, \quad a_3a_1 = a_1a_2{a_3}^2,$$

define a polycyclic group. If $a_1^{\alpha_1}a_2^{\alpha_2}a_3^{\alpha_3}a_1^{\beta_1}a_2^{\beta_2}a_3^{\beta_3} = a_1^{\sigma_1}a_2^{\sigma_2}a_3^{\sigma_3}$ then

$$\begin{aligned}
\sigma_1 &= \alpha_1 + \beta_1,\\
\sigma_2 &= \beta_2 + \big[(\lambda+1)\alpha_2 - \alpha_3\big]\frac{\mu^{\beta_1}}{\sqrt{5}} + \big[(\lambda-2)\alpha_2 - \alpha_3\big]\frac{\lambda^{\beta_1}}{\sqrt{5}},\\
\sigma_3 &= \beta_3 + \big[\alpha_2 + (\lambda-2)\alpha_3\big]\frac{\mu^{\beta_1}}{\sqrt{5}} + \big[-\alpha_2 + (\lambda+1)\alpha_3\big]\frac{\lambda^{\beta_1}}{\sqrt{5}},
\end{aligned}$$

where $\lambda = (1+\sqrt{5})/2$ and $\mu = (1-\sqrt{5})/2$.

As reported in [**5**] and [**10**], quite a bit of work has been done on the rapid computation of products in finite solvable groups, but relatively little research has been devoted to computing efficiently in infinite polycyclic groups. If we only need to compute a few products, there is probably not much we can do. However, if we are going to calculate extensively in some polycyclic group, then it may pay to develop formulas of the type illustrated above. In this case, how can the formulas be found quickly? Every polycyclic group has a faithful representation as a subgroup of $\mathrm{GL}(n, \mathbf{Z})$. Is it possible that for some polycyclic groups G the fastest way to compute with elements in G is to find such a matrix representation and use a fast matrix multiplication algorithm?

References

1. G. Baumslag, F. B. Cannonito, D. J. S. Robinson, and D. Segal, *The algorithmic theory of polycyclic by finite groups*, J. Algebra **142** (1991), 118–149.
2. J. D. Dixon, *The probability of generating the symmertric group*, Math. Z. **110**, 199–295.
3. P. Hall, *Nilpotent Groups*, Notes of lectures given at the Canadian Mathematical Congress, Canadian Math. Soc., 1957; Revised versions issued by Queen Mary College 1969, 1979. See also Collected Works of Philip Hall, 415-462. Oxford: Clarendon Press 1988.
4. D. E. Knuth, *The Art of Computer Programming*, Volume 2, Revised edition, Addison-Wesley, Reading, 1981.
5. C. R. Leedham-Green and L. H. Soicher, *Collection from the left and other strategies*, J. Symbolic Comp. **9** (1990), 665–675.
6. H. W. Lenstra, Jr., *Algorithms in algebraic number theory*, Bull. Amer. Math. Soc. **26** (1992), 211–244.
7. R. Sedgewick, *Algorithms in C*, Addison-Wesley, Reading, 1990.
8. D. Segal, *Decidable properties of polycyclic groups*, Proc London Math. Soc (3) **61** (1990), 497–528.
9. C. C. Sims, *The Knuth-Bendix procedure for strings as a substitute for coset enumeration*, J. Symbolic Comp. **12** (1991), 439-442.
10. M. R. Vaughan-Lee, *Collection from the left*, J. Symbolic Comp. **9** (1990), 725–733.

Department of Mathematics, Rutgers University, New Brunswick, New Jersey 08903

E-mail address: sims@math.rutgers.edu

DIMACS Series in Discrete Mathematics
and Theoretical Computer Science
Volume **11**, 1993

GRAPE: a System for Computing with Graphs and Groups

LEONARD H. SOICHER

1. Introduction

This note briefly describes GRAPE, a computer system for calculations with graphs and groups. GRAPE is designed primarily to study finite graphs arising from group-theoretical and geometrical constructions.

The GRAPE philosophy is that a graph Γ always comes together with a known subgroup G of $\mathrm{Aut}\,\Gamma$ (this is sensible if we assume our graph comes from a group-theoretical or geometrical construction), and that G should be used to reduce the storage and CPU-time requirements for calculations with Γ. The result is that we are routinely able to do calculations with many large symmetric graphs (having 20,000 vertices or more) on a SUN Sparcstation 2. However, even if G is the trivial group, GRAPE tries not to impose an unreasonable overhead for trying to use G.

The GRAPE system consists of the GAP (version 3.1) group theoretical system [**11**] from Aachen, a GAP language library of graph-theoretical functions, and some standalone programs integrated into the system. GAP provides a programming language (the GAP language), a storage manager, objects such as lists, sets, records, permutations, and groups, and a library of group-theoretical routines (written in the GAP language). GAP also allows the execution of operating system commands from within GAP, and the reading and writing of files. This allows us to run various efficient standalone programs under a GAP interface as part of the GRAPE system. At present these standalones are the author's coset enumeration program *enum* (although GAP now has a coset enumerator), B.D. McKay's graph automorphism and isomorphism program *nauty* [**8**], and the author's collapsed adjacency matrix program *coladj*. All computations in the GRAPE system take place via GAP functions, so that the running of standalone programs is invisible to the user. This approach allows us the flexibility

1991 *Mathematics Subject Classification*. Primary 05C25, 05C85.
This paper is in final form and no version of it will be submitted for publication elsewhere.

to add programs written in various languages, but still to have a standard, high-level programming interface via GAP. We have in fact found that most of our graph routines are most easily written in the GAP language, and are reasonably efficient in this form.

GRAPE is being developed by the author at Queen Mary and Westfield College on a SUN Sparcstation 2, but it should be possible to set up GRAPE on most computers running a version of the *UNIX* operating system.

A predecessor to GRAPE, using GAP 2.4 [**9**], was written by the author during visits to the Università di Roma "La Sapienza" and the University of Western Australia. This system has already proved very useful. It was instrumental in the discovery of some new distance-regular graphs [**12**], and in the determination of some of their properties. For some examples in [**5**], the system was used to construct covers of certain graphs, and to determine the automorphism groups of these covers. For [**6**], the system was used to determine when certain 2-arc transitive graphs $QR(q)$ ($q = 7, 9, 17, 23, 25, 31$) are distance-transitive, as these particular $QR(q)$ are not handled by a general result in [**6**] ($QR(q)$ has $2^{(q+1)/2}$ vertices and valency $q + 1$). Also, we are using the *enum* and *coladj* standalone programs to construct (most of) the intersection matrices for the 1-arc transitive graphs of rank ≤ 5 acted on by sporadic simple groups or their automorphism groups (see [**10**]). The biggest cases handled so far (on a University of London Convex supercomputer) are the intersection matrices for the 1-arc transitive graphs on $1,545,600$ vertices arising from the action of Conway's group Co_1 on its $3A$-generated subgroups of order 3 (our notation is that of the ATLAS [**4**]).

We note that the latest version (3.8) of the group theory system *CAYLEY* [**3**] provides graph-theoretical functions and incorporates the *nauty* program. We have found *CAYLEY* useful for much group-theoretical and some graph-theoretical work. The *CAYLEY* approach to graphs is more general than that of GRAPE, but the specialist approach of GRAPE allows it to work with much larger symmetric graphs than *CAYLEY* does.

The reader is referred to [**2**] for graph-theoretical terms not defined here.

2. Representing graphs in GRAPE

In general GRAPE deals with finite directed graphs, possibly with loops, but with no multiple edges. Let $\Gamma = (V, E)$ be such a graph. Let $v \in V$, and denote by $\Gamma_i(v)$ the set of vertices of Γ at distance i from v, with the convention that $\Gamma(v) = \Gamma_1(v)$. We suppose $n = |V|$, $V = \{v_1, \dots, v_n\}$, and discuss how Γ can be represented on a computer. The most common methods are by an *adjaceny matrix* (an $n \times n$ matrix whose (i, j)-entry is 1 if (v_i, v_j) is an edge of Γ, and 0 otherwise), or by a list $\Gamma(v_1), \dots, \Gamma(v_n)$ of adjacency sets.

An adjacency matrix requires n^2 bits of storage, it can be inefficient for computations with sparse graphs, and requires too much store for n greater than about $15,000$ (with present workstation technology). Also, constructing an adjacency matrix requires at least $O(n^2)$ steps. Adjacency matrices are used by *nauty* and *CAYLEY*, but not by (the GAP part) of GRAPE.

A list of adjacency sets requires at least $O(|E|)$ storage, and at least $O(|E|)$ time for its construction. Thus, such a list is inappropriate for a graph with very many edges (if a better way can be found, as described below).

Suppose now we know a group G of automorphisms of Γ. Let $V_1, \ldots, V_k$ be the distinct orbits of G on the vertex set V, with respective representatives $v_1, \ldots, v_k$. Clearly, to store Γ, we need only store permutation generators for G, the list $v_1, \ldots, v_k$ of orbit representatives, and the list $\Gamma(v_1), \ldots, \Gamma(v_k)$. This is how a graph is stored in GRAPE, in the form of a GAP record. This record also contains a Schreier vector (see, for example, [**7**]) to allow the efficient calculation of $\Gamma(v)$ for any $v \in V$. In practice, we often have $k \leq 2$, and so this method of storage can represent a huge saving over conventional means. Even if G is trivial, our method is not significantly worse than a represention by a complete list of adjacency sets.

By default in GRAPE, the vertices of a graph $\Gamma = (V, E)$ are named $1, \ldots, |V|$, and this is how they are always represented internally. Following a suggestion of P.J. Cameron, we allow an (optional) field called `names` for a graph record, so that a user may give his or her names for the vertices (these names can be integers, sets, vectors, or indeed, any GAP type). If a subgraph of Γ is created then this subgraph inherits its vertex names from Γ.

3. Some GRAPE functions

The most common way for a graph to be constructed in GRAPE is to start with a permutation group G of degree n, and a set S of ordered pairs of elements of $\{1, \ldots, n\}$, and then to construct Γ to have vertex set $\{1, \ldots, n\}$, and edge set the union of the G-orbits of the elements of S. We are then able to associate the group G with Γ, and to construct Γ, we only determine the adjacency sets for the orbit representatives $v_1, \ldots, v_k$, say, of G on $\{1, \ldots, n\}$. This calculation is made especially easy if we know (generators for) the stabilizers in G for each of the v_i. Many GRAPE functions allow the user to input such additional useful information, if known, to make the functions run faster.

Now, for ease of exposition, we assume that our graph $\Gamma = (V, E)$ is undirected and has no loops. As before, we suppose $G \leq \operatorname{Aut}\Gamma$ has k orbits on the vertices of Γ, with representatives $v_1, \ldots, v_k$.

Many properties of Γ can be determined by GRAPE functions; for example, connected components, diameter, girth, and various regularity properties, such as whether Γ is distance-regular. Also, given a subset X of V, GRAPE can determine if X is a distance-regular code.

GRAPE can construct induced subgraphs of Γ, such as the induced subgraph Δ on a set $\Gamma_i(v)$. Here, the group associated with Δ is the vertex stabilizer G_v acting on $\Gamma_i(v)$ (if the user does not provide a group). If we do not already know this stabilizer then we can quickly construct an approximation $H \leq G_v$ using random methods (see [**1**] and [**7**]). This is adequate as we need only be sure that H acting on $\Gamma_i(v)$ is a subgroup of Aut Δ, and in practice we will almost always have $H = G_v$. Approximations of this sort are used in some other GRAPE functions.

The most important function in GRAPE is called `LocalInfo`, and is used by many other procedures.

The function `LocalInfo` performs a calculation similar to a breadth-first search in Γ, starting at a given vertex v to successively determine

$$\Gamma_0(v), \Gamma_1(v), \Gamma_2(v), \dots ,$$

among other things. (Here, as you might expect, we first calculate the orbits of (an approximation to) G_v and then we only look at the adjacency sets of representatives of these orbits.) Thus, one thing `LocalInfo` determines is the connected component of Γ containing v.

Now suppose Γ is connected. For $v \in V$ let d_v denote the maximum distance from v to any vertex of Γ. Thus `LocalInfo` determines d_v, and we have that the diameter of Γ is just $\max\{d_{v_1}, \dots , d_{v_k}\}$.

Suppose $v, w \in V$, and $w \in \Gamma_i(v)$. Then `LocalInfo`, in the process of performing its breadth-first search starting at v, determines the numbers

$$\begin{aligned} c_i(v,w) &= |\Gamma_{i-1}(v) \cap \Gamma(w)|, \\ a_i(v,w) &= |\Gamma_i(v) \cap \Gamma(w)|, \\ b_i(v,w) &= |\Gamma_{i+1}(v) \cap \Gamma(w)|. \end{aligned}$$

(Note that these quantities are constant as w varies over an orbit of G_v).

Now define g_v to be the length of the shortest circuit containing v, if v is in a circuit, and define $g_v = \infty$ otherwise. Now let t be the least i such that some $c_i(v, w) \geq 2$ or some $a_i(v, w) \geq 1$, if such an i exists. If no such i exists then $g_v = \infty$. Otherwise, if $c_t(v, w) \geq 2$ for some w, then $g_v = 2t$, else $g_v = 2t + 1$. We also have that the girth of Γ is $\min\{g_{v_1}, \dots , g_{v_k}\}$.

The function `LocalInfo` saves a $c_i(v, w)$, $a_i(v, w)$, or $b_i(v, w)$ just if it depends only on i and v (and not w). Such *parameters* $c_i(v), a_i(v), b_i(v)$, if and when they exist, are used in the determination of many regularity properties of Γ, the strongest of which is distance-regularity. A function very similar to `LocalInfo` can determine if a given subset of the vertices of Γ is a distance-regular code.

Finally, it is possible to use GRAPE functions as an interface to *nauty* to determine the automorphism group of a (vertex-coloured) graph, and to test if two graphs are isomorphic, if these graphs are not too large.

4. Note added: 20 August 1992

The GRAPE system and a beginning user's guide [**13**] are available from the author. The GRAPE library now contains over 50 graph-theoretical functions, many of which were written while the author was visiting the State University of Ghent. All the functionality of the GRAPE system, except for the ability to compute graph automorphism groups and test for graph isomorphism, can now be achieved by GAP code alone.

New features in GRAPE include the ability to analyse the clique structure of a graph, as well as certain other induced subgraph structure, and also functions for constructing quotient graphs, bipartite doubles, point graphs, and edge graphs.

REFERENCES

1. L. Babai, G. Cooperman, L. Finkelstein and A. Seress, *Nearly linear time algorithms for permutation groups with a small base*, ISSAC '91, ACM Press, New York, 1991, pp. 200–209.
2. A.E. Brouwer, A.M. Cohen and A. Neumaier, *Distance-Regular Graphs*, Springer, Berlin and New York, 1989.
3. J.J. Cannon, *An introduction to the group theory language, Cayley*, Computational Group Theory (M.D. Atkinson, ed.), Academic Press, London, 1984, pp. 145–183.
4. J.H. Conway, R.T. Curtis, S.P. Norton, R.A. Parker and R.A. Wilson, *An ATLAS of Finite Groups*, Clarendon Press, Oxford, 1985.
5. D.R. Hughes, *Regular covers of graphs and structures* (to appear).
6. A.A. Ivanov and C.E. Praeger, *On finite affine 2-arc transitive graphs*, University of Western Australia Pure Mathematics Research Report, 1992.
7. J.S. Leon, *On an algorithm for finding a base and strong generating set for a group given by generating permutations*, Math. Comp. **35** (1980), 941–974.
8. B.D. McKay, *nauty user's guide (version 1.5)*, Technical report TR-CS-90-02, Computer Science Department, Australian National University, 1990.
9. A. Niemeyer, W. Nickel and M. Schönert, *GAP: Getting Started and Reference Manual*, Lehrstuhl D für Mathematik, RWTH Aachen, 1988.
10. C.E. Praeger and L.H. Soicher, *Permutation representations and orbital graphs for the sporadic simple groups and their automorphism groups: rank at most five* (to appear).
11. M. Schönert, et. al., *GAP: Groups, Algorithms and Programming*, Lehrstuhl D für Mathematik, RWTH Aachen, 1992.
12. L.H. Soicher, *Three new distance-regular graphs*, Europ. J. Combinatorics (to appear).
13. L.H. Soicher, *Getting started with GRAPE*, preprint, QMW, 1992.

SCHOOL OF MATHEMATICAL SCIENCES, QUEEN MARY AND WESTFIELD COLLEGE, MILE END ROAD, LONDON E1 4NS, U.K.

E-mail: L.H.Soicher@qmw.ac.uk

DIMACS Series in Discrete Mathematics
and Theoretical Computer Science
Volume **11**, 1993

Implications of Parallel Architectures for Permutation Group Computations

BRYANT W. YORK

November 5, 1992

ABSTRACT. Implications of distributed image representation for permutations and the SIMD parallel computer model for permutation group computations are considered. In particular, very practical implementation techniques for basic permutation group computations are developed on a distributed memory, SIMD massively parallel computer, the Connection Machine model CM-2. Sources of parallelism in certain standard algorithms are analyzed and issues involved in mapping them onto the underlying parallel architecture are elucidated. Performance results for parallel implementations on the CM-2 of permutation multiply, permutation inverse, permutation apply, and sifting are presented and used to demonstrate practical scalability considerations.

1. Introduction

The goal of this paper is to elucidate some of the more important issues surrounding parallel permutation group computations on distributed memory parallel computers. Three important factors in the performance of any parallel algorithm on a specific machine are: (1) the speed of the component technology, (2) the system architecture, and (3) how well the algorithm maps to the architecture. In order to avoid generalities, we have chosen a particular architecture and technology, the Connection Machine CM-2; and we have developed

---

1991 *Mathematics Subject Classification.* Primary 20B02, 68Q40; Secondary 20B04.

Connection Machine, CM-2, CM5, *Lisp, C*, and CM Fortran are trademarks of Thinking Machines Corporatio; KSR1 is a trademark of Kendall Square Research; MP-1 is a trademark of MasPar Computer; Sun-3 is a trademark of Sun Microsystems Inc.; DECstation 5000 and Alpha Machine are trademarks of Digital Equipment Corporation.

The author was supported in part by NSF Grant # CCR-9296035.

This paper is in final form and no version of it will be submitted for publication elsewhere.

an implementation of basic permutation group operations to help illustrate our conclusions by way of example. Our main conclusion is that basic permutation group operations are well-suited to parallel systems which have distributed memory, high bandwidth/high function interconnect, a large number of processors, and the data parallel programming model. The data parallel distributed image representation approach was competitive on a price/performance basis for comparable technology generations.

Although there are several serial implementations of systems for permutation group management [**5, 6, 15, 16**], researchers are just beginning to consider the issues of parallel implementation. One of the basic problems addressed in this paper involves determining when it is appropriate to use the interconnection network of a distributed memory parallel machine (DMPM) to implement basic permutation operations. For example, in a serial machine, an n-point permutation multiply requires $3n$ memory references. In a DMPM an n-point permutation multiply may require one or two parallel communications operations and furthermore k such permutation multiplications can be achieved in parallel in the same time (where k is a function of the machine size). Depending upon the network topology and the layout of the permutations across the machine, such a parallel permutation multiply could be achieved in constant time (on a fully-connected network) or in logarithmic time (on a hypercube). We shall see that the network of the CM-2 is implemented by a mixture of fully-connected, mesh-connected, and hypercube-connected structures with different performance characteristics.

Figure 1 depicts a simple example of a single parallel permutation multiply for three pairs of permutations on four points. In this model parallelism is achieved across the points of the permutation as well as across the permutations of the group. Each of the twelve processors in the DMPM has its own local memory and the points of a permutation are distributed over collections of four consecutive processors. Distributed image representation is used; the storage for a point of a permutation begins at the same address within the local memory of each processor and requires $\log_2 n$ bits. In a SIMD DMPM with the notion of a "parallel variable" one parallel variable can designate a vector of permutations (eg. in Figure 1 p, q, and r are each vectors of three permutations on four points). We will refer to one element of the vector (a complete permutation) as a permutation block. In order to execute the vector-parallel permutation multiply, each processor must reference the contents of its p variable (memory reference) and get (communications reference) the contents of the q variable (memory reference) in the processor (of its permutation block) which is at position $Contents(p)$ relative to the beginning of its permutation block. When it receives this value, it stores (memory reference) it into the memory location for the r variable in its own local memory. The parallel variable ID contains the identity permutation in each permutation block of its associated permutation vector and it is used to assist in performing permutation inverse and permutation apply. The

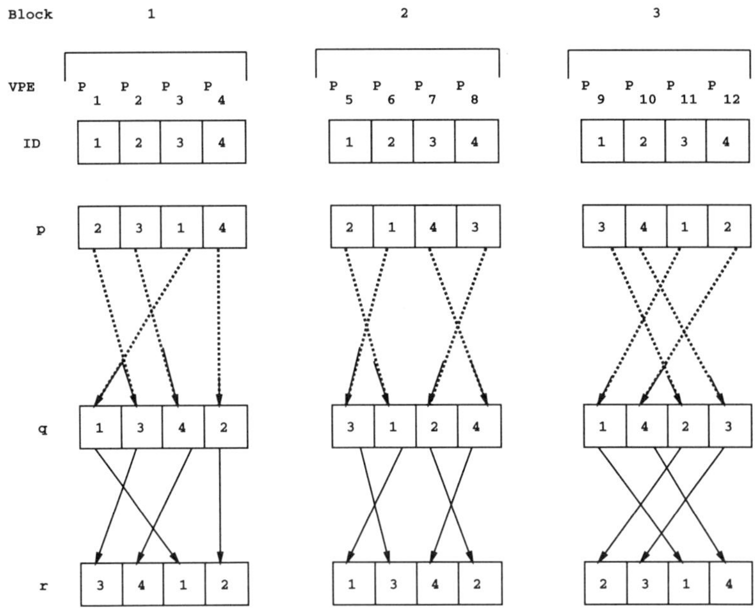

FIGURE 1. Example of Distributed Vector-parallel Perm-multiply

dashed arrows represent the computation of the processor address at the head by the processor at the tail. Solid arrows indicate the communications operation, *get*, which retrieves the value from the processor at the tail and stores it in the parallel variable of the processor at the head.

Clearly three memory references are required to move a point in both serial and parallel permutation multiply; however, in the parallel case the memory references for all points are done in parallel. The performance question becomes "For what size problem is the communications cost of parallel permutation multiply less than the total memory access costs of the serial case?" The answer to this simple question accounts for a large part of the permutation group management problem on a DMPM; however, we shall see that the extent to which the particular DMPM's architecture supports other distributed data structures also has a significant influence.

In the remainder of this paper, we present an overview of the CM-2's architecture (at two levels), implementation descriptions of primitive permutation group operations, and performance results for parallel implementations of permutation multiply, permutation inverse, permutation apply, and sifting. We conclude by considering the scalability implications of the architectural and performance considerations presented in the paper.

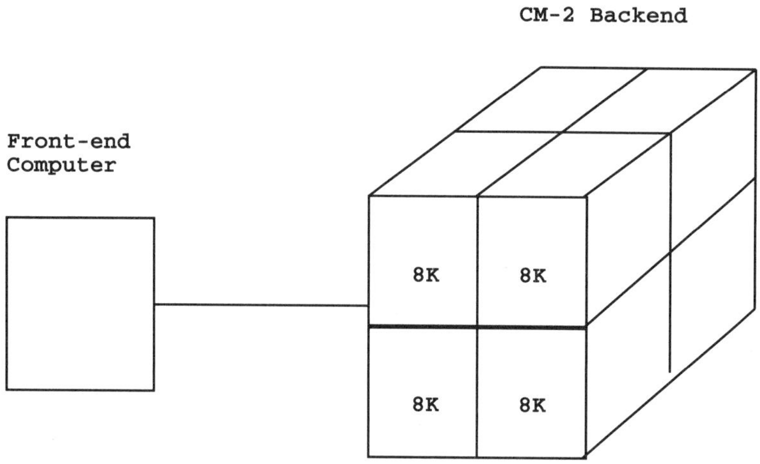

FIGURE 2. Connection Machine, CM-2

2. The Architecture of the Connection Machine model CM-2

The most commonly used classification scheme for parallel computers categorizes machines according to the number of instruction streams, the number of data streams, and whether memory is distributed among or shared by the processors. The MIMD model allows for multiple instruction streams and multiple data streams; the SIMD model supports a single instruction stream and multiple data streams. In either model the memory may be shared or distributed. The CM-2 is a distributed memory (hypercube connected) SIMD computer with software that provides the notions of virtual processors and software reconfiguration of the machine. We present two views of the CM-2's architecture and explore the implications of each for parallel group computations.

2.1. Programmer's View of the CM-2. The CM-2 is basically an SIMD system consisting of a front-end serial computer which drives a back-end multicomputer (see Figure 2). A full back-end contains 64K processing elements (PEs) connected in a 16-dimensional hypercube network. Each PE consists of a bit serial processor with 64Kbits (256Kbits or 1Mbits on the newer models) of bit-addressable local memory. The CM-2 software allows the programmer to impose various grid (mesh) structures on the hypercube of PEs. Only n-dimensional grids ($1 \leq n \leq 31$) in which the length of each dimension is a power of two and the product of the lengths of all dimensions is a power of two multiple of the number of PEs are permitted. These grids may be thought of as discrete finite rectangular geometries. This facility allows the programmer to view the fixed set of PEs (connected as a hypercube in the hardware) as if it simultaneously represented one or more multi-dimensional arrays of *virtual* PEs (VPEs). The set of VPEs is referred to as the "VP set" and the ratio of VPEs to PEs is called the "VP Ratio", abbreviated VPR. A programmer may define multiple VP sets for a given problem and associate several different geometries with the same VP set. Communication is supported within and between VP sets; however, only one geometry may be active for a VP set a time.

High level languages provide instructions for performing general routing in the hypercube as well as fast NEWS grid communications within the discrete geometries. NEWS communication allows a VPE to send messages to its nearest neighbors in the North, East, West, and South directions along a hardware grid more quickly (for short distances) than it could through the general routing network. NEWS communication generalizes to n-dimensions with predictable performance.

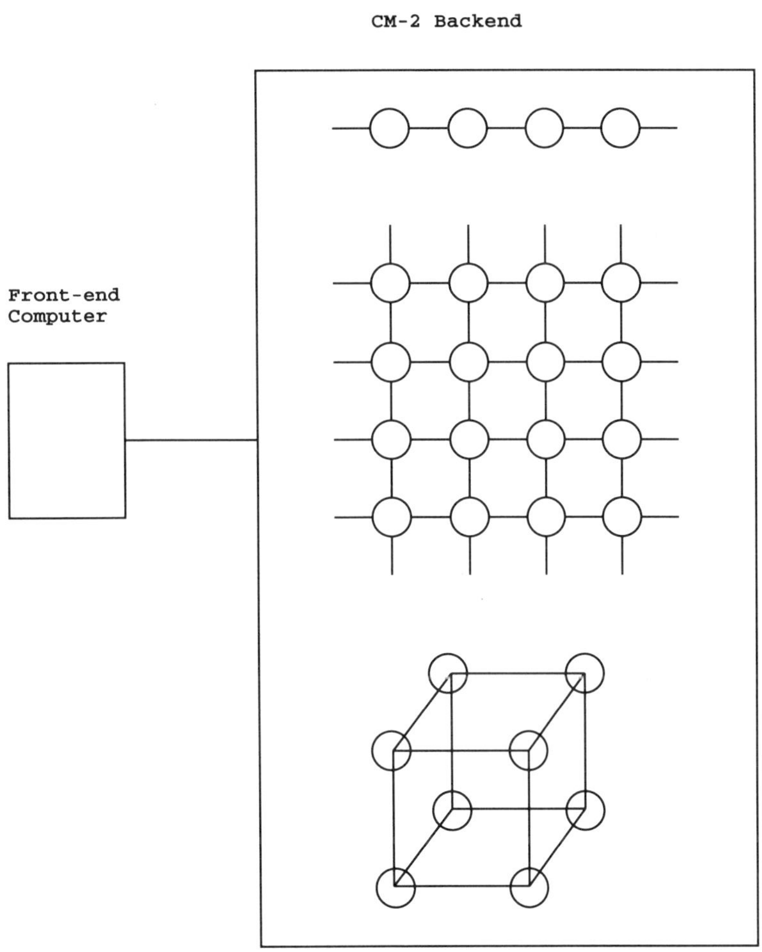

FIGURE 3. Programmer's View of CM-2

2.2. Machine Designer's View of the CM-2. The programmer's view of the CM-2 presented above is the standard view for anyone who codes programs

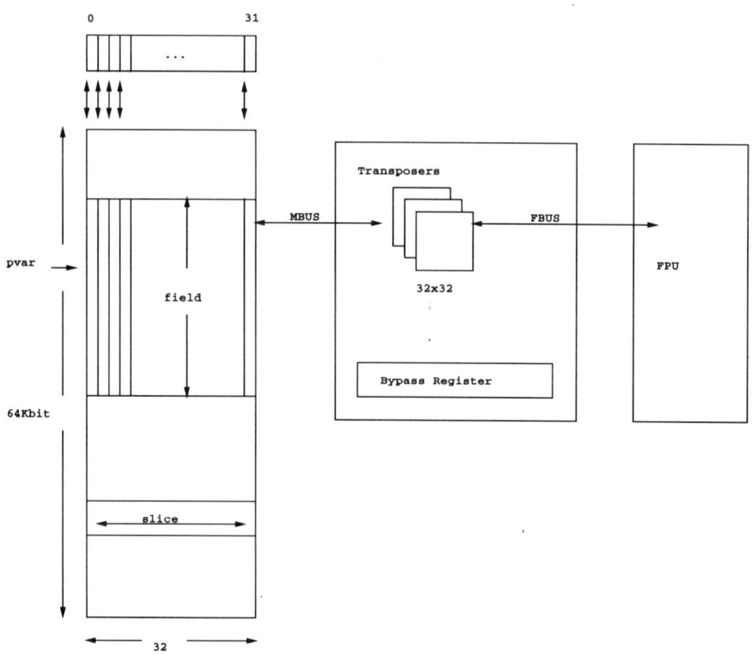

FIGURE 4. CM-2 SPRINT node

for the machine in assembly or any higher level language. However, this is not the only view of the CM-2 and an understanding of the micro-level machine can aid in the design of efficient data representations and algorithms. Thus, our purpose in describing the micro-level machine is to show that the implementation of a given architecture does not always coincide with the high level programmer's view and that non-trivial performance increases can be achieved through knowledge of the micro-level structure. Essentially the following detailed description of the micro machine shows that the network of the CM-2 is actually several networks with different performance characteristics.

At the micro level the CM-2 back-end is actually an 11-dimensional hypercube of so-called SPRINT nodes. Each SPRINT node consists of a serial/parallel interface (hence S PR INT), a floating point unit (FPU), two processor chips (each chip contains 16 bit-serial processors), and a 32x64Kbit (32x256Kbit or 32x1Mbit) memory. Each of the 32 bit-serial processors on a SPRINT node sees its local memory as a 64Kbit array (vertical strip in Figure 4); the SPRINT allows the FPU to view the memory as 64K 32-bit words (horizontal strips). These two views of memory are called "fieldwise" and "slicewise" respectively. A sequencer controls each block of 256 SPRINT nodes and it is driven by instructions which it receives from the front-end computer. The sequencer has a large register file, a scratch RAM, and microcode for directly controlling the FPUs, the SPRINTs, and the bit-serial processors. The SPRINTs may be used to control communication across the hypercube wires of the network. Figure 4 depicts a SPRINT node of the back-end.

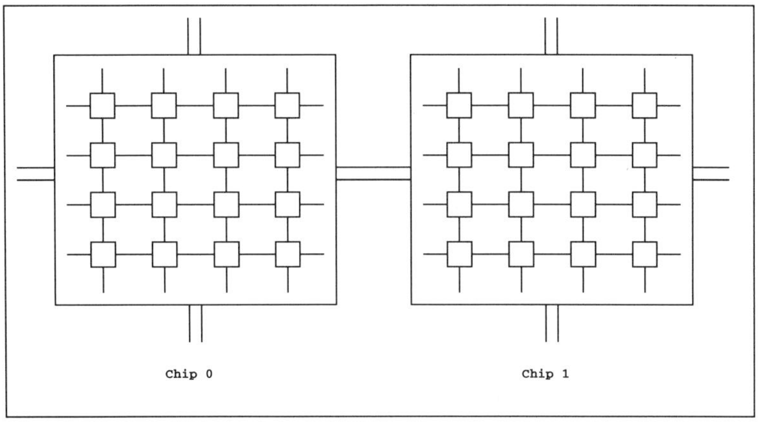

FIGURE 5. Processor Chips and NEWS wires

Within a processor chip the 16 bit-serial PEs are connected in a 4x4 2D (NEWS) mesh. This fast physical interconnection is supported by a 16x16 butterfly network (the Flipper) which allows data to be quickly moved from a field in one processor's memory to a different field in the memory of another processor on the same chip. A pair of bidirectional NEWS wires connects the two chips on the same SPRINT node, and NEWS wires connect chips on adjacent SPRINT nodes in the NEWS grid. Associated with each processor chip is a router which handles messages for the 16 bit-serial processors on the chip. Thus, in a fully configured 64K PEs CM-2, there are 4096 routers connected by 24,576 bidirectional wires. The routers are wired in the pattern of a 12-dimensional binary hypercube.

Finally, each transposer of a SPRINT node is a collection of 32 32-bit registers. There are three transposers per SPRINT node and their primary responsibility is to provide an interface between the fieldwise data of the bit-serial processors and the slicewise data required by the FPU. Slicewise data may be indirectly addressed by the bit-serial processors in a parallel fashion by using the transposers. In general, we have not directly exploited this feature to implement permutation group operations because it is very wasteful of storage. It is utilized by microcode during communications operations.

Besides providing a mechanism for transposing fieldwise data to slicewise data and a mechanism for indirect addressing of shared slicewise arrays of 32-bit values, the transposers may be used for fast communications of large amounts of data. 22 of the registers of one of the transposers are connected to the 11 wires of the 11-dimensional SPRINT node hypercube. 64-bit data can be swapped along all 11 dimensions of the SPRINT hypercube in about 25 microseconds by a primitive operation called *cube-swap*. Although we do not currently make use of these architectural features, they are being analyzed for their use in algorithms which require table lookup.

Network bandwidth and latency are two critical factors in the implementation of distributed permutation operations. It should be noted that it is very difficult to optimize a network for both high bandwidth and low latency; thus, the CM-2 represents a set of tradeoffs which reflect the designer's anticipated use of the machine. A fully-configured 64K PEs CM-2 can perform approximately $250 \mathrm{x} 10^6$ 32-bit sends per second for regular communications patterns; this implies a peak aggregate bandwidth of 8 Gbits/sec. The basic clock rate is 7 Mhz which translates into a 143 ns cycle time; and the memory access time is 100 ns. Each of these parameters may be modified within certain limits for this architecture. For example, the CM-200 has a 10 Mhz clock rate (100 ns cycle time) so it is clear that faster memory is not required.

The purpose of the preceding description is to show that there are different kinds of communications available in the hardware architecture of the CM-2 and they have very different performance characteristics. In this paper we are concerned with the following questions: How does one take advantage of these mechanisms to get good performance in basic permutation group operations? What are the tradeoffs between the management of distributed permutations and local permutations? What are the implications for other data structures required in group computations, such as tables and trees of permutations? What are the *practical* scalability relationships?

2.3. Geometries and Communication Revisited. The CM-2 software allows the programmer to specify certain mesh configurations (geometries) of virtual processing elements in the back-end. The detailed layout of the user's mesh on the processors of the hypercube may be defined by the user, and the choice of layout can dramatically affect performance. In general, NEWS communication is significantly faster than general routing up to a NEWS distance of about 6 hops. For general routing, regular communication patterns are faster than irregular patterns and permutation routing (one-to-one) can be quite fast with the proper data layout.

3. Implementation of Permutation Group Computations

On a traditional serial computer, permutations (in image representation) are generally represented by arrays and permutation operations are implemented as array references. In general, for a permutation on N points, permutation multiply will require $O(N)$ steps. One approach to parallelization of permutation operations would be to distribute an entire permutation to each processor of the parallel computer. Permutation operations would still require $O(N)$ steps as in the serial case, but the N steps would be done in parallel by K processors for a speedup of K. This approach essentially requires a MIMD architecture in which the processors may execute different instructions during the same cycle.

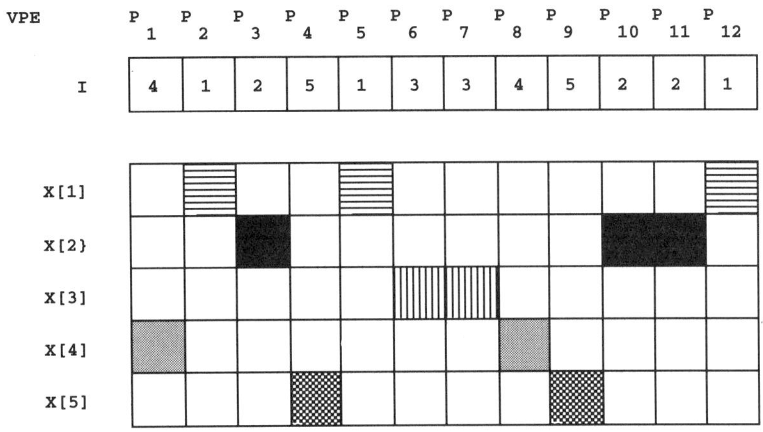

FIGURE 6. Serialized SIMD Parallel Array Reference

At the time this work was begun MIMD machines were not available with more than a few hundred PEs (generally $8 \leq K \leq 128$); the individual PEs were not so powerful; and the total power tradeoff favored massively parallel SIMD over coarse-grained MIMD. More recently, scalable MIMD architectures (CM-5 and KSR1) have been introduced in which the individual PEs are extremely powerful and the total power tradeoff has shifted to more coarse-grained MIMD; however, it is still instructive to consider the implications of the massively parallel SIMD solution.

The permutation-per-processor representation is not well-suited to the CM-2 because it essentially requires that each processor potentially access a different memory address on the same cycle. Since the CM-2 is a SIMD machine in which every bit-serial processor is executing the same instruction (and referencing the same memory address) during a given cycle, parallel (fieldwise) indirect addressing in which different processors reference different memory addresses is precluded. For example, in Figure 6 any attempt to reference a parallel array ($X[5]$) by a parallel variable (I) having different values in different processors would result in a serialization of the parallel reference. In this case, all processors in which the variable, I, has the value 1 would be referenced during the first cycle; all processors in which I has the value 2 in the second cycle, and so on. Thus, it requires N cycles to reference a parallel array by a parallel variable; however, it requires only one cycle to reference a parallel array by a parallel constant. We use the term, cycle, loosely here to refer to the time required to do a single memory reference. Of course, on a bit-addressable, bit-serial machine C cycles are required to reference a bit field of length C. In order to avoid this type of serialization, we have chosen to experiment with distributing the points of a permutation across the processors of the CM-2 and to utilize the communications network to implement the basic permutation operations.

The MasPar MP-1 series of SIMD machines has an indirect addressing capability. The MP-1 represents a very different point in the design space and more experimentation would be required to determine if its choice of communications bandwidth, network topology, processor and memory bandwidth are more suitable to permutation group computations. We did not have access to an MP-1 at this time; however, our experiments with the CM-2 will serve to illustrate our points. The CM-2 also provides indirect addressing on 32-bit slicewise quantities within a SPRINT node. Although this addressing technique is about 4 times faster than the general communications network, it has three major drawbacks: (1) several points must be packed into a 32-bit word and unpacked for use (packing/unpacking can be expensive operations), (2) if we avoid packing/unpacking then there is substantial waste of storage, and (3) the amount of parallelism that could be achieved would be limited to the number of SPRINTs in the machine (1/32 the number of PEs).

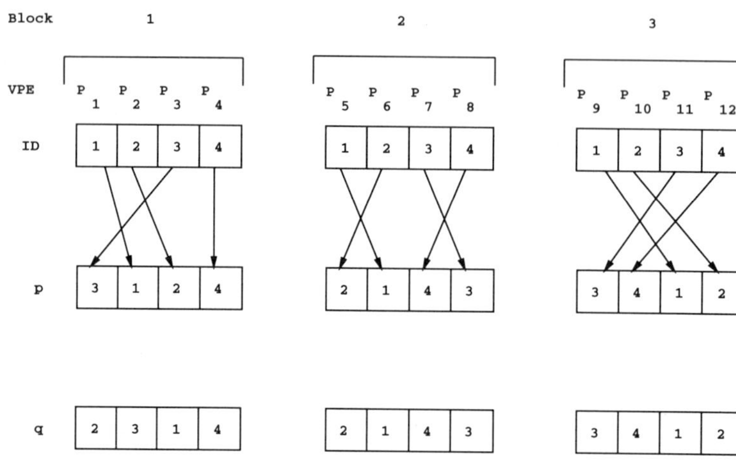

FIGURE 7. Vector-Parallel Perm-inverse, $p = q^{-1}$

Figures 7 and 8 graphically depict the basic operations, *permutation inverse* and *permutation apply*. In Figure 7, q is a vector of three permutations and we wish to compute its inverse in parallel. The procedure is quite simple. Each processor uses the contents of its q variable as the destination processor address and *sends* the contents of its ID variable to the destination processor, where the received value is stored in the parallel variable, p. Notice that permutation multiply is implemented as a *get* and permutation inverse is implemented as a *send*. In general, a get requires *two* sends causing permutation multiply to be almost twice as expensive as permutation inverse.

In Figure 8 we see a graphical depiction of the permutation apply operation. By permutation apply we simply mean the process of determining the image of a point under a permutation. In this case we have a parallel variable s which denotes a vector of three permutations and we wish to apply all three in parallel. There are three distinct phases to this operation. First, we broadcast to all

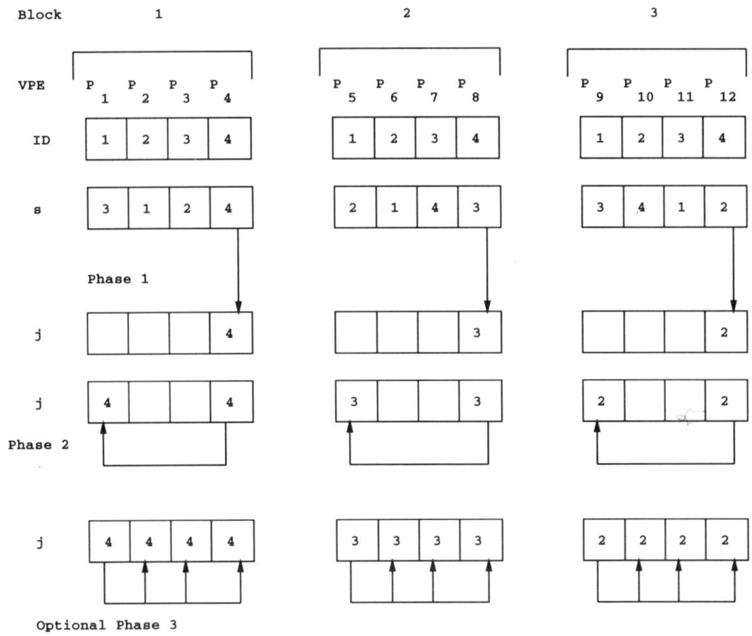

FIGURE 8. Vector-Parallel Perm-apply, $j = 4^s$

processors the point whose images we desire. Secondly, we select only those processors whose ID-value equals the broadcasted point. Only those processors will execute the next operation. All the selected processors store their $s-$value in the result variable, j. In phase 2 each selected processor sends its $j-$value to the first processor in its permutation block. The first processor always acts as a representative for the permutation whenever the entire permutation must act as a unit. Finally, there is an optional third phase in which the $j-$value is spread to all of the processors of the permutation block. Spreading the result is required in those applications where each processor of the permutation will access some other structure in parallel. In Figure 8 the dashed arrows indicate data movement within the memory of a given processor; the single-arrow solid lines of phase 2 indicate an interprocessor *send*; and the multi-arrow solid lines of phase 3 indicate a *spread* operation. It should be noted that, for 2D geometries, phase 2 may be avoided completely.

To begin to understand the performance of basic parallel permutation operations in the distributed representation, we coded the basic routines for permutation multiply, permutation inverse, and permutation apply in the *Lisp language on an 8K PEs CM-2a. We tested them on a range of point sets for $N = 4$ through $N = 256$ at various VP ratios. It should be noted that the size of the point set need not be a power of two; however, non-power-of-two point sets reduce processor utilization.

In the tables in this section we show performance results for permutation

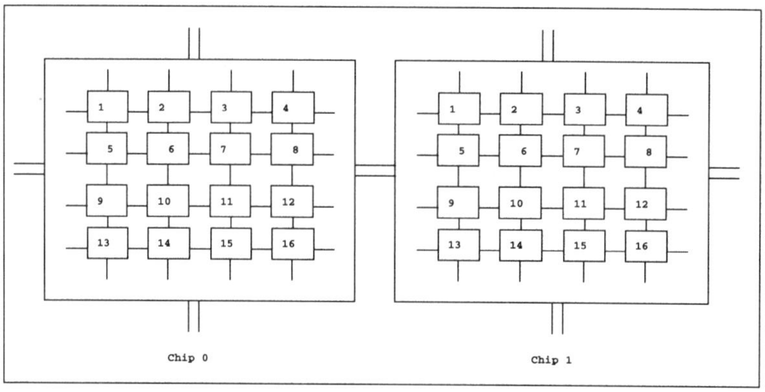

FIGURE 9. Permutation vector layout for $N = 16$

N	*Vector*	*CM Time*	*Elapsed Time*	*% Active*	*Sun-3*	*DEC5000*
4	2048	0.861865	1.472104	58	58.12	6.75
8	1024	0.866733	1.466928	59	53.70	6.33
16	512	0.872113	1.543285	56	52.05	6.13
32	256	0.876625	1.473793	59	50.87	6.03
64	128	0.881334	1.455496	60	50.13	5.98
128	64	0.886452	1.493635	59	50.08	5.95
256	32	1.035927	1.508619	68	50.38	5.95

TABLE 1. Comparison of Time in Seconds for Serial and Parallel (VPR=1) Permutation multiply for 1000 parallel multiplies vs. 1000xVector serial multiplies

multiply, permutation inverse, and permutation apply for 1D geometries and $VPR = 1$. The 1D geometry is one mechanism which allows the programmer to take full advantage of "on-chip" communications performance for $N \leq 16$. In the cases where $N \leq 16$ permutation multiply was achieved with only "on-chip" communication. For $16 < N \leq 32$ permutation multiply was achieved with "on-SPRINT" communication. For $N > 32$ permutation multiply required communication across the hardware hypercube. Figure 9 shows the two processor chips of a SPRINT node. Each chip contains 16 PEs. For $N = 16$ each processor on chip 0 corresponds to one point position in distributed image representation. In fact, this is true for chip 1 as well and for all 512 processor chips in the 8K PEs machine. A parallel variable is defined by an address and a length (in bits); i.e. it begins at the same address in the memory of every processor. Thus, a parallel variable, say p, denotes a vector of 512 16-point permutations distributed one permutation per chip. The field denoted by p need only be $\log_2 N$ (in this case 4) bits long. A second parallel variable q denotes another vector of 512 16-point permutations distributed in the same fashion. A parallel permutation multiply of p by q requires only "on-chip" communication. On-chip communication is achieved through the 16x16 hardware butterfly network called the flipper which moves fields quickly between the memories of processors which share the same chip.

N	*Vector*	*CM Time*	*Elapsed Time*	*% Active*	*Sun-3*	*DEC5000*
4	2048	0.533278	2.549707	20	48.27	5.77
8	1024	0.538978	2.456145	21	44.02	5.35
16	512	0.545108	2.470489	22	41.35	5.17
32	256	0.550906	2.542070	21	39.67	5.05
64	128	0.556737	2.443373	22	39.30	5.00
128	64	0.562647	2.468188	22	39.70	4.97
256	32	0.639169	2.414917	26	39.23	4.97

TABLE 2. Comparison of Time in Seconds for Serial and Parallel (VPR=1) Permutation inverse for 1000 parallel inverses vs. 1000xVector serial inverses

Phase 1						
N	*Vector*	*CM Time*	*Elapsed Time*	*% Active*	*Sun-3*	*DEC5000*
4	2048	0.082394	0.560857	14	40.87	4.47
8	1024	0.087221	0.578320	15	36.88	4.03
16	512	0.099498	0.620232	16	34.23	3.83
32	256	0.095154	0.587993	16	32.72	3.73
64	128	0.096960	0.623969	15	31.72	3.67
128	64	0.099268	0.598347	16	32.25	3.65
256	32	0.101289	0.602863	16	31.77	3.63
Phases 1,2,3						
N	*Vector*	*CM Time*	*Elapsed Time*	*% Active*	*Sun-3*	*DEC5000*
4	2048	0.955525	3.299578	28	40.87	4.47
8	1024	1.041453	4.131721	25	36.88	4.03
16	512	1.032659	3.305256	31	34.23	3.83
32	256	1.081772	3.369904	32	32.72	3.73
64	128	1.053693	3.383366	31	31.72	3.67
128	64	1.090881	3.342725	32	32.25	3.65
256	32	1.055532	3.385909	31	31.77	3.63

TABLE 3. Comparison of Time in Seconds for Serial and Parallel (VPR=1) Permutation apply for 1000 parallel applies vs. 1000xVector serial applies

Tables 1, 2, and 3 show a comparison of run times for serial and parallel permutation multiply, permutation inverse, and permutation apply for $N = 4$ up through $N = 256$. In the parallel case the vector size was chosen to keep the VPR equal to 1 on the 8K PEs CM-2 available for experiments. For example, row 1 of table 1 shows that a pair of vectors of 2,048 4-point permutations was multiplied 1000 times in 0.861865 seconds of CM time. Elapsed time is the total time the code was running on the front-end; it is not an accurate measure of CM performance since the front-end is usually running other processes under timesharing. The % Active column lists the percent of total elapsed time that the CM was actually busy. For these test cases, the test program simply ran a loop of 1000 parallel permutation multiplications with essentially exclusive use of the front-end computer. For such a simple program the front-end code was not fast enough to keep the CM busy 100% of the time. For $VPR = 2$, a 99% Active figure is achieved with low front-end load. The serial time is the time required to perform 2,048,000 4-point permutation multiplications serially on a Sun-3/140 (approximately a 2 MIPS CISC workstation with 8 MB of memory, vintage 1985) and a DECstation 5000/200 (approximately a 24 MIPS RISC workstation with 16MB of memory, vintage 1990). Times for the Sun-3 are shown for comparison with a serial machine from the same component technology generation. In both serial cases, computations were done within physical memory; thus there was no significant paging and in many cases very high cache hit rates were achieved.

Let's briefly consider communications in a little more detail. The CM-2 is based upon the CM-1 which was originally designed to model the human brain – i.e. a richly connected large number of very simple (and slow) processing elements [**9**]. For an 8K PEs CM-2 the network has a peak aggregate bandwidth of about 1 Gbits/sec. Since permutation multiply utilizes the *get* operation which is approximately as costly as two *send* operations, we should not expect to achieve more than 500 Mbits/sec for permutation multiply. The maximum achieved aggregate bandwidth ($N = 4$) for permutation multiply is approximately 20 Mbits/sec (2 bits per point x 4 points per permutation x 2048 permutations = 16,384 bits which get moved 1000 times in .861865 seconds). This is quite substantially less than the peak aggregate bandwidth simply because the front-end can feed *send* instructions to the back-end only at the basic clock rate (7 Mhz). When the basic clock rate is increased to 8.5 Mhz, the corresponding time for 1000 permutation multiplies drops to .711833 and the maximum achieved bandwidth increases to approximately 23 Mbits/sec in precise proportion to the increase in clock speed. The corresponding numbers for permutation inverse are 30 Mbits/sec at 7 Mhz and 37 Mbits/sec at 8.5 Mhz. The point we wish to make here is that the CM-2 network architecture is capable of supporting permutation operations at approximately 25 times the achieved rates; however, modifications in the basic processor and router technology, the front-end interface, and sequencer microcode would be required.

Our experiments were arranged to keep the total number of points moved the same in both the parallel and serial cases. Compared to the Sun-3 the CM-2 speedup factor ranged from 67 for $N = 4$ down to 48 for $N = 256$. It should be noted that on a fully configured 64K PEs CM-2 the vector size could be eight times as large without any substantial additional communications cost penalty. Thus, the speedup factors would range between 384 and 536. The cost to move the same number of points in the parallel case is slowly increasing with N primarily due to the arithmetic involved in computing the destination address by the bit-serial processors. The large increase in CM Time between $N = 128$ and $N = 256$ is due to communications cost; in this case a single permutation spans 8 SPRINTs and *send order* causes some congestion. Although the DECstation 5000 uses newer technology and represents about a factor of 8 performance improvement over the Sun-3, the 8K PEs CM-2 with distributed image representation still provides superior performance by a factor of about 6. It should be mentioned at this time that theoretical "linear speedup" is not the issue when studying the perfomance of practical algorithms on real machines. Since the technologies differ dramatically, the appropriate issue is the ratio of price to performance. For example, in 1987 an 8K PEs CM-2 was less expensive than 48 Sun-3's with the appropriate amount of memory.

Table 2 shows similar results for parallel permutation inverse. The speedup factors compared to the Sun-3 range from 90 at $N = 4$ down to 61 at $N = 256$; for the DECstation 5000 they range from 10.8 at $N = 4$ down to 7.8 at $N = 256$. Table 3 shows very substantial speedups when only phase 1 of parallel permutation apply is required (496 for $N = 4$ down to 313 for $N = 256$ against the Sun-3). When all three phases are required the performance improvement is not so great (ranging from 43 for $N = 4$ down to 30 for $N = 256$ against the Sun-3). It is clear that applications which do not require all three phases of parallel permutation apply will experience substantial performance improvement.

In order to appropriately test the geometries, random permutations were generated for these experiments so that approximately $\frac{1}{2}$ of the test permutations for $N \geq 32$ required "off-chip" communication and $\frac{1}{2}$ for $N \geq 64$ required "off-SPRINT" communication.

Extensive experiments for 2D geometries showed the running time for $VPR = 1$ to be about 1.82 times the running times for the 1D geometries. In addition, the times for 2D geometries grew linearly with the VP ratio as expected. Thus, tailoring the data layout based upon architectural considerations can have a significant impact on performance.

4. Parallel Sifting

In order to test the utility of these primitive operations, a straightforward algorithm for parallel sifting was implemented. It is accomplished by sifting a vector of permutations through a table of coset representatives consisting of the complete right transversals of the point stabilizer chain for a group stored in an $N x N^2$ geometry. In the remainder of this paper we shall refer to this table as the "coset table", although it should not be confused with other uses of this term in different contexts such as coset enumeration. The coset table can be thought of as NxN matrix in which each row corresponds to a complete right transversal.

Input: Complete right transversals T_i for $G^{(i)}$ in $G^{(i+1)}$ and a permutation $\sigma \epsilon S_N$

Output: **true** if $\sigma \epsilon G$, **false** otherwise.

1. $i \leftarrow 0$;
2. $in \leftarrow$**true**;
3. **while** ($i < N$ **and** in) {
4. $i \leftarrow i + 1$;
5. $j \leftarrow i^{\sigma}$;
6.   **if** $\exists \tau \epsilon T_i$ such that $i^{\tau} = j$
7.   **then** $\sigma \leftarrow \sigma\tau^{-1}$;
8.   **else** $in \leftarrow$**false**;
9.   }
10. return(in);

FIGURE 10. Serial Sifting Algorithm

Of course, only the upper triangular portion of the matrix is non-empty and the identity permutation is stored in each entry along the diagonal. Admittedly this geometry is somewhat wasteful of processors and storage, but for $N < 100$ it is not unreasonable.

Figure 10 shows a standard serial algorithm for sifting (adapted from [**17**]). In the parallel case, σ represents a vector of permutations, the point, i, is broadcasted to all processors, and j is a parallel variable which may hold a different value in each processor. The T_i are sets of distributed permutations. Each complete right transversal, T_i, has shape NxK. The N points of the permutation are distributed along the first dimension of the geometry and K such permutations are stored along the second dimension where K is the index of $G^{(i)}$ in $G^{(i+1)}$.

For example, in the case of $N = 4$ σ is a vector of 2048 permutations. When σ is applied to the point 1 (line 5 in figure 10), each permutation of the vector produces a value (via *parallel permutation apply*) in the parallel variable, j. The index, $(i, j) = (1, 1^{\sigma_p}), p = 0, ...2047$, is used to retrieve an entry from the 4x4 the coset table. Each permutation probes the position defined by its index into the coset table with a *get* operation (line 6 in figure 10). In this case several permutations are probing the same position simultaneously. Since the upper triangle of this coset table contains only 6 positions, some congestion in the network results when 2048 processors attempt to access 6 processors. Those permutations whose probes return "empty" are *not representable* by this coset table and they set a bit indicating this status. They are prevented from participating in the remaining steps of the sift (removed from the active context set). All permutations whose probes return "full" retrieve their coset table entry in parallel and perform a parallel permutation multiply (line 7 in figure 10) to produce the permutations to be used in the next step of the sift. (Note: we preprocessed our coset table by precomputing and storing the inverses of the table entries). The algorithm proceeds in this fashion until the only permutations remaining in the active set at the end of N steps are the ones which are representable. Of course, if the active set becomes empty before the N-th step the sift is terminated. Testing for emptiness is easily achieved with one of the "global reduction instructions" of the CM-2 (in this case,CM_global_logior).

N	*Vector*	*CM Time*	*Elapsed Time*	*% Active*	*Sun-3*	*DEC5000*
4	2048	0.008489	0.131483	6	0.420	0.047
8	1024	0.018841	0.214742	8	0.892	0.101
16	512	0.040270	0.384719	10	1.808	0.207
32	256	0.082931	0.721516	11	3.606	0.421
64	128	0.164476	1.390555	11	7.155	0.839

TABLE 4. Parallel Sifting Performance

Table 4 shows parallel sifting times using this technique for $N = 4$ through $N = 64$. Row 1 of the table indicates that a vector of 2048 random permutations on 4 points was sifted through a coset table for S_4 in .008489 seconds on the CM-2, .42 seconds on a Sun-3, and 0.047 seconds on a DECstation 5000/200 for speedup factors of 49 and 5.5 respectively. It should be noted that the parallel sifting time is 1.56 times the amount of time required to perform 3x(permutation apply + permutation multiply) – i.e. $1.56 = .008489/.00545$ where $3(.000956 + .000862) = .005454$. Note that permutation inverses for coset table entries were pre-computed. The difference is directly attributable to the communication associated with determining whether a coset table entry is empty or full. In the serial cases, the sifting time is approximately:
$(N-1)\times$ ($2\times$ permutation apply + permutation multiply),
since a reference into the coset table is the same cost as a permutation apply.

One of the challenges which must be met in order to make the CM-2 a viable vehicle for permutation group management is to develop communications efficient representations for coset tables and other related structures. It should be noted that some computations have already been demonstrated to map well to the restricted geometries currently provided by the CM-2. For example, computing orbits on a 3D grid (NxNxS) where $N =$ the number of points and $S =$ the number of generators can be done very quickly using phase 1 of parallel permutation apply followed by a *send-with-combiner* operation. This technique was utilized in our implementation of group membership testing by the method of random subproducts [**7**] and we are continuing our research efforts in this area.

5. Implications

The main question to be answered in this section is "Over the next decade what kinds of performance improvements can be expected in the area of group computations due to parallelism?" In order to answer this question, we need to look at the implications of general performance trends with regard to the characteristics of parallel group computations. Before we begin to discuss the implications of the experiments, we shall briefly summarize some of the features of the CM-2 which were utilized:

- We made use of the bit-serial processors of the CM-2 to perform arithmetic. These processors are (by design) very slow; they run at 7 Mhz and a minimum of 3 cycles is required to process two 1-bit operands. Thus, a 32-bit add instruction requires at least 96 cycles. Permutation multiply requires the computation of a processor address by adding the value of the processor's stored point (offset) to a base address (the address of the first processor of the permutation block). This is a not significant portion of the time required to execute a parallel permutation operation.
- The bit-addressable memory of the CM-2 allows permutations to be efficiently represented in $\log_2 N$ bits. When $\log_2 N$ is much less than the word length on a standard machine (eg. 32), the CM-2 affords considerable savings in storage. Of course in a word-addressable machine, several permutation points could be packed to a word, but then the cost of packing/unpacking is incurred.
- The CM-2 has special instructions which allow the network to perform computations on data as it passes through the network such as the send-with-combiner instructions.
- Intra-chip communication is very high bandwidth and low latency; intra-SPRINT communication is higher bandwidth and higher latency; inter-SPRINT is even higher bandwidth and higher latency. Whenever possible we utilized the highest bandwidth and lowest latency available for the computation.

In a recent article [**4**] Bell took a candid look at the history of supercomputing and made some bold prognostications. Although Bell is focussed primarily on large scale numerical computing, most of the concepts are relevant to the analysis of parallel group computations. For the most part we shall ignore the technology trends (assuming they benefit all parallel applications) and we will concentrate on the architectural and algorithm mapping issues. A brief assessment of parallel group computations vis-a-vis each issue follows:

SIMD vs. MIMD

The SIMD versus MIMD controversy is being resolved in terms of high-level programming models instead of low-level architectural considerations. The integrating model is called SPMD (Single Program Multiple Data Streams) and it subsumes the form of data parallelism most exploited in our permutation group computations implemented on the SIMD CM-2. This model extends to the CM-5 which is basically a MIMD/MSIMD architecture. Our conclusion is that basic parallel permutation group computations admit a high degree of data parallelism and are consistent with the SPMD model.

Shared Memory vs. Distributed Memory

Our experiments show that a distributed memory parallel computer with distributed image representation for permutations is practical. Shared global memory is unlikely to provide better performance for these local permutation operations and it will certainly detract for high degree permutation groups. The distributed array representation for the coset table experienced some congestion when multiple processors accessed the same table entry; however, there is certainly less congestion than would be experienced under the shared memory model. In the distributed case, table entries may be replicated and and distributed about the machine so as to reduce congestion. In our recent CM-5 implementation [**13**] we used local image representation for permutations and saw speedups of 5 – 10 over the CM-2 attributable almost totally to technology improvement. Finally a single address space is not a requirement since basic permutation group computations partition nicely into blocks of size equal to the degree of the group.

Space Locality and Time Locality

Space locality can offer tremendous benefits as we saw in our implementation on the CM-2 which maintained entire data structures on a SPRINT node. We saw similar results on the CM-5; the processor caches are so large that often many permutations can be stored in cache. Thus for applications which reuse the same set of permutations many times, all the accesses are cache accesses.

Time locality (the last memory referenced will be repeatedly referenced) is not an issue for distributed image representation on the CM-2 since all the points of the permutation are referenced in parallel for permutation multiply and permutation inverse. For permutation apply only a single point is referenced per permutation, but it is unlikely to be repeatedly referenced. It is clear that time locality is an issue on the CM-5 when all the required permutations do not fit in cache simultaneously.

Communications/Computation Balance

A guiding principle in the design of parallel computers is that the bandwidth of the interconnect should increase linearly with the processing power of the individual nodes. A similar principle applies to the corresponding decrease in memory access time and network latency. The problem is that increasing bandwidth generally increases latency. If we consider latency to be the amount of time (or number of instructions) a processor must wait for data to become available once it has been requested, it is a measure of the opportunity cost of lost parallelism.

A careful analysis of the CM-2 architecture and the design tradeoffs of its implementation show that it could have been designed to support parallel permutation group computations more efficiently. Within its architectural paradigm memory bandwidth, communications bandwidth, and processor bandwidth could have been chosen to optimize these operations. For example, 32 Mhz standard cell 8-bit processors, 8-bit wide memories, 50 ns DRAMs, a match between the number of processors per chip and the number of router channels, and some additional parallelism in the router hardware might allow permutation operations to achieve the aggregate bandwidth of the communications network. These technological alternatives were available in 1986; however, these were not the goals of the machine's designers [**20**].

Problem Scalability

From our experiments it is clear that permutation group computations admit a high degree of parallelism and in most cases it is possible to scale the problem to the size of the machine. In terms of practical scalability we are concerned with whether, for a given problem size, adding more processors will allow additional parallelism. Permutation group computations have this character, since, for suitable degree, the number of permutations to be multiplied is generally much larger than the machine size. In our CM-2 implementation we achieved parallelism across the elements (permutations) of the group as well as across the points of each individual permutation. For example, for a group of degree 32 we were able to use a vector size of 256 on an 8K PEs CM-2 and a vector size of 2048 on a 64K PEs CM-2. In a coarser-grained implementation such as on the CM-5, we get parallelism only across the group elements.

Finally, the overriding concern in the development of software for a parallel machine is the ease of programming. The CM-2, as well as many of the other machines mentioned above, admits high–level programming languages such as *Lisp, C*, and CM Fortran. Once the basic permutation operations have been coded for optimal performance in a low–level language, general codes can easily be developed in one of the higher level languages.

References

1. Babai L., Luks E., and Seress A. *On Managing Permutation Groups in* $O(n^4 \log^c n)$, Proc. 28th IEEE Symp. FOCS, (1988), 272-282.
2. Babai L., Luks E., and Seress A. *Permutation Groups in NC*, Proc. XXth ACM STOC, (1987), 409-420.
3. Bhatt S. N., Chung F. R. K., Leighton T., Rosenberg A. L., *Optimal Simulations of Tree Machines*, in Proc. 27th IEEE Symp. FOCS, (1986), 274-282.
4. Bell, G., *Ultracomputers: A Teraflop Before Its Time*, in *Communications of the ACM*, 35(8):26-47, 1992.
5. Cannon J. J., *An Introduction to the Group Theory Language, Cayley*, in *Computational Group Theory*, M. D. Atkinson (ed.), Academic Press, New York, 1984, 145-184.
6. Cooperman G., Finkelstein L., and Purdom P. W. *Fast Group Membership Using a Strong Generating Test for Permutation Groups, Computers and Mathematics*, (E. Kaltofen and S.M. Watt, eds.) Springer-Verlag, New York, 27-36, 1989.
7. Cooperman G., Finkelstein L., and York B. W., *A Parallel Implementation of Group Membership and the Method of Random Subproducts*, in Proc. of the First Summer Institute on Issues and Obstacles in the Practical Implementation of Parallel Algorithms and the Use of Parallel Machines, Dartmouth Institute for Advanced Graduate Studies in Parallel Computation (DAGS/PC), June 23-27, 1992, Dartmouth College, Hanover, NH.
8. Furst M., Hopcroft J., and Luks E. M., *Polynomial-time Algorithms for Permutation Groups, Proc.* **21***st IEEE Symp. FOCS*, (1980), 36-41.
9. Hillis, W. W., *The Connection Machine*, MIT Press, Cambridge, 1985.
10. Hoffman C. M., *Group-Theoretic Algorithms and Graph Isomorphism*, Springer-Verlag, New York, 1982.
11. Johnsson S. L., *Communications Efficient Basic Linear Algebra Computations on Hypercube Architectures*, Journal of Parallel and Distributed Computing, 4, 133-172 (1987).
12. Knuth D. E., *Notes on Efficient Representation of Permutation Groups*, 1981, unpublished manuscript.
13. Kvech K. and York B. W., *Permutation Group Computations on the Connection Machine, CM-5*, Northeastern University Technical Report NU-CCS-92-7.
14. Luks E. M., McKenzie P. *Fast Parallel Computation with Permutation Groups*, Proc. IEEE Symp. FOCS, (1985), 505-514.
15. Niemeyer A., Nickel W., and Schonert M., *GAP: Getting started and Reference Manual*, Lehrstuhl D fur Mathematik, RWTH Aachen, W. Germany.
16. Richardson D., *GROUP: A Computer System for Group-Theoretical Calculations*, M.Sc. Thesis, University of Sidney, 1973.
17. Sims C. C., *Computation with Permutation Groups*, in Proc. of Second Symposium on Symbolic and Algebraic Manipulation, edited by S. R. Petrick, ACM, New York, 1971.
18. Thinking Machines Corporation, *Connection Machine Model CM-2 Technical Summary*, Version 6.0, November 1990, Thinking Machines Corporation, Cambridge, MA.
19. Wu A. Y., *Embedding of Tree Networks in Hypercubes*, Journal of Parallel and Distributed Computing, 2(3):238-249, 1985.
20. private communication with D. Dahl, Thinking Machines Corp. Cambridge, MA

College of Computer Science, Northeastern University, Boston, Massachusetts 02115

E-mail address: york@ccs.northeastern.edu